Applied Mathematics: Body and Soul

Springer
Berlin
Heidelberg
New York
Hong Kong
London
Milan
Paris
Tokyo

K. Eriksson • D. Estep • C. Johnson

Applied Mathematics: Body and Soul

[VOLUME 3]

Calculus in Several Dimensions

Springer

Kenneth Eriksson
Claes Johnson
Chalmers University of Technology
Department of Mathematics
41296 Göteborg, Sweden
e-mail: kenneth|claes@math.chalmers.se

Donald Estep
Colorado State University
Department of Mathematics
Fort Collins, CO 80523-1874
USA
e-mail: estep@math.colostate.edu

Cataloging-in-Publication Data applied for

A catalog record for this book is available from the Library of Congress.

Bibliographic information published by Die Deutsche Bibliothek
Die Deutsche Bibliothek lists this publication in the Deutsche Nationalbibliografie;
detailed bibliographic data is available in the Internet at <http://dnb.ddb.de>.

Mathematics Subject Classification (2000): 15-01, 34-01, 35-01, 49-01, 65-01, 70-01, 76-01

ISBN   978-3-642-05660-4

Springer-Verlag Berlin Heidelberg New York
a member of BertelsmannSpringer Science+Business Media GmbH
springeronline.com
© Springer-Verlag Berlin Heidelberg 2010
Printed in Germany

Cover design: *design & production,* Heidelberg and Anders Logg, Department of Computational Mathematics, Chalmers University of Technology

Printed on acid-free paper   46/3142ck-5 4 3 2 1 0

To the students of Chemical Engineering at Chalmers during 1998–2002, who enthusiastically participated in the development of the reform project behind this book.

Preface

I admit that each and every thing remains in its state until there is reason for change. (Leibniz)

The Need of Reform of Mathematics Education

Mathematics education needs to be reformed as we now pass into the new millennium. We share this conviction with a rapidly increasing number of researchers and teachers of both mathematics and topics of science and engineering based on mathematical modeling. The reason is of course the computer revolution, which has fundamentally changed the possibilities of using mathematical and computational techniques for modeling, simulation and control of real phenomena. New products and systems may be developed and tested through computer simulation on time scales and at costs which are orders of magnitude smaller than those using traditional techniques based on extensive laboratory testing, hand calculations and trial and error.

At the heart of the new simulation techniques lie the new fields of Computational Mathematical Modeling (CMM), including Computational Mechanics, Physics, Fluid Dynamics, Electromagnetics and Chemistry, all based on solving systems of differential equations using computers, combined with geometric modeling/Computer Aided Design (CAD). Computational modeling is also finding revolutionary new applications in biology, medicine, environmental sciences, economy and financial markets.

Education in mathematics forms the basis of science and engineering education from undergraduate to graduate level, because engineering and science are largely based on mathematical modeling. The level and the quality of mathematics education sets the level of the education as a whole. The new technology of CMM/CAD crosses borders between traditional engineering disciplines and schools, and drives strong forces to modernize engineering education in both content and form from basic to graduate level.

Our Reform Program

Our own reform work started some 20 years ago in courses in CMM at advanced undergraduate level, and has through the years successively penetrated through the system to the basic education in calculus and linear algebra. Our aim has become to develop a complete program for mathematics education in science and engineering from basic undergraduate to graduate education. As of now our program contains the series of books:

1. *Computational Differential Equations*, (*CDE*)

2. *Applied Mathematics: Body & Soul I–III*, (*AM I–III*)

3. *Applied Mathematics: Body & Soul IV–*, (*AM IV–*).

AM I–III is the present book in three volumes I–III covering the basics of calculus and linear algebra. *AM IV–* offers a continuation with a series of volumes dedicated to specific areas of applications such as *Dynamical Systems (IV)*, *Fluid Mechanics (V)*, *Solid Mechanics (VI)* and *Electromagnetics (VII)*, which will start appearing in 2003. *CDE* published in 1996 may be be viewed as a first version of the whole *Applied Mathematics: Body & Soul* project.

Our program also contains a variety of software (collected in the *Mathematics Laboratory*), and complementary material with step-by step instructions for self-study, problems with solutions, and projects, all freely available on-line from the web site of the book. Our ambition is to offer a "box" containing a set of books, software and additional instructional material, which can serve as a basis for a full applied mathematics program in science and engineering from basic to graduate level. Of course, we hope this to be an on-going project with new material being added gradually.

We have been running an applied mathematics program based on *AM I–III* from first year for the students of chemical engineering at Chalmers since the Fall 99, and we have used parts of the material from *AM IV–* in advanced undergraduate/beginning graduate courses.

Main Features of the Program:

- The program is based on a synthesis of mathematics, computation and application.

- The program is based on new literature, giving a new unified presentation from the start based on constructive mathematical methods including a computational methodology for differential equations.

- The program contains, as an integrated part, software at different levels of complexity.

- The student acquires solid skills of implementing computational methods and developing applications and software using Matlab.

- The synthesis of mathematics and computation opens mathematics education to applications, and gives a basis for the effective use of modern mathematical methods in mechanics, physics, chemistry and applied subjects.

- The synthesis building on constructive mathematics gives a synergetic effect allowing the study of complex systems already in the basic education, including the basic models of mechanical systems, heat conduction, wave propagation, elasticity, fluid flow, electro-magnetism, reaction-diffusion, molecular dynamics, as well as corresponding multiphysics problems.

- The program increases the motivation of the student by applying mathematical methods to interesting and important concrete problems already from the start.

- Emphasis may be put on problem solving, project work and presentation.

- The program gives theoretical and computational tools and builds confidence.

- The program contains most of the traditional material from basic courses in analysis and linear algebra

- The program includes much material often left out in traditional programs such as constructive proofs of all the basic theorems in analysis and linear algebra and advanced topics such as nonlinear systems of algebraic/differential equations.

- Emphasis is put on giving the student a solid understanding of basic mathematical concepts such as real numbers, Cauchy sequences, Lipschitz continuity, and constructive tools for solving algebraic/differential equations, together with an ability to utilize these tools in advanced applications such as molecular dynamics.

- The program may be run at different levels of ambition concerning both mathematical analysis and computation, while keeping a common basic core.

AM I–III in Brief

Roughly speaking, *AM I–III* contains a synthesis of calculus and linear algebra including computational methods and a variety of applications. Emphasis is put on constructive/computational methods with the double aim of making the mathematics both understandable and useful. Our ambition is to introduce the student early (from the perspective of traditional education) to both advanced mathematical concepts (such as Lipschitz continuity, Cauchy sequence, contraction mapping, initial-value problem for systems of differential equations) and advanced applications such as Lagrangian mechanics, n-body systems, population models, elasticity and electrical circuits, with an approach based on constructive/computational methods.

Thus the idea is that making the student comfortable with both advanced mathematical concepts and modern computational techniques, will open a wealth of possibilities of applying mathematics to problems of real interest. This is in contrast to traditional education where the emphasis is usually put on a set of analytical techniques within a conceptual framework of more limited scope. For example: we already lead the student in the second quarter to write (in Matlab) his/her own solver for general systems of ordinary differential equations based on mathematically sound principles (high conceptual and computational level), while traditional education at the same time often focuses on training the student to master a bag of tricks for symbolic integration. We also teach the student some tricks to that purpose, but our overall goal is different.

Constructive Mathematics: Body & Soul

In our work we have been led to the conviction that the constructive aspects of calculus and linear algebra need to be strengthened. Of course, constructive and computational mathematics are closely related and the development of the computer has boosted computational mathematics in recent years. Mathematical modeling has two basic dual aspects: one symbolic and the other constructive-numerical, which reflect the duality between the infinite and the finite, or the continuous and the discrete. The two aspects have been closely intertwined throughout the development of modern science from the development of calculus in the work of Euler, Lagrange, Laplace and Gauss into the work of von Neumann in our time. For example, Laplace's monumental *Mécanique Céleste* in five volumes presents a symbolic calculus for a mathematical model of gravitation taking the form of Laplace's equation, together with massive numerical computations giving concrete information concerning the motion of the planets in our solar system.

However, beginning with the search for rigor in the foundations of calculus in the 19th century, a split between the symbolic and constructive aspects gradually developed. The split accelerated with the invention of the electronic computer in the 1940s, after which the constructive aspects were pursued in the new fields of numerical analysis and computing sciences, primarily developed outside departments of mathematics. The unfortunate result today is that symbolic mathematics and constructive-numerical mathematics by and large are separate disciplines and are rarely taught together. Typically, a student first meets calculus restricted to its symbolic form and then much later, in a different context, is confronted with the computational side. This state of affairs lacks a sound scientific motivation and causes severe difficulties in courses in physics, mechanics and applied sciences which build on mathematical modeling.

New possibilies are opened by creating from the start a synthesis of constructive and symbolic mathematics representing a synthesis of Body & Soul: with computational techniques available the students may become familiar with nonlinear systems of differential equations already in early calculus, with a wealth of applications. Another consequence is that the basics of calculus, including concepts like real number, Cauchy sequence, convergence, fixed point iteration, contraction mapping, is lifted out of the wardrobe of mathematical obscurities into the real world with direct practical importance. In one shot one can make mathematics education both deeper and broader and lift it to a higher level. This idea underlies the present book, which thus in the setting of a standard engineering program, contains all the basic theorems of calculus including the proofs normally taught only in special honors courses, together with advanced applications such as systems of nonlinear differential equations. We have found that this seemingly impossible program indeed works surprisingly well. Admittedly, this is hard to believe without making real life experiments. We hope the reader will feel encouraged to do so.

Lipschitz Continuity and Cauchy Sequences

The usual definition of the basic concepts of *continuity* and *derivative*, which is presented in most Calculus text books today, build on the concept of *limit*: a real valued function $f(x)$ of a real variable x is said to be continuous at $\bar{x}$ if $\lim_{x \to \bar{x}} f(x) = f(\bar{x})$, and $f(x)$ is said to be differentiable at

$\bar{x}$ with derivative $f'(\bar{x})$ if

$$\lim_{x \to \bar{x}} \frac{f(x) - f(\bar{x})}{x - \bar{x}}$$

exists and equals $f'(\bar{x})$. We use different definitions, where the concept of limit does not intervene: we say that a real-valued function $f(x)$ is Lipschitz continuous with Lipschitz constant L_f on an interval $[a, b]$ if for all $x, \bar{x} \in [a, b]$, we have

$$|f(x) - f(\bar{x})| \le L_f |x - \bar{x}|.$$

Further, we say that $f(x)$ is differentiable at $\bar{x}$ with derivative $f'(\bar{x})$ if there is a constant $K_f(\bar{x})$ such that for all x close to $\bar{x}$

$$|f(x) - f(\bar{x}) - f'(\bar{x})(x - \bar{x})| \le K_f(\bar{x})|x - \bar{x}|^2.$$

This means that we put somewhat more stringent requirements on the concepts of continuity and differentiability than is done in the usual definitions; more precisely, we impose *quantitative* measures in the form of the constants L_f and $K_f(\bar{x})$, whereas the usual definitions using limits are *purely qualitative*.

Using these more stringent definitions we avoid pathological situations, which can only be confusing to the student (in particular in the beginning) and, as indicated, we avoid using the (difficult) concept of limit in a setting where in fact no limit processes are really taking place. Thus, we do not lead the student to definitions of continuity and differentiability suggesting that all the time the variable x is tending to some value $\bar{x}$, that is, all the time some kind of (strange?) limit process is taking place. In fact, continuity expresses that the difference $f(x) - f(\bar{x})$ is small if $x - \bar{x}$ is small, and differentiability expresses that $f(x)$ locally is close to a linear function, and to express these facts we do not have to invoke any limit processes.

These are examples of our leading philosophy of giving Calculus a *quantitative form*, instead of the usual purely qualitative form, which we believe helps both understanding and precision. We believe the price to pay for these advantages is usually well worth paying, and the loss in generality are only some pathological cases of little interest. We can in a natural way relax our definitions, for example to Hölder continuity, while still keeping the quantitative aspect, and thereby increase the pathology of the exceptional cases.

The usual definitions of continuity and differentiability strive for maximal generality, typically considered to be a virtue by a pure mathematician, which however has pathological side effects. With a constructive point of view the interesting world is the constructible world and maximality is not an important issue in itself.

Of course, we do not stay away from limit processes, but we concentrate on issues where the concept of limit really is central, most notably in defining the concept of a *real number* as the limit of a Cauchy sequence of rational numbers, and a solution of an algebraic or differential equation as the limit of a Cauchy sequence of approximate solutions. Thus, we give the concept of *Cauchy sequence* a central role, while maintaining a constructive approach seeking constructive processes for generating Cauchy sequences.

In standard Calculus texts, the concepts of Cauchy sequence and Lipschitz continuity are not used, believing them to be too difficult to be presented to freshmen, while the concept of real number is left undefined (seemingly believing that a freshman is so familiar with this concept from early life that no further discussion is needed). In contrast, in our constructive approach these concepts play a central role already from start, and in particular we give a good deal of attention to the fundamental aspect of the constructibility of real numbers (viewed as possibly never-ending decimal expansions).

We emphasize that taking a constructive approach does not make mathematical life more difficult in any important way, as is often claimed by the ruling mathematical school of formalists/logicists: All theorems of interest in Calculus and Linear Algebra survive, with possibly some small unessential modifications to keep the quantitative aspect and make the proofs more precise. As a result we are able to present basic theorems such as Contraction Mapping Principle, Implicit Function theorem, Inverse Function theorem, Convergence of Newton's Method, in a setting of several variables with complete proofs as a part of our basic Calculus, while these results in the standard curriculum are considered to be much too difficult for this level.

Proofs and Theorems

Most mathematics books including Calculus texts follow a theorem-proof style, where first a theorem is presented and then a corresponding proof is given. This is seldom appreciated very much by the students, who often have difficulties with the role and nature of the proof concept.

We usually turn this around and first present a line of thought leading to some result, and then we state a corresponding theorem as a summary of the hypothesis and the main result obtained. We thus rather use a proof-theorem format. We believe this is in fact often more natural than the theorem-proof style, since by first presenting the line of thought the different ingredients, like hypotheses, may be introduced in a logical order. The proof will then be just like any other line of thought, where one successively derives consequences from some starting point using different hypothesis as one goes along. We hope this will help to eliminate the often perceived mystery of proofs, simply because the student will not be aware of the fact that a proof is being presented; it will just be a logical line of thought, like any logical line of thought in everyday life. Only when the line of thought is finished, one may go back and call it a proof, and in a theorem collect the main result arrived at, including the required hypotheses. As a consequence, in the Latex version of the book we do use a theorem-environment, but not any proof-environment; the proof is just a logical line of thought preceding a theorem collecting the hypothesis and the main result.

The Mathematics Laboratory

We have developed various pieces of software to support our program into what we refer to as the *Mathematics Laboratory*. Some of the software serves the purpose of illustrating mathematical concepts such as roots of equations, Lipschitz continuity, fixed point iteration, differentiability, the definition of the integral and basic calculus for functions of several variables; other pieces are supposed to be used as models for the students own computer realizations; finally some pieces are aimed at applications such as solvers for differential equations. New pieces are being added continuously. Our ambition is to also add different multi-media realizations of various parts of the material.

In our program the students get a training from start in using Matlab as a tool for computation. The development of the constructive mathematical aspects of the basic topics of real numbers, functions, equations, derivatives and integrals, goes hand in hand with experience of solving equations with fixed point iteration or Newton's method, quadrature, and numerical methods or differential equations. The students see from their own experience that abstract symbolic concepts have roots deep down into constructive computation, which also gives a direct coupling to applications and physical reality.

Go to http://www.phi.chalmers.se/bodysoul/

The *Applied Mathematics: Body & Soul* project has a web site containing additional instructional material and the *Mathematics Laboratory*. We hope that the web site for the student will be a good friend helping to (independently) digest and progress through the material, and that for the teacher it may offer inspiration. We also hope the web site may serve as a forum for exchange of ideas and experience related the project, and we therefore invite both students and teachers to submit material.

Acknowledgment

The authors of this book want to thank sincerely the following colleagues and graduate students for contributing valuable material, corrections and suggestions for improvement: Rickard Bergström, Niklas Eriksson, Johan Hoffman, Mats Larson, Stig Larsson, Mårten Levenstam, Anders Logg, Klas Samuelsson and Nils Svanstedt, all actively participating in the development of our reform project. And again, sincere thanks to all the students of chemical engineering at Chalmers who carried the burden of being exposed to new material often in incomplete form, and who have given much enthusiastic criticism and feed-back.

The source of mathematicians pictures is the MacTutor History of Mathematics archive, and some images are copied from old volumes of Deadalus, the yearly report from The Swedish Museum of Technology.

My heart is sad and lonely
for you I sigh, dear, only
Why haven't you seen it
I'm all for you body and soul
(Green, Body and Soul)

Contents Volume 3

Contents Volume 1

21 Analytic Geometry in $\mathbb{R}^3$ **313**

Contents Volume 2

Volume 3

Calculus in Several Dimensions

$$\Delta u = \nabla \cdot \nabla u$$

$$\frac{\partial u}{\partial t} - \Delta u = f$$

$$\frac{\partial^2 u}{\partial t^2} - \Delta u = f$$

$$i\frac{\partial u}{\partial t} = -\frac{1}{2}\Delta u + Vu$$

$$\frac{\partial u}{\partial t} + u \cdot \nabla u - \nu\Delta u = f, \quad \nabla \cdot u = 0$$

$$\nabla \times H = J, \quad \nabla \cdot B = 0, \quad B = \mu H.$$

54

Vector-Valued Functions of Several Real Variables

Auch die Chemiker müssen sich allmählich an den Gedanken gewöhnen, dass ihnen die theoretische Chemie ohne die Beherrschung der Elemente der höheren Analysis ein Buch mit sieben Siegeln bleiben wird. Ein Differential- oder Integralzeichen muss aufhören, für den Chemiker eine unverständliche Hieroglyphe zu sein,... wenn er sich nicht der Gefahr aussetzen will, für die Entwicklung der theoretischen Chemie jedes Verständnis zu verlieren. (H. Jahn, Grundriss der Elektrochemie, 1895)

54.1 Introduction

We now turn to the extension of the basic concepts of real-valued functions of one real variable, such as Lipschitz continuity and differentiability, to vector-valued functions of several variables. We have carefully prepared the material so that this extension will be as natural and smooth as possible. We shall see that the proofs of the basic theorems like the Chain rule, the Mean Value theorem, Taylor's theorem, the Contraction Mapping theorem and the Inverse Function theorem, extend almost word by word to the more complicated situation of vector valued functions of several real variables.

We consider functions $f : \mathbb{R}^n \to \mathbb{R}^m$ that are vector valued in the sense that the value $f(x) = (f_1(x), \ldots, f_m(x))$ is a vector in $\mathbb{R}^m$ with components $f_i : \mathbb{R}^n \to \mathbb{R}$ for $i = 1, \ldots, m$, where with $f_i(x) = f_i(x_1, \ldots, x_n)$ and $x = (x_1, \ldots, x_n) \in \mathbb{R}^n$. As usual, we view $x = (x_1, \ldots, x_n)$ as a n-column vector and $f(x) = (f_1(x), \ldots, f_m(x))$ as a m-column vector.

As particular examples of vector-valued functions, we first consider *curves*, which are functions $g : \mathbb{R} \to \mathbb{R}^n$, and *surfaces*, which are functions $g : \mathbb{R}^2 \to \mathbb{R}^n$. We then discuss composite functions $f \circ g : \mathbb{R} \to \mathbb{R}^m$, where $g : \mathbb{R} \to \mathbb{R}^n$ is a curve and $f : \mathbb{R}^n \to \mathbb{R}^m$, with $f \circ g$ again being a curve. We recall that $f \circ g(t) = f(g(t))$.

The inputs to the functions reside in the n dimensional vector space $\mathbb{R}^n$ and it is worthwhile to consider the properties of $\mathbb{R}^n$. Of particular importance is the notion of Cauchy sequence and convergence for sequences $\{x^{(j)}\}_{j=1}^\infty$ of vectors $x^{(j)} = (x_1^{(j)}, \ldots, x_n^{(j)}) \in \mathbb{R}^n$ with coordinates $x_k^{(j)}$, $k = 1, \ldots, n$. We say that the sequence $\{x^{(j)}\}_{j=1}^\infty$ is a *Cauchy sequence* if for all $\epsilon > 0$, there is a natural number N so that

$$\|x^{(i)} - x^{(j)}\| \leq \epsilon \quad \text{for } i, j > N.$$

Here $\| \cdot \|$ denotes the Euclidean norm in $\mathbb{R}^n$, that is, $\|x\| = (\sum_{i=1}^n x_i^2)^{1/2}$. Sometimes, it is convenient to work with the norms $\|x\|_1 = \sum_{i=1}^n |x_i|$ or $\|x\|_\infty = \max_{i=1,\ldots,n} |x_i|$. We say that the sequence $\{x^{(j)}\}_{j=1}^\infty$ of vectors in $\mathbb{R}^n$ *converges* to $x \in \mathbb{R}^n$ if for all $\epsilon > 0$, there is a natural number N so that

$$\|x - x^{(i)}\| \leq \epsilon \quad \text{for } i > N.$$

It is easy to show that a convergent sequence is a Cauchy sequence and conversely that a Cauchy sequence converges. We obtain these results applying the corresponding results for sequences in $\mathbb{R}$ to each of the coordinates of the vectors in $\mathbb{R}^n$.

Example 54.1. The sequence $\{x^{(i)}\}_{i=1}^\infty$ in $\mathbb{R}^2$, $x^{(i)} = (1 - i^{-2}, \exp(-i))$, converges to $(1, 0)$.

54.2   Curves in $\mathbb{R}^n$

A function $g : I \to \mathbb{R}^n$, where $I = [a, b]$ is an interval of real numbers, is a *curve* in $\mathbb{R}^n$, see Fig. 54.1. If we use t as the independent variable ranging over I, then we say that the curve $g(t)$ is *parametrized* by the variable t. We also refer to the set of points $\Gamma = \{g(t) \in \mathbb{R}^n : t \in I\}$ as the curve Γ parameterized by the function $g : I \to \mathbb{R}^n$.

Example 54.2. The simplest example of a curve is a straight line. The function $g : \mathbb{R} \to \mathbb{R}^2$ given by

$$g(t) = \bar{x} + tz,$$

where $z \in \mathbb{R}^2$ and $\bar{x} \in \mathbb{R}^2$, is a straight line in $\mathbb{R}^2$ through the point $\bar{x}$ with direction z, see Fig. 54.2.

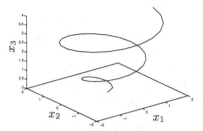

Fig. 54.1. The curve $g : [0, 4] \to \mathbb{R}^3$ with $g(t) = \left(t^{1/2} \cos(\pi t),\, t^{1/2} \sin(\pi t), t\right)$

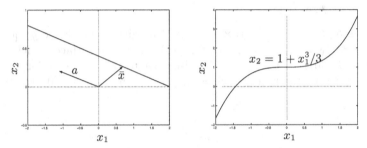

Fig. 54.2. On the *left*: the curve $g(t) = \bar{x} + ta$. On the *right*: a curve $g(t) = (t, f(t))$

Example 54.3. Let $f : [a, b] \to \mathbb{R}$ be given, and define $g : [a, b] \to \mathbb{R}^2$ by $g(t) = (g_1(t), g_2(t)) = (t, f(t))$. This curve is simply the graph of the function $f : [a, b] \to \mathbb{R}$, see Fig. 54.2.

54.3 Different Parameterizations of a Curve

It is possible to use different parametrizations for the set of points forming a curve. If $h : [c, d] \to [a, b]$ is a one-to-one mapping, then the composite function $f = g \circ h : [c, d] \to \mathbb{R}^2$ is a *reparameterization* of the curve $\{g(t) : t \in [a, b]\}$ given by $g : [a, b] \to \mathbb{R}^2$.

Example 54.4. The function $f : [0, \infty) \to \mathbb{R}^3$ given by

$$f(\tau) = (\tau \cos(\pi \tau^2), \tau \sin(\pi \tau^2), \tau^2),$$

is a reparameterization of the curve $g : [0, \infty) \to \mathbb{R}^3$ given by

$$g(t) = (\sqrt{t} \cos(\pi t), \sqrt{t} \sin(\pi t), t),$$

obtained setting $t = h(\tau) = \tau^2$. We have $f = g \circ h$.

54.4   Surfaces in $\mathbb{R}^n$, $n \geq 3$

A function $g : Q \to \mathbb{R}^n$, where $n \geq 3$ and Q is a subdomain of $\mathbb{R}^2$, may be viewed to be a *surface* S in $\mathbb{R}^n$, see Fig. 54.3. We write $g = g(y)$ with $y = (y_1, y_2) \in Q$ and say that S is parameterized by $y \in Q$. We may also identify the surface S with the set of points $S = \{g(y) \in \mathbb{R}^n : y \in Q\}$, and reparameterize S by $f = g \circ h : \tilde{Q} \to \mathbb{R}^n$ if $h : \tilde{Q} \to Q$ is a one-to-one mapping of a domain $\tilde{Q}$ in $\mathbb{R}^2$ onto Q.

Example 54.5. The simplest example of a surface $g : \mathbb{R}^2 \to \mathbb{R}^3$ is a plane in $\mathbb{R}^3$ given by

$$g(y) = g(y_1, y_2) = \bar{x} + y_1 b_1 + y_2 b_2, \quad y \in \mathbb{R}^2,$$

where $\bar{x}$, b_1, $b_2 \in \mathbb{R}^3$.

Example 54.6. Let $f : [0, 1] \times [0, 1] \to \mathbb{R}$ be given, and define $g : [0, 1] \times [0, 1] \to \mathbb{R}^3$ by $g(y_1, y_2) = (y_1, y_2, f(y_1, y_2))$. This is a surface, which is the graph of $f : [0, 1] \times [0, 1] \to \mathbb{R}$. We also refer to this surface briefly as the surface given by the function $x_3 = f(x_1, x_2)$ with $(x_1, x_2) \in [0, 1] \times [0, 1]$.

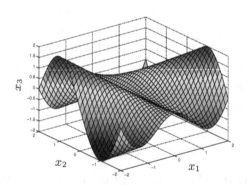

Fig. 54.3. The surface $s(y_1, y_2) = (y_1, y_2, y_1 \sin((y_1 + y_2)\pi/2))$ with $-1 \leq y_1, y_2 \leq 1$, or briefly the surface $x_3 = x_1 \sin((x_1 + x_2)\pi/2)$ with $-1 \leq x_1, x_2 \leq 1$

54.5   Lipschitz Continuity

We say that $f : \mathbb{R}^n \to \mathbb{R}^m$ is Lipschitz continuous on $\mathbb{R}^n$ if there is a constant L such that

$$\|f(x) - f(y)\| \leq L\|x - y\| \quad \text{for all } x, y \in \mathbb{R}^n. \tag{54.1}$$

This definition extends easily to functions $f : A \to \mathbb{R}^m$ with the domain $D(f) = A$ being a subset of $\mathbb{R}^n$. For example, A may be the unit n-cube $[0,1]^n = \{x \in \mathbb{R}^n : 0 \le x_i \le 1, i = 1, \dots, n\}$ or the unit n-disc $\{x \in \mathbb{R}^n : \|x\| \le 1\}$.

To check if a function $f : A \to \mathbb{R}^m$ is Lipschitz continuous on some subset A of $\mathbb{R}^n$, it suffices to check that the component functions $f_i : A \to \mathbb{R}$ are Lipschitz continuous. This is because

$$|f_i(x) - f_i(y)| \le L_i \|x - y\| \quad \text{for } i = 1, \dots, m,$$

implies

$$\|f(x) - f(y)\|^2 = \sum_{i=1}^m |f_i(x) - f_i(y)|^2 \le \sum_{i=1}^m L_i^2 \|x - y\|^2,$$

which shows that $\|f(x) - f(y)\| \le L\|x - y\|$ with $L = (\sum_i L_i^2)^{\frac{1}{2}}$.

Example 54.7. The function $f : [0,1] \times [0,1] \to \mathbb{R}^2$ defined by $f(x_1, x_2) = (x_1 + x_2, x_1 x_2)$, is Lipschitz continuous with Lipschitz constant $L = 2$. To show this, we note that $f_1(x_1, x_2) = x_1 + x_2$ is Lipschitz continuous on $[0,1] \times [0,1]$ with Lipschitz constant $L_1 = \sqrt{2}$ because $|f_1(x_1, x_2) - f_1(y_1, y_2)| \le |x_1 - y_1| + |x_2 - y_2| \le \sqrt{2}\|x - y\|$ by Cauchy's inequality. Similarly, $f_2(x_1, x_2) = x_1 x_2$ is Lipschitz continuous on $[0,1] \times [0,1]$ with Lipschitz constant $L_2 = \sqrt{2}$ since $|x_1 x_2 - y_1 y_2| = |x_1 x_2 - y_1 x_2 + y_1 x_2 - y_1 y_2| \le |x_1 - y_1| + |x_2 - y_2| \le \sqrt{2}\|x - y\|$.

Example 54.8. The function $f : \mathbb{R}^n \to \mathbb{R}^n$ defined by

$$f(x_1, \dots, x_n) = (x_n, x_{n-1}, \dots, x_1),$$

is Lipschitz continuous with Lipschitz constant $L = 1$.

Example 54.9. A linear transformation $f : \mathbb{R}^n \to \mathbb{R}^m$ given by an $m \times n$ matrix $A = (a_{ij})$, with $f(x) = Ax$ and x a n-column vector, is Lipschitz continuous with Lipschitz constant $L = \|A\|$. We made this observation in Chapter *Analytic geometry in $\mathbb{R}^n$*. We repeat the argument:

$$L = \max_{x \ne y} \frac{\|f(x) - f(y)\|}{\|x - y\|} = \max_{x \ne y} \frac{\|Ax - Ay\|}{\|x - y\|}$$

$$= \max_{x \ne y} \frac{\|A(x - y)\|}{\|x - y\|} = \max_{x \ne 0} \frac{\|Ax\|}{\|x\|} = \|A\|.$$

Concerning the definition of the matrix norm $\|A\|$, we note that the function $F(x) = \|Ax\|/\|x\|$ is homogeneous of degree zero, that is, $F(\lambda x) = F(x)$ for all non-zero real numbers λ, and thus $\|A\|$ is the maximum value of $F(x)$ on the closed and bounded set $\{x \in \mathbb{R}^n : \|x\| = 1\}$, which is a finite real number.

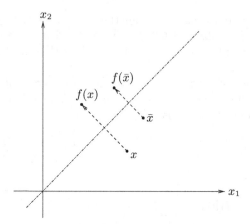

Fig. 54.4. Illustration of the mapping $f(x_1, x_2) = (x_2, x_1)$, which is clearly Lipschitz continuous with $L = 1$

We recall that if A is a diagonal $n \times n$ matrix with diagonal elements λ_i, then $\|A\| = \max_i |\lambda_i|$.

54.6   Differentiability: Jacobian, Gradient and Tangent

We say that $f : \mathbb{R}^n \rightarrow \mathbb{R}^m$ is *differentiable at* $\bar{x} \in \mathbb{R}^n$ if there is a $m \times n$ matrix $M(\bar{x}) = (m_{ij}(\bar{x}))$, called the *Jacobian* of the function $f(x)$ at $\bar{x}$, and a constant $K_f(\bar{x})$ such that for all x close to $\bar{x}$,

$$f(x) = f(\bar{x}) + M(\bar{x})(x - \bar{x}) + E_f(x, \bar{x}), \qquad (54.2)$$

where $E_f(x, \bar{x}) = (E_f(x, \bar{x})_i)$ is an m-vector satisfying $\|E_f(x, \bar{x})\| \leq K_f(\bar{x})\|x - \bar{x}\|^2$. We also denote the Jacobian by $Df(\bar{x})$ or $f'(\bar{x})$ so that $M(\bar{x}) = Df(\bar{x}) = f'(\bar{x})$. Since $f(x)$ is a m-column vector, or $m \times 1$ matrix, and x is a n-column vector, or $n \times 1$ matrix, $M(\bar{x})(x-\bar{x})$ is the product of the $m \times n$ matrix $M(\bar{x})$ and the $n \times 1$ matrix $x - \bar{x}$ yielding a $m \times 1$ matrix or a m-column vector.

We say that $f : A \rightarrow \mathbb{R}^m$, where A is a subset of $\mathbb{R}^n$, is *differentiable on* A if $f(x)$ is differentiable at $\bar{x}$ for all $\bar{x} \in A$. We say that $f : A \rightarrow \mathbb{R}^m$ is *uniformly differentiable on* A if the constant $K_f(\bar{x}) = K_f$ can be chosen independently of $\bar{x} \in A$.

We now show how to determine a specific element $m_{ij}(\bar{x})$ of the Jacobian using the relation (54.2). We consider the coordinate function $f_i(x_1, \ldots, x_n)$ and setting $x = \bar{x} + se_j$, where e_j is the jth standard basis vector and s is a small real number, we focus on the variation of $f_i(x_1, \ldots, x_n)$ as the

Fig. 54.5. Carl Jacobi (1804–51): "It is often more convenient to possess the ashes of great men than to possess the men themselves during their lifetime" (on the return of Descarte's remains to France)

variable x_j varies in a neighborhood of $\bar{x}_j$. The relation (54.2) states that for small non-zero real numbers s,

$$f_i(\bar{x} + se_j) = f_i(\bar{x}) + m_{ij}(\bar{x})s + E_f(\bar{x} + se_j, \bar{x})_i, \qquad (54.3)$$

where $\|x - \bar{x}\|^2 = \|se_j\|^2 = s^2$ implies

$$|E_f(\bar{x} + se_j, \bar{x})_i| \le K_f(\bar{x})s^2.$$

Note that by assumption $\|E_f(x, \bar{x})\| \le K_f(\bar{x})\|x - \bar{x}\|^2$, and so each coordinate function $E_f(\bar{x} + se_j, \bar{x})_i$ satisfies $|E_f(x, \bar{x})_i| \le K_f(\bar{x})\|x - \bar{x}\|^2$.

Now, dividing by s in (54.3) and letting s tend to zero, we find that

$$m_{ij}(\bar{x}) = \lim_{s \to 0} \frac{f_i(\bar{x} + se_j) - f_i(\bar{x})}{s}, \qquad (54.4)$$

which we can also write as

$$m_{ij}(\bar{x}) = \qquad\qquad\qquad\qquad\qquad\qquad\qquad\qquad\qquad (54.5)$$
$$\lim_{x_j \to \bar{x}_j} \frac{f_i(\bar{x}_1, \ldots, \bar{x}_{j-1}, x_j, \bar{x}_{j+1}, \ldots, \bar{x}_n) - f_i(\bar{x}_1, \ldots, \bar{x}_{j-1}, \bar{x}_j, \bar{x}_{j+1}, \ldots, \bar{x}_n)}{x_j - \bar{x}_j}.$$

We refer to $m_{ij}(\bar{x})$ as the *partial derivative* of f_i with respect to x_j at $\bar{x}$, and we use the alternative notation $m_{ij}(\bar{x}) = \frac{\partial f_i}{\partial x_j}(\bar{x})$. To compute $\frac{\partial f_i}{\partial x_j}(\bar{x})$ we freeze all coordinates at $\bar{x}$ but the coordinate x_j and then let x_j vary in a neighborhood of $\bar{x}_j$. The formula

$$\frac{\partial f_i}{\partial x_j}(\bar{x}) = \tag{54.6}$$

$$\lim_{x_j \to \bar{x}_j} \frac{f_i(\bar{x}_1, \ldots, \bar{x}_{j-1}, x_j, \bar{x}_{j+1}, \ldots, \bar{x}_n) - f_i(\bar{x}_1, \ldots, \bar{x}_{j-1}, \bar{x}_j, \bar{x}_{j+1}, \ldots, \bar{x}_n)}{x_j - \bar{x}_j},$$

states that we compute the partial derivative with respect to the variable x_j by keeping all the other variables $x_1, \ldots, x_{j-1}, x_{j+1}, \ldots, x_n$ constant. Thus, computing partial derivatives should be a pleasure using our previous expertise of computing derivatives of functions of one real variable!

We may express the computation alternatively as follows:

$$\frac{\partial f_i}{\partial x_j}(\bar{x}) = m_{ij}(\bar{x}) = g'_{ij}(0) = \frac{dg_{ij}}{ds}(0), \tag{54.7}$$

where $g_{ij}(s) = f_i(\bar{x} + se_j)$.

Example 54.10. Let $f : \mathbb{R}^3 \to \mathbb{R}$ be given by $f(x_1, x_2, x_3) = x_1 e^{x_2} \sin(x_3)$. We compute

$$\frac{\partial f}{\partial x_1}(\bar{x}) = e^{\bar{x}_2} \sin(\bar{x}_3), \quad \frac{\partial f}{\partial x_2}(\bar{x}) = \bar{x}_1 e^{\bar{x}_2} \sin(\bar{x}_3),$$

$$\frac{\partial f}{\partial x_3}(\bar{x}) = \bar{x}_1 e^{\bar{x}_2} \cos(\bar{x}_3),$$

and thus

$$f'(\bar{x}) = (e^{\bar{x}_2} \sin(\bar{x}_3), \bar{x}_1 e^{\bar{x}_2} \sin(\bar{x}_3), \bar{x}_1 e^{\bar{x}_2} \cos(\bar{x}_3))$$

Example 54.11. If $f : \mathbb{R}^3 \to \mathbb{R}^2$ is given by $f(x) = \begin{pmatrix} \exp(x_1^2 + x_2^2) \\ \sin(x_2 + 2x_3) \end{pmatrix}$, then

$$f'(x) = \begin{pmatrix} 2x_1 \exp(x_1^2 + x_2^2) & 2x_2 \exp(x_1^2 + x_2^2) & 0 \\ 0 & \cos(x_2 + 2x_3) & 2\cos(x_2 + 2x_3) \end{pmatrix}.$$

We have now shown how to compute the elements of a Jacobian using the usual rules for differentiation with respect to one real variable. This opens a whole new world of applications to explore. The setting is thus a differentiable function $f : \mathbb{R}^n \to \mathbb{R}^m$ satisfying for suitable $x, \bar{x} \in \mathbb{R}^n$:

$$f(x) = f(\bar{x}) + f'(\bar{x})(x - \bar{x}) + E_f(x, \bar{x}), \tag{54.8}$$

with $\|E_f(x, \bar{x})\| \le K_f(\bar{x})\|x - \bar{x}\|^2$, where $f'(\bar{x}) = Df(\bar{x})$ is the Jacobian $m \times n$ matrix with elements $\frac{\partial f_i}{\partial x_j}$:

$$f'(\bar{x}) = Df(\bar{x}) = \begin{pmatrix} \frac{\partial f_1}{\partial x_1}(\bar{x}) & \frac{\partial f_1}{\partial x_2}(\bar{x}) & \cdots & \frac{\partial f_1}{\partial x_n}(\bar{x}) \\ \frac{\partial f_2}{\partial x_1}(\bar{x}) & \frac{\partial f_2}{\partial x_2}(\bar{x}) & \cdots & \frac{\partial f_2}{\partial x_n}(\bar{x}) \\ \cdots & \cdots & \cdots & \\ \frac{\partial f_m}{\partial x_1}(\bar{x}) & \frac{\partial f_m}{\partial x_2}(\bar{x}) & \cdots & \frac{\partial f_m}{\partial x_n}(\bar{x}) \end{pmatrix}.$$

Sometimes we use the following notation for the Jacobian $f'(x)$ of a function $y = f(x)$ with $f : \mathbb{R}^n \to \mathbb{R}^m$:

$$f'(x) = \frac{dy_1, \dots, dy_m}{dx_1, \dots, dx_n}(x) \tag{54.9}$$

The function $x \to \hat{f}(x) = f(\bar{x}) + f'(\bar{x})(x - \bar{x})$ is called the *linearization* of the function $x \to f(x)$ at $x = \bar{x}$. We have

$$\hat{f}(x) = f'(\bar{x})x + f(\bar{x}) - f'(\bar{x})\bar{x} = Ax + b,$$

with $A = f'(\bar{x})$ a $m \times n$ matrix and $b = f(\bar{x}) - f'(\bar{x})\bar{x}$ a m-column vector. We say that $\hat{f}(x)$ is an *affine transformation*, which is a transformation of the form $x \to Ax + b$, where x is a n-column vector, A is a $m \times n$ matrix and b is a m-column vector. The Jacobian $\hat{f}'(x)$ of the linearization $\hat{f}(x) = Ax + b$ is a constant matrix equal to the matrix A, because the partial derivatives of Ax with respect to x are simply the elements of the matrix A.

If $f : \mathbb{R}^n \to \mathbb{R}$, that is $m = 1$, then we also denote the Jacobian f' by ∇f, that is,

$$f'(\bar{x}) = \nabla f(\bar{x}) = \left(\frac{\partial f}{\partial x_1}(\bar{x}), \dots, \frac{\partial f}{\partial x_n}(\bar{x}) \right).$$

In words, $\nabla f(\bar{x})$ is the n-row vector or $1 \times n$ matrix of partial derivatives of $f(x)$ with respect to $x_1, x_2, \dots, x_n$ at $\bar{x}$. We refer to $\nabla f(\bar{x})$ as the *gradient* of $f(x)$ at $\bar{x}$. If $f : \mathbb{R}^n \to \mathbb{R}$ is differentiable at $\bar{x}$, we thus have

$$f(x) = f(\bar{x}) + \nabla f(\bar{x})(x - \bar{x}) + E_f(x, \bar{x}), \tag{54.10}$$

with $|E_f(x, \bar{x})| \le K_f(\bar{x})\|x - \bar{x}\|^2$, and $\hat{f}(x) = f(\bar{x}) + \nabla f(\bar{x})(x - \bar{x})$ is the linearization of $f(x)$ at $x = \bar{x}$. We may alternatively express the product $\nabla f(\bar{x})(x - \bar{x})$ of the n-row vector ($1 \times n$ matrix) $\nabla f(\bar{x})$ with the n-column vector ($n \times 1$ matrix) $(x - \bar{x})$ as the scalar product $\nabla f(\bar{x}) \cdot (x - \bar{x})$ of the n-vector $\nabla f(\bar{x})$ with the n-vector $(x - \bar{x})$. We thus often write (54.10) in the form

$$f(x) = f(\bar{x}) + \nabla f(\bar{x}) \cdot (x - \bar{x}) + E_f(x, \bar{x}). \tag{54.11}$$

Example 54.12. If $f : \mathbb{R}^3 \to \mathbb{R}$ is given by $f(x) = x_1^2 + 2x_2^3 + 3x_3^4$, then

$$\nabla f(x) = (2x_1, 6x_2^2, 12x_3^3).$$

Example 54.13. The equation $x_3 = f(x)$ with $f : \mathbb{R}^2 \to \mathbb{R}$ and $x = (x_1, x_2)$ represents a surface in $\mathbb{R}^3$ (the graph of the function f). The linearization

$$x_3 = f(\bar{x}) + \nabla f(\bar{x}) \cdot (x - \bar{x})$$

$$= f(\bar{x}) + \frac{\partial f}{\partial x_1}(\bar{x})(x_1 - \bar{x}_1) + \frac{\partial f}{\partial x_2}(\bar{x})(x_2 - \bar{x}_2)$$

with $\bar{x} = (\bar{x}_1, \bar{x}_2)$, represents the *tangent plane* at $x = \bar{x}$, see Fig. 54.6.

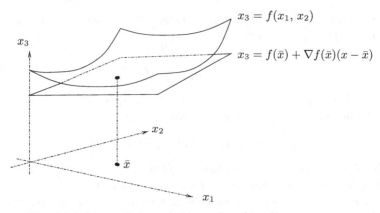

Fig. 54.6. The surface $x_3 = f(x_1, x_2)$ and its tangent plane

Example 54.14. Consider now a curve $f : \mathbb{R} \to \mathbb{R}^m$, that is, $f(t) = (f_1(t), \ldots, f_m(t))$ with $t \in \mathbb{R}$ and we have a situation with $n = 1$. The linearization $t \to \hat{f}(t) = f(\bar{t}) + f'(\bar{t})(t - \bar{t})$ at $\bar{t}$ represents a straight line in $\mathbb{R}^m$ through the point $f(\bar{t})$ and the Jacobian $f'(\bar{t}) = (f_1'(\bar{t}), \ldots, f_m'(\bar{t}))$ gives the direction of the *tangent* to the curve $f : \mathbb{R} \to \mathbb{R}^m$ at $f(\bar{t})$, see Fig. 54.7.

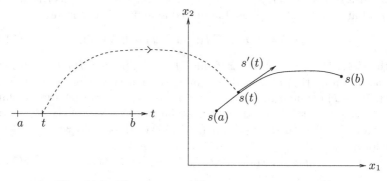

Fig. 54.7. The tangent $s'(t)$ to a curve given by $s(t)$

54.7 The Chain Rule

Let $g : \mathbb{R}^n \to \mathbb{R}^m$ and $f : \mathbb{R}^m \to \mathbb{R}^p$ and consider the composite function $f \circ g : \mathbb{R}^n \to \mathbb{R}^p$ defined by $f \circ g(x) = f(g(x))$. Under suitable assumptions of differentiability and Lipschitz continuity, we shall prove a *Chain rule* generalizing the Chain rule of Chapter *Differentiation rules* in the case

$n = m = p = 1$. Using linearizations of f and g, we have

$$f(g(x)) = f(g(\bar{x})) + f'(g(\bar{x}))(g(x) - g(\bar{x})) + E_f(g(x), g(\bar{x}))$$
$$= f(g(\bar{x})) + f'(g(\bar{x}))g'(\bar{x})(x - \bar{x}) + f'(g(\bar{x}))E_g(x, \bar{x}) + E_f(g(x), g(\bar{x})),$$

where we may naturally assume that

$$\|E_f(g(x), g(\bar{x}))\| \leq K_f \|g(x) - g(\bar{x})\|^2 \leq K_f L_g^2 \|x - \bar{x}\|^2,$$

and $\|f'(g(\bar{x}))E_g(x, \bar{x})\| \leq \|f'(g(\bar{x}))\| K_g \|x - \bar{x}\|^2$, with suitable constants of differentiability K_f and K_g and Lipschitz constant L_g. We have now proved:

Theorem 54.1 (The Chain rule) *If $g : \mathbb{R}^n \to \mathbb{R}^m$ is differentiable at $\bar{x} \in \mathbb{R}^n$, and $f : \mathbb{R}^m \to \mathbb{R}^p$ is differentiable at $g(\bar{x}) \in \mathbb{R}^m$ and further $g : \mathbb{R}^n \to \mathbb{R}^m$ is Lipschitz continuous, then the composite function $f \circ g : \mathbb{R}^n \to \mathbb{R}^p$ is differentiable at $\bar{x} \in \mathbb{R}^n$ with Jacobian*

$$(f \circ g)'(\bar{x}) = f'(g(\bar{x}))g'(\bar{x}).$$

The Chain rule has a wealth of applications and we now turn to harvest a couple of the most basic examples.

54.8   The Mean Value Theorem

Let $f : \mathbb{R}^n \to \mathbb{R}$ be differentiable on $\mathbb{R}^n$ with a Lipschitz continuous gradient, and for given $x, \bar{x} \in \mathbb{R}^n$ consider the function $h : \mathbb{R} \to \mathbb{R}$ defined by

$$h(t) = f(\bar{x} + t(x - \bar{x})) = f \circ g(t),$$

with $g(t) = \bar{x} + t(x - \bar{x})$ representing the straight line through $\bar{x}$ and x. We have

$$f(x) - f(\bar{x}) = h(1) - h(0) = h'(\bar{t}),$$

for some $\bar{t} \in [0, 1]$, where we applied the usual Mean Value theorem to the function $h(t)$. By the Chain rule we have

$$h'(t) = \nabla f(g(t)) \cdot g'(t) = \nabla f(g(t)) \cdot (x - \bar{x}),$$

and we have now proved:

Theorem 54.2 (Mean Value theorem) *Let $f : \mathbb{R}^n \to \mathbb{R}$ be differentiable on $\mathbb{R}^n$ with a Lipschitz continuous gradient ∇f. Then for given x and $\bar{x}$ in $\mathbb{R}^n$, there is $y = x + \bar{t}(x - \bar{x})$ with $\bar{t} \in [0, 1]$, such that*

$$f(x) - f(\bar{x}) = \nabla f(y) \cdot (x - \bar{x}).$$

With the help of the Mean Value theorem we express the difference $f(x) - f(\bar{x})$ as the scalar product of the gradient $\nabla f(y)$ with the difference $x - \bar{x}$, where y is a point somewhere on the straight line between x and $\bar{x}$.

We may extend the Mean Value theorem to a function $f : \mathbb{R}^n \to \mathbb{R}^m$ to take the form

$$f(x) - f(\bar{x}) = f'(y)(x - \bar{x}),$$

where y is a point on the straight line between x and $\bar{x}$, which may be different for different rows of $f'(y)$. We may then estimate:

$$\|f(x) - f(\bar{x})\| = \|f'(y) \cdot (x - \bar{x})\| \leq \|f'(y)\|\|x - \bar{x}\|,$$

and we may thus estimate the Lipschitz constant of f by $\max_y \|f'(y)\|$ with $\|f'(y)\|$ the (Euclidean) matrix norm of $f'(y)$.

Example 54.15. Let $f : \mathbb{R}^n \to \mathbb{R}$ be given by $f(x) = \sin(\sum_{j=1}^n x_j)$. We have

$$\frac{\partial f}{\partial x_i}(\bar{x}) = \cos\left(\sum_{j=1}^n \bar{x}_j\right) \quad \text{for } i = 1, \ldots, n,$$

and thus $|\frac{\partial f}{\partial x_i}(\bar{x})| \leq 1$ for $i = 1, \ldots, n$, and therefore

$$\|\nabla f(\bar{x})\| \leq \sqrt{n}.$$

We conclude that $f(x) = \sin(\sum_{j=1}^n x_j)$ is Lipschitz continuous with Lipschitz constant $\sqrt{n}$.

54.9   Direction of Steepest Descent and the Gradient

Let $f : \mathbb{R}^n \to \mathbb{R}$ be a given function and suppose we want to study the variation of $f(x)$ in a neighborhood of a given point $\bar{x} \in \mathbb{R}^n$. More precisely, let x vary on the line through $\bar{x}$ in a given direction $z \in \mathbb{R}^n$, that is assume that $x = \bar{x} + tz$ where t is a real variable varying in a neighborhood of 0. Assuming f to be differentiable, the linearization formula (54.8) implies

$$f(x) = f(\bar{x}) + t\nabla f(\bar{x}) \cdot z + E_f(x, \bar{x}), \tag{54.12}$$

where $|E_f(x, \bar{x})| \leq t^2 K_f \|z\|^2$ and $\nabla f(\bar{x}) \cdot z$ is the scalar product of the gradient $\nabla f(\bar{x}) \in \mathbb{R}^n$ and the vector $z \in \mathbb{R}^n$. If $\nabla f(\bar{x}) \cdot z \neq 0$, then the linear term $t\nabla f(\bar{x}) \cdot z$ will dominate the quadratic term $E_f(x, \bar{x})$ for small t. So the linearization

$$\hat{f}(x) = f(\bar{x}) + t\nabla f(\bar{x}) \cdot z$$

will be a good approximation of $f(x)$ for $x = \bar{x} + tz$ close to $\bar{x}$. Thus if $\nabla f(\bar{x}) \cdot z \neq 0$, then we get good information on the variation of $f(x)$ along the line $x = \bar{x} + tz$ by studying the linear function $t \to f(\bar{x}) + t\nabla f(\bar{x}) \cdot z$ with slope $\nabla f(\bar{x}) \cdot z$. In particular, if $\nabla f(\bar{x}) \cdot z > 0$ and $x = \bar{x} + tz$ then $\hat{f}(x)$ increases as we increase t and decreases as we decrease t. Similarly, if $\nabla f(\bar{x}) \cdot z < 0$ and $x = \bar{x} + tz$ then $\hat{f}(x)$ decreases as we increase t and increases as we decrease t.

Alternatively, we may consider the composite function $F_z : \mathbb{R} \to \mathbb{R}$ defined by $F_z(t) = f(g_z(t))$ with $g_z : \mathbb{R} \to \mathbb{R}^n$ given by $g_z(t) = \bar{x} + tz$. Obviously, $F_z(t)$ describes the variation of $f(x)$ on the straight line through $\bar{x}$ with direction z, with $F_z(0) = f(\bar{x})$. Of course, the derivative $F_z'(0)$ gives important information on this variation close to $\bar{x}$. By the Chain rule we have

$$F_z'(0) = \nabla f(\bar{x})z = \nabla f(\bar{x}) \cdot z,$$

and we retrieve $\nabla f(\bar{x}) \cdot z$ as a quantity of interest. In particular, the sign of $\nabla f(\bar{x}) \cdot z$ determines if $F_z(t)$ is increasing or decreasing at $t = 0$.

We may now ask how to choose the direction z to get maximal increase or decrease. We assume $\nabla f(\bar{x}) \neq 0$ to avoid the trivial case with $F_z'(0) = 0$ for all z. It is then natural to normalize z so $\|z\| = 1$ and we study the quantity $F_z'(0) = \nabla f(\bar{x}) \cdot z$ as we vary $z \in \mathbb{R}^n$ with $\|z\| = 1$. We conclude that the scalar product $\nabla f(\bar{x}) \cdot z$ is maximized if we choose z in the direction of the gradient $\nabla f(\bar{x})$,

$$z = \frac{\nabla f(\bar{x})}{\|\nabla f(\bar{x})\|},$$

which is called the direction of *steepest ascent*. For this gives

$$\max_{\|z\|=1} F_z'(0) = \nabla f(\bar{x}) \cdot \frac{\nabla f(\bar{x})}{\|\nabla f(\bar{x})\|} = \|\nabla f(\bar{x})\|.$$

Similarly, the scalar product is minimized if we choose z in the opposite direction of the gradient $\nabla f(\bar{x})$,

$$z = -\frac{\nabla f(\bar{x})}{\|\nabla f(\bar{x})\|},$$

which is called the direction of *steepest descent*, see Fig. 54.8. For then

$$\min_{\|z\|=1} F_z'(0) = -\nabla f(\bar{x}) \cdot \frac{\nabla f(\bar{x})}{\|\nabla f(\bar{x})\|} = -\|\nabla f(\bar{x})\|.$$

If $\nabla f(\bar{x}) = 0$, then $\bar{x}$ is said to be a *stationary point*. If $\bar{x}$ is a stationary point, then evidently $\nabla f(\bar{x}) \cdot z = 0$ for any direction z and

$$f(x) = f(\bar{x}) + E_f(x, \bar{x}).$$

The difference $f(x) - f(\bar{x})$ is then quadratically small in the distance $\|x - \bar{x}\|$, that is $|f(x) - f(\bar{x})| \leq K_f \|x - \bar{x}\|^2$, and $f(x)$ is very close to the constant value $f(\bar{x})$ for x close to $\bar{x}$.

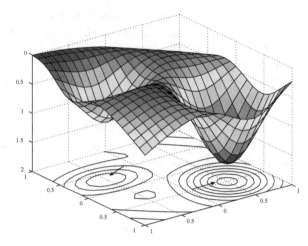

Fig. 54.8. Directions of steepest descent on a "hiking map"

54.10   A Minimum Point Is a Stationary Point

Suppose $\bar{x} \in \mathbb{R}^n$ is a *minimum point* for the function $f : \mathbb{R}^n \to \mathbb{R}$, that is

$$f(x) \geq f(\bar{x}) \quad \text{for } x \in \mathbb{R}^n. \tag{54.13}$$

We shall show that if $f(x)$ is differentiable at a minimum point $\bar{x}$, then

$$\nabla f(\bar{x}) = 0. \tag{54.14}$$

For if $\nabla f(\bar{x}) \neq 0$, then we could move in the direction of steepest descent from $\bar{x}$ to a point x close to $\bar{x}$ with $f(x) < f(\bar{x})$, contradicting (54.13). Consequently, in order to find minimum points of a function $f : \mathbb{R}^n \to \mathbb{R}$, we are led to try to solve the equation $g(x) = 0$, where $g = \nabla f : \mathbb{R}^n \to \mathbb{R}^n$. Here, we interpret $\nabla f(x)$ as a n-column vector.

A whole world of applications in mechanics, physics and other areas may be formulated as solving equations of the form $\nabla f(x) = 0$, that is as finding stationary points. We shall meet many applications below.

54.11   The Method of Steepest Descent

Let $f : \mathbb{R}^n \to \mathbb{R}$ be given and consider the problem of finding a minimum point $\bar{x}$. To do so it is natural to try a *method of Steepest Descent*: Given an approximation $\bar{y}$ of $\bar{x}$ with $\nabla f(\bar{y}) \neq 0$, we move from $\bar{y}$ to a new point y in the direction of steepest descent:

$$y = \bar{y} - \alpha \frac{\nabla f(\bar{y})}{\|\nabla f(\bar{y})\|},$$

where $\alpha > 0$ is a step length to be chosen. We know that $f(y)$ decreases as α increases from 0 and the question is just to find a reasonable value of α. This can be done by increasing α in small steps until $f(y)$ doesn't decrease anymore. The procedure is then repeated with $\bar{y}$ replaced by y. Evidently, the method of Steepest Descent is closely connected to Fixed Point Iteration for solving the equation $\nabla f(x) = 0$ in the form

$$x = x - \alpha \nabla f(x)$$

where we let $\alpha > 0$ include the normalizing factor $1/\|\nabla f(\bar{y})\|$.

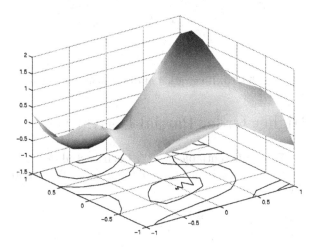

Fig. 54.9. The method of Steepest Descent for $f(x_1, x_2) = x_1 \sin(x_1 + x_2) + x_2 \cos(2x_1 - 3x_2)$ starting at $(.5, .5)$ with $\alpha = .3$

54.12   Directional Derivatives

Consider a function $f : \mathbb{R}^n \to \mathbb{R}$, let $g_z(t) = \bar{x} + tz$ with $z \in \mathbb{R}^n$ a given vector normalized to $\|z\| = 1$, and consider the composite function $F_z(t) = f(\bar{x} + tz)$. The Chain rule implies

$$F_z'(0) = \nabla f(\bar{x}) \cdot z,$$

and

$$\nabla f(\bar{x}) \cdot z$$

is called the *derivative of $f(x)$ in the direction z at $\bar{x}$*, see Fig. 54.10.

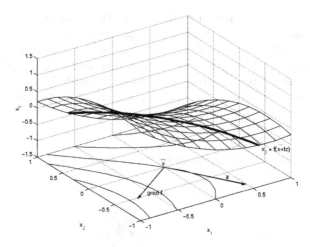

Fig. 54.10. Illustration of directional derivative

54.13 Higher Order Partial Derivatives

Let $f : \mathbb{R}^n \to \mathbb{R}$ be differentiable on $\mathbb{R}^n$. Each partial derivative $\frac{\partial f}{\partial x_i}(\bar{x})$ is a function of $\bar{x} \in \mathbb{R}^n$ may be itself be differentiable. We denote its partial derivatives by

$$\frac{\partial}{\partial x_j}\frac{\partial f}{\partial x_i}(\bar{x}) = \frac{\partial^2 f}{\partial x_j \partial x_i}(\bar{x}), \quad i, j = 1, \ldots, n, \ \bar{x} \in \mathbb{R}^n,$$

which are called the *partial derivatives of f of second order at $\bar{x}$*. It turns out that under appropriate continuity assumptions, the order of differentiation does not matter. In other words, we shall prove that

$$\frac{\partial^2 f}{\partial x_j \partial x_i}(\bar{x}) = \frac{\partial^2 f}{\partial x_i \partial x_j}(\bar{x}).$$

We carry out the proof in the case $n = 2$ with $i = 1$ and $j = 2$. We rewrite the expression

$$A = f(x_1, x_2) - f(\bar{x}_1, x_2) - f(x_1, \bar{x}_2) + f(\bar{x}_1, \bar{x}_2), \qquad (54.15)$$

as

$$A = f(x_1, x_2) - f(x_1, \bar{x}_2) - f(\bar{x}_1, x_2) + f(\bar{x}_1, \bar{x}_2), \qquad (54.16)$$

by shifting the order of the two mid terms. First, we set $F(x_1, x_2) = f(x_1, x_2) - f(\bar{x}_1, x_2)$ and use (54.15) to write

$$A = F(x_1, x_2) - F(x_1, \bar{x}_2).$$

The Mean Value theorem implies

$$A = \frac{\partial F}{\partial x_2}(x_1, y_2)(x_2 - \bar{x}_2) = \left(\frac{\partial f}{\partial x_2}(x_1, y_2) - \frac{\partial f}{\partial x_2}(\bar{x}_1, y_2) \right)(x_2 - \bar{x}_2)$$

for some $y_2 \in [\bar{x}_2, x_2]$. We use the Mean value theorem once again to get

$$A = \frac{\partial^2 f}{\partial x_1 \partial x_2}(y_1, y_2)(x_1 - \bar{x}_1)(x_2 - \bar{x}_2),$$

with $y_1 \in [\bar{x}_1, x_1]$. We next rewrite A using (54.16) in the form

$$A = G(x_1, x_2) - G(\bar{x}_1, x_2),$$

where $G(x_1, x_2) = f(x_1, x_2) - f(x_1, \bar{x}_2)$. Using the Mean Value theorem twice as above, we obtain

$$A = \frac{\partial^2 f}{\partial x_2 \partial x_1}(z_1, z_2)(x_1 - \bar{x}_1)(x_2 - \bar{x}_2),$$

where $z_i \in [\bar{x}_i, x_i]$, $i = 1, 2$. Assuming the second partial derivatives are Lipschitz continuous at $\bar{x}$ and letting x_i tend to $\bar{x}_i$ for $i = 1, 2$ gives

$$\frac{\partial^2 f}{\partial x_1 \partial x_2}(\bar{x}) = \frac{\partial^2 f}{\partial x_2 \partial x_1}(\bar{x}).$$

We have proved the following fundamental result:

Theorem 54.3 *If the partial derivatives of second order of a function $f :$ $\mathbb{R}^n \to \mathbb{R}$ are all Lipschitz continuous, then the order of application of the derivatives of second order is irrelevant.*

The result directly generalizes to higher order partial derivatives: if the derivatives are Lipschitz continuous, then the order of application doesn't matter. What a relief!

54.14   Taylor's Theorem

Suppose $f : \mathbb{R}^n \to \mathbb{R}$ has Lipschitz continuous partial derivatives of order 2. For given $x, \bar{x} \in \mathbb{R}^n$, consider the function $h : \mathbb{R} \to \mathbb{R}$ defined by

$$h(t) = f(\bar{x} + t(x - \bar{x})) = f \circ g(t),$$

where $g(t) = \bar{x} + t(x - \bar{x})$ is the straight line through $\bar{x}$ and x. Clearly $h(1) = f(x)$ and $h(0) = f(\bar{x})$, so the Taylor's theorem applied to $h(t)$ gives

$$h(1) = h(0) + h'(0) + \frac{1}{2}h''(\bar{t}),$$

for some $\bar{t} \in [0, 1]$. We compute using the Chain rule:

$$h'(t) = \nabla f(g(t)) \cdot (x - \bar{x}) = \sum_{i=1}^{n} \frac{\partial f}{\partial x_i}(g(t))(x_i - \bar{x}_i),$$

and similarly by a further differentiation with respect to t:

$$h''(t) = \sum_{i=1}^{n} \sum_{j=1}^{n} \frac{\partial^2 f}{\partial x_i \partial x_j}(g(t))(x_i - \bar{x}_i)(x_j - \bar{x}_j).$$

We thus obtain

$$f(x) = f(\bar{x}) + \nabla f(\bar{x}) \cdot (x - \bar{x}) + \frac{1}{2} \sum_{i,j=1}^{n} \frac{\partial^2 f}{\partial x_i \partial x_j}(y)(x_i - \bar{x}_i)(x_j - \bar{x}_j), \quad (54.17)$$

for some $y = \bar{x} + \bar{t}(x - \bar{x})$ with $t \in [0, 1]$. The $n \times n$ matrix $H(\bar{x}) = (h_{ij}(\bar{x}))$ with elements $h_{ij}(\bar{x}) = \frac{\partial^2 f}{\partial x_i \partial x_j}(\bar{x})$ is called the *Hessian* of $f(x)$ at $x = \bar{x}$. The Hessian is the matrix of all second partial derivatives of $f : \mathbb{R}^n \to \mathbb{R}$. With matrix vector notation with x a n-column vector, we can write

$$\sum_{i,j=1}^{n} \frac{\partial^2 f}{\partial x_i \partial x_j}(y)(x_i - \bar{x}_i)(x_j - \bar{x}_j) = (x - \bar{x})^\top H(y)(x - \bar{x}).$$

We summarize:

Theorem 54.4 (Taylor's theorem) *Let* $f : \mathbb{R}^n \to \mathbb{R}$ *be twice differentiable with Lipschitz continuous Hessian* $H = (h_{ij})$ *with elements* $h_{ij} = \frac{\partial^2 f}{\partial x_i \partial x_j}$. *Then, for given* x *and* $\bar{x} \in \mathbb{R}^n$, *there is* $y = x + \bar{t}(x - \bar{x})$ *with* $\bar{t} \in [0, 1]$, *such that*

$$f(x) = f(\bar{x}) + \nabla f(\bar{x}) \cdot (x - \bar{x}) + \frac{1}{2} \sum_{i,j=1}^{n} \frac{\partial^2 f}{\partial x_i \partial x_j}(y)(x_i - \bar{x}_i)(x_j - \bar{x}_j)$$

$$= f(\bar{x}) + \nabla f(\bar{x}) \cdot (x - \bar{x}) + \frac{1}{2}(x - \bar{x})^\top H(y)(x - \bar{x}).$$

54.15 The Contraction Mapping Theorem

We shall now prove the following generalization of the Contraction Mapping theorem.

Theorem 54.5 *If* $g : \mathbb{R}^n \to \mathbb{R}^n$ *is Lipschitz continuous with Lipschitz constant* $L < 1$, *then the equation* $x = g(x)$ *has a unique solution* $\bar{x} = \lim_{i \to \infty} x^{(i)}$, *where* $\{x^{(i)}\}_{i=1}^{\infty}$ *is a sequence in* $\mathbb{R}^n$ *generated by Fixed Point Iteration:* $x^{(i)} = g(x^{(i-1)})$, $i = 1, 2, \ldots$, *starting with any initial value* $x^{(0)}$.

The proof is word by word the same as in the case $g : \mathbb{R} \to \mathbb{R}$ considered in Chapter *Fixed Points and Contraction Mappings*. We repeat the proof for the convenience of the reader. Subtracting the equation $x^{(k)} = g(x^{(k-1)})$ from $x^{(k+1)} = g(x^{(k)})$, we get

$$x^{(k+1)} - x^{(k)} = g(x^{(k)}) - g(x^{(k-1)}),$$

and using the Lipschitz continuity of g, we thus have

$$\|x^{(k+1)} - x^{(k)}\| \le L\|x^{(k)} - x^{(k-1)}\|.$$

Repeating this estimate, we find that

$$\|x^{(k+1)} - x^{(k)}\| \le L^k\|x^{(1)} - x^{(0)}\|,$$

and thus for $j > i$

$$\|x^{(i)} - x^{(j)}\| \le \sum_{k=i}^{j-1} \|x^{(k)} - x^{(k+1)}\|$$

$$\le \|x^{(1)} - x^{(0)}\| \sum_{k=i}^{j-1} L^k = \|x^{(1)} - x^{(0)}\| L^i \frac{1 - L^{j-i}}{1 - L}.$$

Since $L < 1$, $\{x^{(i)}\}_{i=1}^{\infty}$ is a Cauchy sequence in $\mathbb{R}^n$, and therefore converges to a limit $\bar{x} = \lim_{i \to \infty} x^{(i)}$. Passing to the limit in the equation $x^{(i)} = g(x^{(i-1)})$ shows that $\bar{x} = g(\bar{x})$ and thus $\bar{x}$ is a fixed point of $g : \mathbb{R}^n \to \mathbb{R}^n$. Uniqueness follows from the fact that if $\bar{y} = g(\bar{y})$, then $\|\bar{x} - \bar{y}\| = \|g(\bar{x}) - g(\bar{y})\| \le L\|\bar{x} - \bar{y}\|$ which is impossible unless $\bar{y} = \bar{x}$, because $L < 1$.

Example 54.16. Consider the function $g : \mathbb{R}^2 \to \mathbb{R}^2$ defined by $g(x) = (g_1(x), g_2(x))$ with

$$g_1(x) = \frac{1}{4 + |x_1| + |x_2|}, \quad g_2(x) = \frac{1}{4 + |\sin(x_1)| + |\cos(x_2)|}.$$

We have

$$\left|\frac{\partial g_i}{\partial x_j}\right| \le \frac{1}{16},$$

and thus by simple estimates

$$\|g(x) - g(y)\| \le \frac{1}{4}\|x - y\|,$$

which shows that $g : \mathbb{R}^2 \to \mathbb{R}^2$ is Lipschitz continuous with Lipschitz constant $L_g \le \frac{1}{4}$. The equation $x = g(x)$ thus has a unique solution.

54.16 Solving $f(x) = 0$ with $f : \mathbb{R}^n \to \mathbb{R}^n$

The Contraction Mapping theorem can be applied as follows. Suppose $f : \mathbb{R}^n \to \mathbb{R}^n$ is given and we want to solve the equation $f(x) = 0$. Introduce

$$g(x) = x - Af(x),$$

where A is some non-singular $n \times n$ matrix with constant coefficients to be chosen. The equation $x = g(x)$ is then equivalent to the equation $f(x) = 0$. If $g : \mathbb{R}^n \to \mathbb{R}^n$ is Lipschitz continuous with Lipschitz constant $L < 1$, then $g(x)$ has a unique fixed point $\bar{x}$ and thus $f(\bar{x}) = 0$. We have

$$g'(x) = I - Af'(x),$$

and thus we are led to choose the matrix A so that

$$\|I - Af'(x)\| \leq 1$$

for x close to the root $\bar{x}$. The ideal choice seems to be:

$$A = f'(\bar{x})^{-1},$$

assuming that $f'(\bar{x})$ is non-singular, since then $g'(\bar{x}) = 0$. In applications, we may seek to choose A close to $f'(\bar{x})^{-1}$ with the hope that the corresponding $g'(x) = I - Af'(x)$ will have $\|g'(x)\|$ small for x close to the root $\bar{x}$, leading to a quick convergence. In Newton's method we choose $A = f'(x)^{-1}$, see below.

Example 54.17. Consider the initial value problem $\dot{u}(t) = f(u(t))$ for $t > 0$, $u(0) = u_0$, where $f : \mathbb{R}^n \to \mathbb{R}^n$ is a given Lipschitz continuous function with Lipschitz constant L_f, and as usual $\dot{u} = \frac{du}{dt}$. Consider the backward Euler method

$$U(t_i) = U(t_{i-1}) + k_i f(U(t_i)), \qquad (54.18)$$

where $0 = t_0 < t_1 < t_2 \ldots$ is a sequence of increasing discrete time levels with time steps $k_i = t_i - t_{i-1}$. To determine $U(t_i) \in \mathbb{R}^n$ satisfying (54.18) having already determined $U(t_{i-1})$, we have to solve the nonlinear system of equations

$$V = U(t_{i-1}) + k_i f(V) \qquad (54.19)$$

in the unknown $V \in \mathbb{R}^n$. This equation is of the form $V = g(V)$ with $g(V) = U(t_{i-1}) + k_i f(V)$ and $g : \mathbb{R}^n \to \mathbb{R}^n$.

Therefore, we use the Fixed Point Iteration

$$V^{(m)} = U(t_{i-1}) + k_i f(V^{(m-1)}), \qquad m = 1, 2, \ldots,$$

choosing say $V^{(0)} = U(t_{i-1})$ to try to solve for the new value. If L_f denotes the Lipschitz constant of $f : \mathbb{R}^n \to \mathbb{R}^n$, then

$$\|g(V) - g(W)\| = \|k_i(f(V) - f(W))\| \leq k_i L_f \|V - W\|, \qquad V, W \in \mathbb{R}^n,$$

and thus $g : \mathbb{R}^n \to \mathbb{R}^n$ is Lipschitz continuous with Lipschitz constant $L_g = k_i L_f$. Now $L_g < 1$ if the time step k_i satisfies $k_i < 1/L_f$ and thus the Fixed Point Iteration to determine $U(t_i)$ in (54.18) converges if $k_i < 1/L_f$. This gives a method for numerical solution of a very large class of initial value problems of the form $\dot{u}(t) = f(u(t))$ for $t > 0$, $u(0) = u_0$. The only restriction is to choose sufficiently small time steps, which however can be a severe restriction if the Lipschitz constant L_f is very large in the sense of requiring massive computational work (very small time steps). Thus, caution for large Lipschitz constants L_f!!

54.17 The Inverse Function Theorem

Suppose $f : \mathbb{R}^n \to \mathbb{R}^n$ is a given function and let $\bar{y} = f(\bar{x})$, where $\bar{x} \in \mathbb{R}^n$ is given. We shall prove that if $f'(\bar{x})$ is non-singular, then for $y \in \mathbb{R}^n$ close to $\bar{y}$, the equation

$$f(x) = y \qquad (54.20)$$

has a unique solution x. Thus, we can define x as a function of y for y close to $\bar{y}$, which is called the inverse function $x = f^{-1}(y)$ of $y = f(x)$. To show that (54.20) has a unique solution x for any given y close to $\bar{y}$, we consider the Fixed Point iteration for $x = g(x)$ with $g(x) = x - (f'(\bar{x}))^{-1}(f(x) - y)$, which has the fixed point x satisfying $f(x) = y$ as desired. The iteration is

$$x^{(j)} = x^{(j-1)} - (f'(\bar{x}))^{-1}(f(x^{(j-1)}) - y), \quad j = 1, 2, \dots,$$

with $x^{(0)} = \bar{x}$. To analyze the convergence, we subtract

$$x^{(j-1)} = x^{(j-2)} - (f'(\bar{x}))^{-1}(f(x^{(j-2)}) - y)$$

and write $e^j = x^{(j)} - x^{(j-1)}$ to get

$$e^j = e^{j-1} - (f'(\bar{x}))^{-1}(f(x^{(j-1)}) - f(x^{(j-2)})) \quad \text{for } j = 1, 2, \dots .$$

The Mean Value theorem implies

$$f_i(x^{(j-1)}) - f_i(x^{(j-2)}) = f'(z)e^{j-1},$$

where z lies on the straight line between $x^{(j-1)}$ and $x^{(j-2)}$. Note there might be possibly different z for different rows of $f'(z)$. We conclude that

$$e^j = \left(I - (f'(\bar{x}))^{-1}f'(z)\right)e^{j-1}.$$

Assuming now that

$$\|I - (f'(\bar{x}))^{-1}f'(z)\| \le \theta, \qquad (54.21)$$

where $\theta < 1$ is a positive constant, we have

$$\|e^j\| \le \theta\|e^{j-1}\|.$$

As in the proof of the Contraction Mapping theorem, this shows that the sequence $\{x^{(j)}\}_{j=1}^{\infty}$ is a Cauchy sequence and thus converges to a vector $x \in \mathbb{R}^n$ satisfying $f(x) = y$.

The condition for convergence is obviously (54.21). This condition is satisfied if the coefficients of the Jacobian $f'(x)$ are Lipschitz continuous close to $\bar{x}$ and $f'(\bar{x})$ is non-singular so that $(f'(\bar{x}))^{-1}$ exists, and we restrict y to be sufficiently close to $\bar{y}$.

We summarize in the following (very famous):

Theorem 54.6 (Inverse Function theorem) *Let $f : \mathbb{R}^n \to \mathbb{R}^n$ and assume the coefficients of $f'(x)$ are Lipschitz continuous close to $\bar{x}$ and $f'(\bar{x})$ is non-singular. Then for y sufficiently close to $\bar{y} = f(\bar{x})$, the equation $f(x) = y$ has a unique solution x. This defines x as a function $x = f^{-1}(y)$ of y.*

Carl Jacobi (1804–51), German mathematician, was the first to study the role of the determinant of the Jacobian in the inverse function theorem, and also gave important contributions to many areas of mathematics including the budding theory of first order partial differential equations.

54.18   The Implicit Function Theorem

There is an important generalization of the Inverse Function theorem. Let $f : \mathbb{R}^n \times \mathbb{R}^m \to \mathbb{R}^n$ be a given function with value $f(x, y) \in \mathbb{R}^n$ for $x \in \mathbb{R}^n$ and $y \in \mathbb{R}^m$. Assume that $f(\bar{x}, \bar{y}) = 0$ and consider the equation in $x \in \mathbb{R}^n$,

$$f(x, y) = 0,$$

for $y \in \mathbb{R}^m$ close to $\bar{y}$. In the case of the Inverse Function theorem we considered a special case of this situation with $f : \mathbb{R}^n \times \mathbb{R} \to \mathbb{R}^n$ defined by $f(x, y) = g(x) - y$ with $g : \mathbb{R}^n \to \mathbb{R}^n$.

We define the Jacobian $f'_x(x, y)$ of $f(x, y)$ with respect to x at (x, y) to be the $n \times n$ matrix with elements

$$\frac{\partial f_i}{\partial x_j}(x, y).$$

Assuming now that $f'_x(\bar{x}, \bar{y})$ is non-singular, we consider the Fixed Point iteration:

$$x^{(j)} = x^{(j-1)} - (f'_x(\bar{x}, \bar{y}))^{-1} f(x^{(j-1)}, y),$$

connected to solving the equation $f(x, y) = 0$. Arguing as above, we can show this iteration generates a sequence $\{x^{(j)}\}_{j=1}\infty$ that converges to $x \in \mathbb{R}^n$ satisfying $f(x, y) = 0$ assuming $f'_x(x, y)$ is Lipschitz continuous for x close to $\bar{x}$ and y close to $\bar{y}$. This defines x as a function $g(y)$ of y for y close to $\bar{y}$. We have now proved the (also very famous):

Theorem 54.7 (Implicit Function theorem) *Let* $f : \mathbb{R}^n \times \mathbb{R}^m \to \mathbb{R}^n$ *with* $f(x, y) \in \mathbb{R}^n$ *and* $x \in \mathbb{R}^n$ *and* $y \in \mathbb{R}^m$, *and assume that* $f(\bar{x}, \bar{y}) = 0$. *Assume that the Jacobian* $f'_x(x, y)$ *with respect to* x *is Lipschitz continuous for* x *close to* $\bar{x}$ *and* y *close to* $\bar{y}$, *and that* $f'_x(\bar{x}, \bar{y})$ *is non-singular. Then for* y *close to* $\bar{y}$, *the equation* $f(x, y) = 0$ *has a unique solution* $x = g(y)$. *This defines* x *as a function* $g(y)$ *of* y.

54.19   Newton's Method

We next turn to *Newton's method* for solving an equation $f(x) = 0$ with $f : \mathbb{R}^n \to \mathbb{R}^n$, which reads:

$$x^{(i+1)} = x^{(i)} - f'(x^{(i)})^{-1} f(x^{(i)}), \quad \text{for } i = 0, 1, 2, \ldots, \tag{54.22}$$

where $x^{(0)}$ is an initial approximation. Newton's method corresponds to Fixed Point iteration for $x = g(x)$ with $g(x) = x - f'(x)^{-1} f(x)$. We shall prove that Newton's method converges quadratically close to a root $\bar{x}$ when $f'(\bar{x})$ is non-singular. The argument is the same is as in the case $n = 1$ considered above. Setting $e^i = \bar{x} - x^{(i)}$, and using $\bar{x} = \bar{x} - f'(x^{(i)})^{-1} f(\bar{x})$ if $f(\bar{x}) = 0$, we have

$$\bar{x} - x^{(i+1)} = \bar{x} - x^{(i)} - f'(x^{(i)})^{-1} (f(\bar{x}) - f(x^{(i)}))$$
$$= \bar{x} - x^{(i)} - f'(x^{(i)})^{-1} (f'(x^{(i)}) + E_f(x^{(i)}, \bar{x})) = f'(x^{(i)})^{-1} E_f(x^{(i)}, \bar{x}).$$

We conclude that

$$\|\bar{x} - x^{(i+1)}\| \leq C \|\bar{x} - x^{(i)}\|^2$$

provided

$$\cdot \ \|f'(x^{(i)})^{-1}\| \leq C,$$

where C is some positive constant. We have proved the following fundamental result:

Theorem 54.8 (Newton's method) *If* $\bar{x}$ *is a root of* $f : \mathbb{R}^n \to \mathbb{R}^n$ *such that* $f(x)$ *is uniformly differentiable with a Lipschitz continuous derivative close to* $\bar{x}$ *and* $f'(\bar{x})$ *is non-singular, then Newton's method for solving* $f(x) = 0$ *converges quadratically if started sufficiently close to* $\bar{x}$.

In concrete implementations of Newton's method we may rewrite (54.22) as

$$f'(x^{(i)}) z = -f(x^{(i)}),$$
$$x^{(i+1)} = x^{(i)} + z,$$

where $f'(x^{(i)}) z = -f(x^{(i)})$ is a system of equations in z that is solved by Gaussian elimination or by some iterative method.

Example 54.18. We return to the equation (54.19), that is,

$$h(V) = V - k_i f(V) - U(t_{i-1}) = 0.$$

To apply Newton's method to solve the equation $h(V) = 0$, we compute

$$h'(v) = I - k_i f'(v),$$

and conclude that $h'(v)$ will be non-singular at v, if $k_i < \|f'(v)\|^{-1}$. We conclude that Newton's method converges if k_i is sufficiently small and we start close to the root. Again the restriction on the time step is connected to the Lipschitz constant L_f of f, since L_f reflects the size of $\|f'(v)\|$.

54.20   Differentiation Under the Integral Sign

Finally, we show that if the limits of integration of an integral are independent of a variable x_1, then the operation of taking the partial derivative with respect x_1 can be moved past the integral sign. Let then $f : \mathbb{R}^2 \to \mathbb{R}$ be a function of two real variables x_1 and x_2 and consider the integral

$$\int_0^1 f(x_1, x_2)\, dx_2 = g(x_1),$$

which is a function $g(x_1)$ of x_1. We shall now prove that

$$\frac{dg}{dx_1}(\bar{x}_1) = \int_0^1 \frac{\partial f}{\partial x_1}(\bar{x}_1, x_2)\, dx_2, \qquad (54.23)$$

which is referred to as "differentiation under the integral sign". The proof starts by writing

$$f(x_1, x_2) = f(\bar{x}_1, x_2) + \frac{\partial f}{\partial x_1}(\bar{x}_1, x_2)(x_1 - \bar{x}_1) + E_f(x_1, \bar{x}_1, x_2),$$

where we assume that

$$|E_f(x_1, \bar{x}_1, x_2)| \le K_f(\bar{x}_1 - x_1)^2.$$

Taylor's theorem implies this is true provided the second partial derivatives of f are bounded. Integration with respect to x_2 yields

$$\int_0^1 f(x_1, x_2)\, dx_2 = \int_0^1 f(\bar{x}_1, x_2)\, dx_2$$

$$+ (x_1 - \bar{x}_1) \int_0^1 \frac{\partial f}{\partial x_1}(\bar{x}_1, x_2)\, dx_2 + \int_0^1 E_f(x_1, \bar{x}_1, x_2)\, dx_2.$$

Since

$$\left| \int_0^1 E_f(x_1, \bar{x}_1, x_2)\, dx_2 \right| \leq K_f(\bar{x}_1 - x_1)^2$$

(54.23) follows after dividing by $(x_1 - \bar{x}_1)$ and taking the limit as x_1 tends to $\bar{x}_1$. We summarize:

Theorem 54.9 (Differentiation under the integral sign) *If the second partial derivatives of $f(x_1, x_2)$ are bounded, then for $x_1 \in \mathbb{R}$,*

$$\frac{d}{dx_1} \int_0^1 f(x_1, x_2)\, dx_2 = \int_0^1 \frac{\partial f}{\partial x_1}(x_1, x_2)\, dx_2 \qquad (54.24)$$

Example 54.19.

$$\frac{d}{dx} \int_0^1 (1 + xy^2)^{-1}\, dy = \int_0^1 \frac{\partial}{\partial x}(1 + xy^2)^{-1}\, dy = - \int_0^1 \frac{y^2}{(1 + xy^2)^2}\, dy.$$

Chapter 54   Problems

54.1. Sketch the following surfaces in $\mathbb{R}^3$: (a) $\Gamma = \{x : x_3 = x_1^2 + x_2^2\}$, (b) $\Gamma = \{x : x_3 = x_1^2 - x_2^2\}$, (c) $\Gamma = \{x : x_3 = x_1 + x_2^2\}$, (d) $\Gamma = \{x : x_3 = x_1^4 + x_2^6\}$. Determine the tangent planes to the surfaces at different points.

54.2. Determine whether the following functions are Lipschitz continuous or not on $\{x : |x| < 1\}$ and determine Lipschitz constants:

- (a) $f : \mathbb{R}^3 \to \mathbb{R}^3$ where $f(x) = x|x|^2$,
- (b) $f : \mathbb{R}^3 \to \mathbb{R}$ where $f(x) = \sin|x|^2$,
- (c) $f : \mathbb{R}^2 \to \mathbb{R}^3$ where $f(x) = (x_1, x_2, \sin|x|^2)$,
- (d) $f : \mathbb{R}^3 \to \mathbb{R}$ where $f(x) = 1/|x|$,
- (e) $f : \mathbb{R}^3 \to \mathbb{R}^3$ where $f(x) = x\sin(|x|)$, (optional)
- (f) $f : \mathbb{R}^3 \to \mathbb{R}$ where $f(x) = \sin(|x|)/|x|$. (optional)

54.3. For the functions in the previous exercise, determine which are contractions in $\{x : |x| < 1\}$ and find their fixed points (optional).

54.4. Linearize the following functions on $\mathbb{R}^3$ at $x = (1, 2, 3)$:

- (a) $f(x) = |x|^2$,
- (b) $f(x) = \sin(|x|^2)$,
- (c) $f(x) = (|x|^2, \sin(x_2))$,
- (d) $f(x) = (|x|^2, \sin(x_2), x_1 x_2 \cos(x_3))$.

54.5. Compute the determinant of the Jacobian of the following functions: (a) $f(x) = (x_1^3 - 3x_1x_2^2, 3x_1x_2^2 - x_2^3)$, (b) $f(x) = (x_1e^{x_2}\cos(x_3), x_1e^{x_2}\sin(x_3), x_1e^{x_2})$.

54.6. Compute the second order Taylor polynomials at $(0, 0, 0)$ of the following functions $f : \mathbb{R}^3 \rightarrow \mathbb{R}$: (a) $f(x) = \sqrt{1 + x_1 + x_2 + x_3}$, (b) $f(x) = (x_1 - 1)x_2x_3$, (c) $f(x) = \sin(\cos(x_1x_2x_3))$, (d) $\exp(-x_1^2 - x_2^2 - x_3^2)$, (e) try to estimate the errors in the approximations in (a)-(d).

54.7. Linearize $f \circ s$, where $f(x) = x_1x_2x_3$ at $t = 1$ with (a) $s(t) = (t, t^2, t^3)$, (b) $s(t) = (\cos(t), \sin(t), t)$, (c) $s(t) = (t, 1, t^{-1})$.

54.8. Evaluate $\int_0^\infty y^n e^{-xy}\, dy$ for $x > 0$ by repeated differentiation with respect to x of $\int_0^\infty e^{-xy}\, dy$.

54.9. Try to minimize the function $u(x) = x_1^2 + x_2^2 + 2x_3^2$ by starting at $x = (1, 1, 1)$ using the method of steepest descent. Seek the largest step length for which the iteration converges.

54.10. Compute the roots of the equation $(x_1^2 - x_2^2 - 3x_1 + x_2 + 4, 2x_1x_2 - 3x_2 - x_1 + 3) = (0, 0)$ using Newton's method.

54.11. Generalize Taylor's theorem for a function $f : \mathbb{R}^n \rightarrow \mathbb{R}$ to third order.

54.12. Is the function $f(x_1, x_2) = \dfrac{x_1^2 - x_2^2}{x_1^2 + x_2^2}$ Lipschitz continuous close to $(0, 0)$?

> Jacobi and Euler were kindred spirits in the way they created their mathematics. Both were prolific writers and even more prolific calculators; both drew a great deal of insight from immense algorithmical work; both laboured in many fields of mathematics (Euler, in this respect, greatly surpassed Jacobi); and both at any moment could draw from the vast armoury of mathematical methods just those weapons which would promise the best results in the attack of a given problem. (Sciba)

55

Level Curves/Surfaces and the Gradient

It would make no sense to overload the student with all kinds of little things that might be of occasional use. Instead, it is important that students become familiar with ways to think mathematically, recognize the need for applying mathematical methods to engineering problems, realize that mathematics is a systematic science built on relatively few principles and get a firm grasp for the interrelation between theory, computing and experiment. (E. Kreyszig, in Preface to Advanced Engineering Mathematics, 1993)

55.1 Level Curves

A *level curve* of a function $u : \mathbb{R}^2 \rightarrow \mathbb{R}$ is a curve $g : [a, b] \rightarrow \mathbb{R}^2$ such that

$$u(g(t)) = c \quad \text{for } t \in [a, b], \tag{55.1}$$

where c is a constant. A level curve is also called an *isoline*. The points x on a level curve $x = g(t)$ satisfying (55.1), all have the same function value $u(x) = u(g(t)) = c$. By plotting the level curves or isolines for a collection of different constants c, we get a *level curve plot* or *contour plot* of the function $u(x)$. The level curves are the projections onto $\mathbb{R}^2$ of the intersections of the planes $x_3 = c$ in $\mathbb{R}^3$ with the graph $\{x \in \mathbb{R}^2 : x_3 = u(x_1, x_2), (x_1, x_2) \in \mathbb{R}^2\}$. We illustrate in Fig. 55.2.

Example 55.1. The level curves of the function $u(x) = x_1^2 + x_2^2$ are the circles $x_1^2 + x_2^2 = c$ with $c \geq 0$ a constant. The level curves of the function

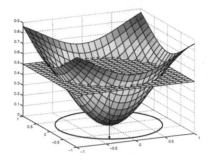

Fig. 55.1. Projection onto $\mathbb{R}^2$ of the intersection of $x_3 = c$ and $x_3 = u(x_1, x_2)$ (with $u(x_1, x_2) = 1 - \exp(-x_1^2 - x_2^2)$ and $c = .5$) gives a level curve

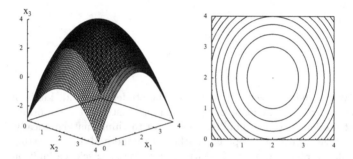

Fig. 55.2. A plot of a surface and the corresponding contour plot with contour curves shown every .7 units starting at the maximum height of 4

$u(x) = 2x_1^2 + x_2^2$ are the ellipses $2x_1^2 + x_2^2 = c$ with $c \geq 0$. The level curves of the function $u(x) = x_1^2 - x_2$ are the parabolas $x_2 = x_1^2 - c$ with c a constant.

Example 55.2. A hiking map indicates the level curves of the function $u : \mathbb{R}^2 \to \mathbb{R}$ that gives the height of a point $x \in \mathbb{R}^2$ above a reference level, like the see level. The difference in height between two nearby level curves is typically 10 meters. The change in height between two points can be obtained by counting the number of contour lines intersected by a line joining the two points. This is useful when planning a hiking trip. Recall Fig. 54.8.

A level curve $u(g(t)) = c$ may be thought of as the shore-lines with the sea level equal to c above the reference level.

55.2   Local Existence of Level Curves

The local existence of level curves follows from the following special case of the Implicit Function theorem, where the level curve is given by $t \to (t, g(t))$ or $t \to (g(t), t)$, where $g : \mathbb{R} \to \mathbb{R}$.

Theorem 55.1 *Assume $u : \mathbb{R}^2 \to \mathbb{R}$ has continuous partial derivatives and $u(\bar{x}_1, \bar{x}_2) = c$. If $\frac{\partial u}{\partial x_2}(\bar{x}_1, \bar{x}_2) \neq 0$, then there is a $\delta > 0$ such that $u(x_1, x_2) = c$ has a unique solution $x_2 = g(x_1)$ for $|x_1 - \bar{x}_1| < \delta$. If $\frac{\partial u}{\partial x_1}(\bar{x}_1, \bar{x}_2) \neq 0$, then there is a $\delta > 0$ such that $u(x_1, x_2) = c$ has a unique solution $x_1 = g(x_2)$ for $|x_2 - \bar{x}_2| < \delta$.*

Notice that if $\frac{\partial u}{\partial x_2}(\bar{x}_1, \bar{x}_2) = 0$, then the level curve is parallel to the x_2-axis, and thus we cannot expect the equation $u(x_1, x_2) = c$ to define x_2 as a function of x_1 (a corresponding function $x_2 = g(x_1)$ would then have infinite slope at $x_1 = \bar{x}_1$).

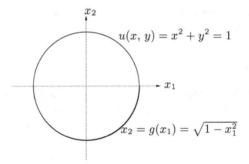

Fig. 55.3. $x_2 = -\sqrt{1 - x_1^2}$ giving one piece of the level curve $u(x_1, x_2) = x_1^2 + x_2^2 = 1$

55.3   Level Curves and the Gradient

Differentiating both sides of (55.1), we get using the Chain rule

$$\frac{d}{dt}u(g(t)) = \nabla u(x) \cdot g'(t) = \frac{\partial u}{\partial x_1}(g(t))g_1'(t) + \frac{\partial u}{\partial x_2}(g(t))g_2'(t) = 0.$$

Since $g'(t) = (g_1'(t), g_2'(t))$ is the direction of the tangent of the curve $g(t)$, this means that the direction $g'(t)$ of a level curve of a function $u : \mathbb{R}^2 \to \mathbb{R}$ is orthogonal to the gradient $\nabla u(g(t))$. Recall that the gradient $\nabla u(x)$ points in the direction of the steepest ascent of the function $u(x)$ at x, and the direction perpendicular to the gradient (the direction of the level curve) is a direction in which u stays constant, see Fig. 55.4. Moving along a level curve the function stays constant, and moving in the direction of the gradient the function increases as quickly as possible!

Since the gradient $\nabla(\bar{x})$ is a normal to the tangent to the level curve through $\bar{x}$, we can write the equation for the tangent to the level curve through $\bar{x}$ in the form $\nabla u(\bar{x}) \cdot (x - \bar{x}) = 0$.

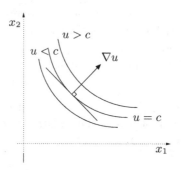

Fig. 55.4. The gradient of $\nabla u(x)$ of a function $u : \mathbb{R}^2 \to \mathbb{R}$ is orthogonal to the level curve of u through x

We summarize:

Theorem 55.2 *The gradient $\nabla u(g(t))$ of a function $u : \mathbb{R}^2 \to \mathbb{R}$ is orthogonal to the tangent $g'(t)$ of a level curve $g : I \to \mathbb{R}$. We can write the equation for the tangent to the level curve through $\bar{x}$ in the form $\nabla u(\bar{x}) \cdot (x - \bar{x}) = 0$.*

Example 55.3. Consider the function $u(x_1, x_2) = x_1^2 + x_2^2$ with circular level curves $g(t) = (g_1(t), g_2(t))$ satisfying $g_1^2(t) + g_2^2(t) = c^2$. We have $\nabla u(x) = (2x_1, 2x_2)$ and differentiating $g_1^2(t) + g_2^2(t) = c^2$ with respect to t we get $0 = 2g_1(t)g_1'(t) + 2g_2(t)g_2'(t) = \nabla u(g(t)) \cdot g'(t)$ as expected. Alternatively, parameterizing a level curve $g(t)$ satisfying $g_1^2(t) + g_2^2(t) = c^2$ by $g(t) = c(\cos(t), \sin(t))$, we have $g'(t) = c(-\sin(t), \cos(t)) = (-x_2(t), x_1(t))$ with $x = g(t)$. We check that $\nabla u(g(t)) \cdot g'(t) = 2(x_1(t), x_2(t)) \times (-x_2(t), x_1(t)) = 0$.

Example 55.4. If $u : \mathbb{R}^2 \to \mathbb{R}$ is of the form $u(x_1, x_2) = f(x_1) - x_2$, where $f : \mathbb{R} \to \mathbb{R}$, then $\nabla u(x) = (f'(x_1), -1)$. A level curve $u(g(t)) = c$ can be parameterized by $g(t) = (t, f(t) - c)$, and $g'(t) = (1, f'(t))$. Clearly, $\nabla u(g(t)) \cdot g'(t) = (f'(t), -1) \cdot (1, f'(t)) = 0$.

55.4   Level Surfaces

A *level surface* of a function $u : \mathbb{R}^3 \to \mathbb{R}$ is a surface $g : Q \to \mathbb{R}^3$, where Q is a subset of $\mathbb{R}^2$, such that

$$u(g(y)) = c \quad \text{for } y \in Q, \tag{55.2}$$

where c is a constant. A level surface is also called an *isosurface*. The points on a level surface $g(t)$ satisfying (55.1) all have the same function value $u(g(y)) = c$.

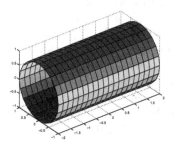

Fig. 55.5. A piece of the level surface $u(x_1, x_2, x_3) = x_1^2 + x_3^2 = 1$

55.5   Local Existence of Level Surfaces

The local existence of level surfaces follows from the following special case of the Implicit Function theorem. We find that the level surface is parameterized as $g(y_1, y_2) = (y_1, y_2, f(y_1, y_2))$, $g(y_1, y_3) = (y_1, f(y_1, y_2), y_3)$ or $g(y_2, y_3) = (f(y_2, y_3), y_2, y_3)$ with some function $f : \mathbb{R}^2 \to \mathbb{R}$, depending on which partial derivative is non-zero.

Theorem 55.3 *Assume* $u : \mathbb{R}^3 \to \mathbb{R}$ *has continuous partial derivatives and* $u(\bar{x}_1, \bar{x}_2, \bar{x}_3) = c$, *where* c *is a constant. If* $\partial u / \partial x_3 \neq 0$, *then there is a* $\delta > 0$ *such that* $u(x_1, x_2, x_3) = c$ *has a unique solution* $x_3 = f(x_1, x_2)$ *for* $\|(x_1, x_2) - (\bar{x}_1, \bar{x}_2)\| < \delta$. *If* $\partial u / \partial x_2 \neq 0$, *then there is a* $\delta > 0$ *such that* $u(x_1, x_2, x_3) = c$ *has a unique solution* $x_2 = g(x_1, x_3)$ *for* $\|(x_1, x_3) - (\bar{x}_1, \bar{x}_3)\| < \delta$. *If* $\partial u / \partial x_1 \neq 0$, *then there is a* $\delta > 0$ *such that* $u(x_1, x_2, x_3) = c$ *has a unique solution* $x_1 = g(x_2, x_3)$ *for* $\|(x_2, x_3) - (\bar{x}_2, \bar{x}_3)\| < \delta$.

55.6   Level Surfaces and the Gradient

Differentiating both sides of (55.2) with respect to y_1 and y_2, where $y = (y_1, y_2)$, we get using the Chain rule

$$\frac{\partial}{\partial y_i} u(g(y)) = \nabla u(g(y)) \cdot g'_{,i}(y) = 0, \quad i = 1, 2,$$

where we use the notation

$$g'_{,i}(y) = \frac{\partial}{\partial y_i} g(y).$$

We use the comma in $g'_{,i}$ to indicate differentiation with respect to x_i, while g_i will denote component i of $g = (g_1, g_2, g_3)$. We recall that the tangent plane (linearization) of $g(y)$ at $\bar{x} = g(\bar{y})$ is given by $(y_1, y_2) \rightarrow g(\bar{y}) + (y_1 - \bar{y}_1)g'_{,1}(\bar{y}) + (y_2 - \bar{y}_2)g'_{,2}(\bar{y})$, and we conclude that $\nabla u(g(\bar{y}))$ is orthogonal to the tangent plane of the level surface through $\bar{x} = g(\bar{y})$. We say that $\nabla u(g(\bar{y}))$ is *orthogonal to the level surface* $u(x) = c$ through $\bar{x} = g(\bar{y})$, or that $\nabla u(g(\bar{y}))$ is a *normal to the level surface* $u(x) = c$ at $\bar{x} = g(\bar{y})$, see Fig. 55.6. Since $\nabla u(\bar{x})$ thus is a normal to the tangent plane at x, the equation for the tangent plane to a level surface through $\bar{x}$ can also be written $\nabla u(\bar{x}) \cdot (x - \bar{x}) = 0$.

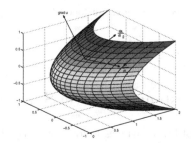

Fig. 55.6. The gradient $\nabla u(x) = (2x_1, -1, 2x_3)$ of $u(x_1, x_2, x_3) = x_1^2 + x_3^2 - x_2$ is orthogonal to a level surface $(x_1, x_3) \rightarrow g(x_1, x_3) = (x_1, x_1^2 + x_3^2 + c, x_3)$ since $g'_1 = (1, 2x_1, 0)$ and $g'_3 = (0, 2x_3, 1)$

We summarize:

Theorem 55.4 *The gradient $\nabla u(\bar{x})$ of a function $u : \mathbb{R}^3 \rightarrow \mathbb{R}$, is orthogonal to the tangent plane $(y_1, y_2) \rightarrow g(\bar{y}) + (y_1 - \bar{y}_1)g'_{,1}(\bar{y}) + (y_2 - \bar{y}_2)g'_{,2}(\bar{y})$ of a level surface $y \rightarrow x = g(y)$, where $\bar{x} = g(\bar{y})$. The equation for the tangent plane of a level surface through $\bar{x}$ can also be written $\nabla u(\bar{x}) \cdot (x - \bar{x}) = 0$.*

Example 55.5. Consider the function $u(x) = x_1^2 + x_2^2 + x_3^2$ with the level surfaces $g(y)$ satisfying $g_1^2(y) + g_2^2(y) + s_3^2(y) = c^2$ representing spheres centered at the origin with radii c. The gradient $\nabla(x) = 2x$ is evidently orthogonal to a tangent plane of a level surface at x.

Example 55.6. If $u : \mathbb{R}^3 \rightarrow \mathbb{R}$ is of the form $u(x_1, x_2, x_3) = f(x_1, x_2) - x_3$, where $f : \mathbb{R} \rightarrow \mathbb{R}$, then $\nabla u(x) = (f'_{,1}(x_1, x_2), f'_{,2}(x_1, x_2), -1)$. A level surface $u(g(y)) = c$ can be parameterized by $g(y) = (y_1, y_2, f(y_1, y_2) - c)$, and $g'_1(y) = (1, 0, f'_{,1}(y))$ and $g'_2(y) = (0, 1, f'_{,2}(y))$. Clearly, $\nabla u(g(y)) \cdot g'_{,i}(y) = 0$ for $i = 1, 2$.

Chapter 55 Problems

55.1. Sketch the following surfaces in $\mathbb{R}^3$: (a) $\Gamma = \{x : x_1^2 + x_2^2 = x_3\}$,
(b) $\Gamma = \{x : x_1^2 + x_2^2 = x_3^2\}$, (c) $\Gamma = \{x : x_1^2 + x_2^2 = -x_3^2\}$,
(d) $\Gamma = \{x : x_1^2 + 2x_2^2 + 3x_3^2 = 6\}$. Determine the tangent planes to the surfaces at various points.

55.2. Find parametrization of the curves for the intersections of the surfaces in the previous exercise with the plane $x_3 = 1$.

55.3. Show that the surface $\Gamma = \{x : x_1^2 + 2x_2^2 + 3x_3^2 + x_1 x_3^3 = 7\}$ can be expressed in the form $x_3 = g(x_1, x_2)$ close to $(1, 1, 1)$.

55.4. Compute the gradients of the following functions $f : \mathbb{R}^3 \rightarrow \mathbb{R}$:
(a) $f(x) = x_1^n(x_2^n + x_3^n)$, (b) $f(x) = |x|$, (c) $f(x) = |x|^2$, (d) $f(x) = 1/|x|$,
(e) $f(x) = \exp(x_1 x_2 x_3)$.

55.5. For each of the functions in the previous exercise, determine the equation for the tangent plane to the level surface $f(x) = f(1, 1, 1)$ at $x = (1, 1, 1)$.

55.6. Determine the equation for the tangent plane at $x = (1, 2, 3)$ for the following surfaces: (a) $x_3 = \frac{3}{2}x_1 x_2$, (b) $x_1^2 + x_2^2 + x_3^2 = 14$, (c) $x_2 = \sin(2\pi x_1) + 2\cos(2\pi x_3)$.

55.7. Determine the tangent plane and normal vector to the ellipse $x_1^2 + 3x_2^2 = 10$ at $x = (1, \sqrt{3})$.

55.8. Let $f : Q \rightarrow \mathbb{R}$, where $Q = [0, 1] \times [0, 1]$ is the unit square, satisfy $f(x) = 0$ for x on the boundary of Q. Prove under convenient assumptions that there is a point $y \in Q$ such that $\nabla f(y) = 0$.

56

Linearization and Stability of Initial Value Problems

The logos of somewome to that base anything, when most characteristically mantissa minus, comes to nullum in the endth: orso, here is nowet badder than the sin of Aha with his cosin Lil, verswaysed on coversvised, and all that's consecants and cotangincies...
(Finnegans Wake, James Joyce)

56.1 Introduction

We continue the study of the general initial value problem (40.1), now focussing on the *stability* of solutions, which is a measure of the *sensitivity of solutions to perturbations in given data*. This is a fundamentally important aspect of the behavior of solutions, which we touched upon in Chapter *The general initial value problem*, and which we now consider more closely.

We consider an autonomous problem of the form

$$\dot{u}(t) = f(u(t)) \quad \text{for } 0 < t \le T, \, u(0) = u^0, \tag{56.1}$$

where $f : \mathbb{R}^d \to \mathbb{R}^d$ is a given bounded Lipschitz continuous function, $u^0 \in \mathbb{R}^d$ is a given initial value, and we seek a solution $u : [0, T] \to \mathbb{R}^d$, where we think of $[0, T]$ as a given time interval. To study the stability of a given solution $u(t)$ to small perturbations in given data, e.g. in the given initial data u^0, we will consider an associated *linearized problem* that arises upon linearizing the function $v \to f(v)$ around the solution $u(t)$.

56.2 Stationary Solutions

We consider first the simplest case of a *stationary solution* $u(t) = \bar{u}$ for $0 \leq t \leq T$, that is a solution $u(t)$ of (56.1) that is independent of time t. Since $\dot{u}(t) = 0$ if $u(t)$ is independent of time, $u(t) = \bar{u}$ is a stationary solution if $f(\bar{u}) = 0$ and $u^0 = \bar{u}$, where $\bar{u} = (\bar{u}_1, \ldots, \bar{u}_d) \in \mathbb{R}^d$. The equation $f(\bar{u}) = 0$ corresponds to a system of d equations $f_i(\bar{u}_1, \ldots, \bar{u}_d) = 0$, $i = 1, \ldots, d$, in d unknowns $\bar{u}_1, \ldots, \bar{u}_d$, where the f_i are the components of f. We studied such systems of equations in Chapter *Vector-valued functions of several real variables*. Here, we assume the existence of a stationary solution $u(t) = \bar{u}$ so that $\bar{u} \in \mathbb{R}^d$ satisfies the equation $f(\bar{u}) = 0$. In general, there may be several roots $\bar{u}$ of the equation $f(v) = 0$ and thus there may be several stationary solutions. We also refer to a stationary solution $u(t) = \bar{u}$ as an *equilibrium solution*.

Example 56.1. The stationary solutions $\bar{u}$ of the Crash model

$$\begin{cases} \dot{u}_1 + \nu u_1 - \kappa u_1 u_2 = \nu & t > 0, \\ \dot{u}_2 + 2\nu u_2 - \nu u_2 u_1 = 0 & t > 0, \end{cases} \tag{56.2}$$

of the form $\dot{u} = f(u)$ with $f(u) = (-\nu u_1 + \kappa u_1 u_2 + \nu, -2\nu u_2 + \nu u_2 u_1)$, are $\bar{u} = (1, 0)$ and $\bar{u} = (2, \frac{\nu}{\kappa})$.

56.3 Linearization at a Stationary Solution

We shall now study perturbations of a given stationary solution under small perturbations of initial data. We thus assume $f(\bar{u}) = 0$ and denote the corresponding equilibrium solution by $\bar{u}(t)$ for $t > 0$, that is $\bar{u}(t) = \bar{u}$ for $t > 0$. We consider the initial value problem (56.1) with $u^0 = \bar{u} + \varphi^0$, where $\varphi^0 \in \mathbb{R}^d$ is a given small perturbation of the initial data $\bar{u}$. We denote the corresponding solution by $u(t)$ and focus attention on the corresponding perturbation in the solution, that is $\psi(t) = u(t) - \bar{u}(t) = u(t) - \bar{u}$. We want to derive a differential equation for the perturbation $\psi(t)$, and to this end we linearize f at $\bar{u}$ and write

$$f(u(t)) = f(\bar{u} + \psi(t)) = f(\bar{u}) + f'(\bar{u})\psi(t) + e(t),$$

where $f'(\bar{u})$ is the Jacobian of $f : \mathbb{R}^d \rightarrow \mathbb{R}^d$ at $\bar{u}$ and the error term $e(t)$ is quadratic in $\psi(t)$ (and thus is very small if $\psi(t)$ is small). Since $f(\bar{u}) = 0$ and $u(t)$ satisfies (56.1), we have

$$\dot{\psi}(t) = \frac{d}{dt}(\bar{u} + \psi(t)) = f(u(t)) = f'(\bar{u})\psi(t) + e(t).$$

Neglecting the quadratic term $e(t)$, we are led to a linear initial value problem,

$$\dot{\varphi}(t) = f'(\bar{u})\varphi(t) \quad \text{for } t > 0, \quad \varphi(0) = \varphi^0, \tag{56.3}$$

where $\varphi(t)$ is an approximation of the perturbation $\psi(t) = u(t) - \bar{u}$ up to a second order term. We refer to (56.3) as the *linearized problem* associated to the stationary solution $\bar{u}$ of (56.1). Since $f'(\bar{u})$ is a constant $d \times d$ matrix, we can express the solution to (56.3) using the matrix exponential as

$$\varphi(t) = \exp(tA)\varphi^0 \quad \text{for } 0 < t \le T, \tag{56.4}$$

where $A = f'(\bar{u})$. We thus have a formula that describes the evolution of perturbation $\varphi(t)$ starting from an initial perturbation $\varphi(0) = \varphi^0$. Depending on the nature of the matrix $\exp(tA)$, the perturbation may increase or decrease with time, reflecting a stronger or lesser sensitivity of the solution $u(t)$ to perturbations in initial data and therefore different stability features of the given problem.

We know that if A is diagonalizable, so that $A = B\Lambda B^{-1}$ where B is a non-singular $d \times d$ matrix and Λ is a diagonal matrix with the eigenvalues $\lambda_1, \ldots, \lambda_d$ of A on the diagonal, then

$$\varphi(t) = B\exp(t\Lambda)B^{-1}\varphi^0 \quad \text{for } t \ge 0. \tag{56.5}$$

We see that each component of $\varphi(t)$ is a linear combination of $\exp(t\lambda_1), \ldots,$ $\exp(t\lambda_d)$ and the sign of the real part $\operatorname{Re} \lambda_i$ of λ_i determines if the corresponding term grows or decays exponentially. If some $\operatorname{Re} \lambda_i > 0$, then we have exponential growth of certain perturbations, which indicates that the corresponding stationary solution $\bar{u}$ is *unstable*. On the other hand, if all $\operatorname{Re} \lambda_i \le 0$, then we would expect $\bar{u}$ to be *stable*.

These considerations are qualitative in nature, and to be more precise we should base judgements of stability or instability on quantitative estimates of perturbation growth. In the diagonalizable case, (56.5) implies in the Euclidean vector and matrix norms that

$$\|\varphi(t)\| \le \|B\|\|B^{-1}\| \max_{i=1,\ldots,d} \exp(t\lambda_i)\|\varphi^0\|. \tag{56.6}$$

We see that the maximal perturbation growth is governed by the maximal exponential factors $\exp(t\lambda_i)$ as well as the factors $\|B\|$ and $\|B^{-1}\|$ related to the transformation matrix B. If the transformation matrix B is orthogonal, then $\|B\| = \|B^{-1}\| = 1$, and the perturbation growth is governed solely by the exponential factors $\exp(t\lambda_i)$. We give this case special attention:

56.4 Stability Analysis when $f'(\bar{u})$ Is Symmetric

If $A = f'(\bar{u})$ is symmetric so that $A = Q\Lambda Q^{-1}$ with Q orthogonal and Λ a diagonal matrix with real diagonal elements λ_i, then

$$\|\varphi(t)\| \le \max_{i=1,\ldots,d} \exp(t\lambda_i)\|\varphi^0\|. \tag{56.7}$$

In particular, if all eigenvalues $\lambda_i \leq 0$ then perturbations $\varphi(t)$ cannot grow with time, and we say that the solution $\bar{u}$ is stable. On the other hand, if some eigenvalue $\lambda_i > 0$ and the corresponding eigenvector is g_i then $\varphi(t) = \exp(t\lambda_i)g_i$ solves the linearized initial value problem (56.3) with $\varphi^0 = g_i$, and evidently the particular perturbation $\varphi(t)$ grows exponentially. We then say that the solution $\bar{u}$ is *unstable*. Of course, the size of the positive eigenvalues influence the perturbation growth, so that if $\lambda_i > 0$ is small, then then growth is slow and the instability is mild. Likewise, if λ_i is small negative, then the exponential decay is slow.

56.5   Stability Factors

We may express the stability features of a particular perturbation φ^0 through a *stability factor* $S(T, \varphi_0)$ defined as follows:

$$S(T, \varphi_0) = \max_{0 \leq t \leq T} \frac{\|\varphi(t)\|}{\|\varphi^0\|}.$$

where $\varphi(t)$ solves the linearized problem (56.3) with initial data φ^0. The stability factor $S(T, \varphi_0)$ measures the maximal growth of the norm of $\varphi(t)$ over the time interval $[0, T]$ versus the norm of the initial value φ_0.

We can now seek to capture the overall stability features of a stationary solution $\bar{u}$ by maximization over all different perturbations:

$$S(T) = \max_{\varphi^0 \neq 0} S(T, \varphi_0).$$

If the stability factor $S(T)$ is large, then some perturbations grow very much over the time interval $[0, T]$, which indicates a strong sensitivity to perturbations or *instability*. On the other hand, if $S(T)$ is of moderate size then the perturbation growth is moderate, which signifies *stability*. Using the Euclidean matrix norm, we can also express $S(T)$ as

$$S(T) = \max_{0 \leq t \leq T} \| \exp(tA) \|.$$

Example 56.2. If $A = f'(\bar{u})$ is symmetric with eigenvalues $\lambda_1, \ldots, \lambda_d$, then

$$S(T) = \max_{i=1,\ldots,d} \max_{0 \leq t \leq T} \exp(t\lambda_i).$$

In particular, if all $\lambda_i \leq 0$, then $S(T) = 1$.

Example 56.3. The initial value problem for a pendulum takes the form

$$\dot{u}_1 = u_2, \quad \dot{u}_2 = -\sin(u_1) \quad \text{for } t > 0,$$
$$u_1(0) = u_{01}, \quad u_2(0) = u_{02},$$

corresponding to $f(u) = (u_2, -\sin(u_1))$ and the equilibrium solutions are $\bar{u} = (0,0)$ and $\bar{u} = (\pi, 0)$. We have

$$f'(\bar{u}) = \begin{pmatrix} 0 & 1 \\ -\cos(\bar{u}_1) & 0 \end{pmatrix},$$

and the linearized problem at $\bar{u} = (0,0)$ thus takes the form

$$\dot{\varphi}(t) = \begin{pmatrix} 0 & 1 \\ -1 & 0 \end{pmatrix} \varphi(t) \equiv A_0 \varphi(t) \quad \text{for } t > 0, \quad \varphi(0) = \varphi^0,$$

with solution

$$\varphi_1(t) = \varphi_1^0 \cos(t) + \varphi_2^0 \sin(t), \quad \varphi_2(t) = -\varphi_1^0 \sin(t) + \varphi_2^0 \cos(t).$$

It follows by a direct computation (or using that $\begin{pmatrix} \cos(t) & \sin(t) \\ -\sin(t) & \cos(t) \end{pmatrix}$ is an orthogonal matrix), that for $t > 0$

$$\|\varphi(t)\|^2 = \|\varphi_0\|^2,$$

and thus the norm $\|\varphi(t)\|$ of a solution $\varphi(t)$ of the linearized equations is constant in time, which means that the stability factor $S(T) = 1$ for all $T > 0$. We conclude that if the norm of a perturbation is small initially, it will stay small for all time. This means that the equilibrium solution $\bar{u} = (0,0)$ is *stable*. More precisely, if the pendulum is perturbed initially a little from its bottom position, the pendulum will oscillate back and forth around the bottom position with constant amplitude. This fits our direct experimental experience of course.

Note that the linearized operator A_0 is non-symmetric; the eigenvalues of A_0 are purely imaginary $\pm i$, which says that $\|\varphi(t)\| = \|\varphi_0\|$, that is a perturbation neither grows nor decays. Another way to derive this fact is to use the fact that A_0 is *antisymmetric*, that is $A_0^\top = -A_0$, which shows that $(A_0\varphi, \varphi) = (\varphi, A_0^\top \varphi) = -(\varphi, A_0\varphi) = -(A_0\varphi, \varphi)$, and thus $(A_0\varphi, \varphi) = 0$, where $(\cdot, \cdot)$ is the $\mathbb{R}^2$ scalar product. It follows from the equation $\dot{\varphi} = A_0\varphi$ upon multiplication by φ that $0 = (\dot{\varphi}, \varphi) = \frac{1}{2}\frac{d}{dt}(\varphi, \varphi) = \frac{1}{2}\frac{d}{dt}\|\varphi\|^2$, which proves that $\|\varphi(t)\|^2 = \|\varphi_0\|^2$.

The linearized problem at $\bar{u} = (\pi, 0)$ reads

$$\dot{\varphi}(t) = \begin{pmatrix} 0 & 1 \\ 1 & 0 \end{pmatrix} \varphi(t) \equiv A_\pi \varphi(t) \quad \text{for } t > 0, \quad \varphi(0) = \varphi^0,$$

with symmetric matrix A_π with eigenvalues ± 1. Since one eigenvalue is positive, the stationary solution $\bar{u} = (\pi, 0)$ is unstable. More precisely, the solution is given by

$$\varphi_1 = \frac{\varphi_1^0}{2}(e^t + e^{-t}) + \frac{\varphi_2^0}{2}(e^t - e^{-t}), \quad \varphi_2 = \frac{\varphi_1^0}{2}(e^t - e^{-t}) + \frac{\varphi_2^0}{2}(e^t + e^{-t}),$$

and due to the exponential factor e^t, perturbations will grow exponentially in time, and thus an initially small perturbation will become large as soon as $t \geq 10$ say. Physically, this means that if the pendulum is perturbed initially a little from its top position, the pendulum will eventually move away from the top position, even if the initial perturbation is very small. This fact of course has direct experimental evidence: to balance a pendulum with the weight in the top position is tricky business. Small perturbations quickly grow to large perturbations and the equilibrium solution $(\pi, 0)$ of the pendulum is unstable.

Example 56.4. The linearization of the Crash model (56.2) at the equilibrium solution $\bar{u} = (1, 0)$, takes the form

$$\dot{\varphi}(t) = \begin{pmatrix} -\nu & \kappa \\ 0 & -\nu \end{pmatrix} \varphi(t) \equiv A_{\nu,\kappa}\varphi(t) \quad \text{for } t > 0, \quad \varphi(0) = \varphi^0, \qquad (56.8)$$

The solution is given by $\varphi_2(t) = \varphi_2^0 \exp(-\nu t)$, and $\varphi_1(t) = t\kappa \exp(-\nu t)\varphi_2^0 + \exp(-\nu t)\varphi_1^0$. Clearly, $\varphi_2(t)$ decays monotonically to zero and so does $\varphi_1(t)$ if $\kappa = 0$. But, if $\kappa \neq 0$ then $\varphi_1(t)$ reaches the following value, assuming for simplicity that $\varphi_{01} = 0$,

$$\varphi_1(\nu^{-1}) = \nu^{-1}\kappa \exp(-1)\varphi_2^0,$$

which contains the factor ν^{-1} that is large if ν is small. In other words, the stability factor $S(\nu^{-1}) \sim \nu^{-1}$, which is large if ν is small. Eventually, however, $\varphi_1(t)$ decays to zero. As a result, the equilibrium solution $(1, 0)$ is stable only to small perturbations, since we saw in the Chapter The Crash model that $(1, 0)$ is unstable to perturbations above a certain threshold depending on λ. Note that here the Jacobian $f'(\bar{u}) = A_{\nu,\kappa}$ has a double eigenvalue $-\nu$, but $A_{\nu,\kappa}$ is non-symmetric and the space of eigenvectors is one-dimensional and is spanned by $(1, 0)$. As a result, the term $t\kappa \exp(-\nu t)\varphi_2^0$ with linear growth in t appears; thus in this highly non-symmetric problem (if ν is small), large perturbation growth $\sim \nu^{-1}$ is possible although all eigenvalues are non-positive.

The matrix $A_{\nu,\kappa}$ is an example of a *non-normal* matrix. A non-normal matrix A is a matrix such that $A^\mathsf{T} A \neq A A^\mathsf{T}$. A non-normal matrix may or may not be diagonalizable, and if diagonalizable so that $A = B\Lambda B^{-1}$, we may have $\|B\|$ or $\|B^{-1}\|$ large, resulting in large stability factors in the corresponding linearized problem, as we just saw (cf. Problem 56.5).

The linearization at the equilibrium solution $\bar{u} = (2, \frac{\nu}{\kappa})$ takes the form

$$\dot{\varphi}(t) = \begin{pmatrix} 0 & 2\kappa \\ \frac{\nu^2}{\kappa} & 0 \end{pmatrix} \varphi(t) \quad \text{for } t > 0, \quad \varphi(0) = \varphi^0. \qquad (56.9)$$

The eigenvalues of the Jacobian are $\pm\sqrt{2}\nu$ and the solution is a linear combination of $\exp(\sqrt{2}\nu t)$ and $\exp(-\sqrt{2}\nu t)$ and thus has one exponentially growing part with growth factor $\exp(\sqrt{2}\nu t)$. The equilibrium solution $u = (2, \frac{\nu}{\kappa})$ is thus unstable.

56.6 Stability of Time-Dependent Solutions

We now seek to extend the scope to linearization and linearized stability for a time-dependent solution $\bar{u}(t)$ of (56.1). We want to study solutions of the form $u(t) = \bar{u}(t) + \psi(t)$, where $\psi(t)$ is a perturbation. Using $\frac{d}{dt}\bar{u} = f(\bar{u})$ and linearizing f at $\bar{u}(t)$, we obtain

$$\frac{d}{dt}(\bar{u} + \psi)(t) = f(\bar{u}(t)) + f'(\bar{u}(t))\psi(t) + e(t),$$

with $e(t)$ quadratic in $\psi(t)$. This leads to the linearized equation

$$\dot{\varphi}(t) = A(t)\varphi(t) \quad \text{for } t > 0, \varphi(0) = \varphi_0, \tag{56.10}$$

where $A(t) = f'(\bar{u}(t))$ is an $d \times d$ matrix that now depends on t if $\bar{u}(t)$ depends on t. We have no analytical solution formula to this general problem and thus although the stability properties of the given solution $\bar{u}(t)$ are expressed through the solutions $\varphi(t)$ of the linearized problem (56.10), it may be difficult to analytically assess these properties. We may define stability factors $S(T, \varphi_0)$ and $S(T)$ just as above, and we may say that a solution $\bar{u}(t)$ is stable if $S(T)$ is moderately large, and unstable if $S(T)$ is large. To determine $S(T)$ in general, we have to use numerical methods and solve (56.10) with different initial data φ^0. We return to the computation of stability factors in the next chapter on adaptive solvers for initial value problems.

56.7 Sum Up

The question of stability of solutions to initial value problems is of fundamental importance. We can give an affirmative answer in the case of a stationary solution with corresponding symmetric Jacobian. In this case a positive eigenvalue signifies instability, with the instability increasing with increasing eigenvalue, and all eigenvalues non-positive means stability. The case of an anti-symmetric Jacobian also signifies stability with the norm of perturbations being constant in time. If the Jacobian is non-normal we have to watch out and remember that just looking at the sign of the real part of eigenvalues may be misleading: in the non-normal case algebraic growth may in fact dominate slow exponential decay for finite time. In these cases and also for time-dependent solutions, an analytical stability analysis may be out of reach and the desired information about stability may be obtained by numerical solution of the associated linearized problem.

Chapter 56 Problems

56.1. Determine the stationary solutions to the system

$$\dot{u}_1 = u_2(1 - u_1^2),$$
$$\dot{u}_2 = 2 - u_1 u_2,$$

and study the stability of these solutions.

56.2. Determine the stationary solutions to the following system (Minea's equation) for different values of $\delta > 0$ and γ,

$$\dot{u}_1 = -u_1 - \delta(u_2^2 + u_3^2) + \gamma,$$
$$\dot{u}_2 = -u_2 - \delta u_1 u_2,$$
$$\dot{u}_3 = -u_3 - \delta u_1 u_3,$$

and study the stability of these solutions.

56.3. Determine the stationary solutions of the system (56.1) with (a) $f(u) = (u_1(1-u_2), u_2(1-u_1))$, (b) $f(u) = (-2(u_1-10)+u_2 \exp(u_1), -2u_2 - u_2 \exp(u_1))$, (c) $f(u) = (u_1 + u_1 u_2^2 + u_1 u_3^2, -u_1 + u_2 - u_2 u_3 + u_1 u_2 u_3, u_2 + u_3 - u_1^2)$, and study the stability of these solutions.

56.4. Determine the stationary solutions of the system (56.1) with (56.1) with (a) $f(u) = (-1001u_1 + 999u_2, 999u_1 - 1001u_2)$, (b) $f(u) = (-u_1 + 3u_2 + 5u_3, -4u_2 + 6u_3, u_3)$, (c) $f(u) = (u_2, -u_1 - 4u_2)$, and study the stability of these solutions.

56.5. Analyze the stability of the following variant of the linearized problem (56.8) with $\epsilon > 0$ small,

$$\dot{\varphi}(t) = \begin{pmatrix} -\nu & \kappa \\ \epsilon & -\nu \end{pmatrix} \varphi(t) \equiv A_{\nu,\kappa,\epsilon} \varphi(t) \quad \text{for } t > 0, \quad \varphi(0) = \varphi^0, \qquad (56.11)$$

by diagonalizing the matrix $\equiv A_{\nu,\kappa,\epsilon}$. Note that the diagonalization degenerates as ϵ tends to zero (that is, the two eigenvectors become parallel). Check if $A_{\nu,\kappa,\epsilon}$ is a normal or non-normal matrix.

57
Adaptive Solvers for IVPs

> On two occasions I have been asked (by members of Parliament), "Pray, Mr Babbage, if you put into the machine wrong figures, will the right answer come out?". I am not able rightly to apprehend the kind of confusion of ideas that could provoke such a question. (Babbage (1792–1871))

57.1   Introduction

In this chapter, we discuss the important issue of *adaptive error control* for numerical methods for initial value problems. This is the subject of automated choice of the time step with the purpose of controlling the numerical error to within a given tolerance level. The basic idea is to combine *feed-back* information from the computation concerning the *residual* of the computed solution and the results of auxiliary computations of *stability factors*. We focus first on on the cG(1) method and then comment on the backward Euler method, also referred to as dG(0), the discontinuous Galerkin method with piecewise constants.

We also discuss the application of cG(1) and dG(0) to a class of so-called *stiff* IVPs typically arising in chemical reaction modeling.

57.2   The cG(1) Method

We recall that cG(1), the continuous Galerkin method with polynomials of order 1, for the initial value problem $\dot{u}(t) = f(u(t))$ for $t > 0$, $u(0) = u^0$, with $f : \mathbb{R}^d \to \mathbb{R}^d$, takes the form

$$U(t_n) = U(t_{n-1}) + \int_{t_{n-1}}^{t_n} f(U(t))\, dt, \quad n = 1, 2, \ldots, \tag{57.1}$$

where $U(t)$ is continuous piecewise linear with nodal values $U(t_n) \in \mathbb{R}^d$ at an increasing sequence of discrete time levels $0 = t_0 < t_1 < \ldots$, and $U(0) = u^0$. If we evaluate the integral in (57.1) with the midpoint quadrature rule, we obtain the Midpoint method:

$$U(t_n) = U(t_{n-1}) + k_n f\left(\frac{U(t_n) + U(t_{n-1})}{2}\right), \quad n = 1, 2, \ldots, \tag{57.2}$$

where $k_n = t_n - t_{n-1}$ is the time step. The cG(1)-method is the first in a family of cG(q)-methods with $q = 1, 2, \ldots,$, where the solution is approximated by continuous piecewise polynomials of order q. The Galerkin "orthogonality" of cG(1) is expressed by the fact that the method can be formulated

$$\int_{t_{n-1}}^{t_n} (\dot{U}(t) - f(U(t))) \cdot v\, dt = 0, \quad n = 1, 2, \ldots, \tag{57.3}$$

for all $v \in \mathbb{R}^d$. This says that the *residual*

$$R(U(t)) = \dot{U}(t) - f(U(t)), \quad t \in [0, T], \tag{57.4}$$

of the continuous piecewise linear approximate solution $U(t)$ is *orthogonal* to the constant functions $v(t) = v \in \mathbb{R}^d$ on each subinterval (t_{n-1}, t_n). The residual $\dot{u}(t) - f(u(t))$ of the exact solution is zero since $\dot{u}(t) = f(u(t))$, while the residual of $R(U(t))$ of the approximate solution $U(t)$ is non-zero in general. Similarly, in cG(q) the residual is orthogonal on (t_{n-1}, t_n) to polynomials of degree $q-1$. Note that (57.1) is a vector equation that reads

$$U_i(t_n) = U_i(t_{n-1}) + \int_{t_{n-1}}^{t_n} f_i(U(t))\, dt, \quad n = 1, 2, \ldots, i = 1, \ldots, d,$$

as can be seen from (57.3) upon setting $v = e_i$, $i = 1, \ldots, d$.

We will now study the problem of *automatic step-size control* with the purpose of keeping the error

$$\|u(T) - U(T)\| \le TOL,$$

where $T = t_N$ is a final time and TOL is a given tolerance, while using as few time steps as possible. The objective is the same as that of computing an integral over an interval $[0, T]$ using numerical quadrature to a certain tolerance using as few quadrature points as possible. This is exactly the problem we meet in the case of a scalar initial value problem $\dot{u}(t) = f(u(t), t)$ with $f(u(t), t) = f(t)$.

We shall derive an *a posteriori error estimate* in which the final error $\|u(T) - U(T)\|$ is estimated in terms of the residual $R(U(t) = \dot{U}(t) - f(U(t))$ and certain *stability factors* that measure the *accumulation* of the numerical errors introduced in each time step.

The a posteriori error estimate takes the form

$$\|u(T) - U(T)\| \leq S_c(T) \max_{0 \leq t \leq T} \|k(t)R(U(t)\|, \qquad (57.5)$$

where $k(t) = k_n = t_n - t_{n-1}$ for $t \in [t_{n-1}, t_n)$ and where the stability factor $S_c(T)$ is defined as follows. We consider the linearized problem

$$-\dot{\varphi}(t) = A^\top(t)\varphi(t) \quad \text{for } 0 < t < T, \ \varphi(T) = \varphi^0, \qquad (57.6)$$

where

$$A(t) = \int_0^1 f'(su(t) + (1-s)U(t)) \, ds.$$

We note that replacing $u(t)$ by $U(t)$ gives the following approximate formula for $A(t)$,

$$A(t) \approx f'(U(t)),$$

assuming $U(t)$ is close to $u(t)$. We conclude that $A(t)$ is close to the Jacobian $f'(u(t))$ of $f(v)$ at $v = u(t)$ if $U(t)$ is a reasonable approximation of $u(t)$ Note that the dual $A^\top(t)$ of $A(t)$ occurs in (57.6). Note further that the linearized dual problem (57.6) runs *backward* in time since the initial value $\varphi(T) = \varphi^0$ is specified at time $t = T$. We are now ready to introduce the following stability factors:

$$S_d(T) = \max_{\varphi^0 \in \mathbb{R}^d} \frac{\|\varphi(t)\|}{\|\varphi^0\|},$$

$$S_c(T) = \max_{\varphi^0 \in \mathbb{R}^d} \frac{\int_0^t \|\dot{\varphi}(s)\| \, ds}{\|\varphi^0\|}, \qquad (57.7)$$

where φ solves (57.6). We note that the stability factors measure different features of the the dual solution φ. The stability factor $S_d(t)$ measures the maximal perturbation growth over the time interval $[0, T]$. We met this factor in the previous chapter. We shall see that this factor is tailored to measure the effect of an error in the initial data u^0 and the "d" in S_d refers to "data". The stability factor $S_c(t)$ measures the integral of $\|\dot{\varphi}\|$ over $[0, T]$ and is geared to evaluate the error in cG(1) and the "c" in S_c refers to "computation".

We shall give the proof of (57.5) below, first in a very simple case with $n = 1$ and $f(u(t)) = au(t)$ with a a constant and then in the general case. The proofs are very similar. Before plunging into the proofs, we shall try to digest the a posteriori error estimate, and see how it can be used to design an adaptive algorithm aiming at controlling the final error $\|u(T) - U(T)\|$ on a given tolerance level with as few time steps as possible.

The stability factors $S_c(T)$ and $S_d(T)$ can be computed by numerically solving the linearized dual problem (57.6) with $\varphi^0 = e_i$ for $i = 1, \ldots, d$. If d is large, then we may reduce the variation of the initial data by limiting the error control to certain components only, or by trying to choose φ^0 parallel to $u(T) - U(T)$, which we approximate as $U_h(T) - U_H(T)$ with $U_h(T)$ and $U_H(T)$ being approximations computed with two different tolerances.

57.3   Adaptive Time Step Control for cG(1)

We recall the basic error estimate (57.5):

$$\|u(T) - U(T)\| \le S_c(T) \max_{0 \le t \le T} \|k(t))R(t)\|, \qquad (57.8)$$

where $R(t) = \dot{U}(t) - f(U(t))$ and we assume that the stability factor $S(T)$ has been computed or estimated. We will return to this issue below. To achieve $\|u(T) - U(T)\| \le TOL$, we use (57.5) to choose the time steps $k_n = t_n - t_{n-1}$ so that

$$k(t) = k_n \approx \frac{TOL}{S_c(T)R_n} \quad \text{for } t \in [t_{n-1}, t_n), \qquad (57.9)$$

where

$$R_n = \max_{t_{n-1} \le t \le t_n} \|\dot{U}(t) - f(U(t))\|$$

is the residual on the time interval $[t_{n-1}, t_n)$. Note that the residual R_n is computable from the computed solution $U(t)$ and if $S_c(T)$ is known, timestepalg gives an equation for the time step $k_n = t_n - t_{n-1}$, where t_{n-1} already known. As with adaptive numerical quadrature, (57.9) yields a nonlinear equation for the time step $k_n = t_n - t_{n-1}$ that we can seek to solve using some form of trial-and-error strategy or by prediction, e.g. replacing R_n by R_{n-1}.

57.4   Analysis of cG(1) for a Linear Scalar IVP

We shall now prove an a posteriori error estimate for cG(1) for a a linear scalar IVP of the form

$$\dot{u}(t) = au(t) + f(t) \quad \text{for } t > 0, u(0) = u^0, \qquad (57.10)$$

where a is a constant and $f(t)$ is a given function. The analysis is based on representing the error in terms of the solution $\varphi(t)$ of the following dual problem:

$$\begin{cases} -\dot{\varphi} = a\varphi & \text{for } T > t \geq 0, \\ \varphi(T) = e(T), \end{cases} \tag{57.11}$$

where $e = u - U$. Note again that (57.11) runs "backwards" in time starting at time t_N and that the time derivative term $\dot{\varphi}$ has a minus sign. We start from the identity

$$|e(T)|^2 = |e(T)|^2 + \int_0^T e\left(-\dot{\varphi} - a\varphi\right) dt,$$

and integrate by parts to get the following representation of $|e(T)|^2$,

$$|e(T)|^2 = \int_0^T (\dot{e} - ae)\varphi \, dt + e(0)\varphi(0),$$

where we allow $U(0)$ to be different from $u(0)$, corresponding to an error in the initial value $u(0)$. Since u solves the differential equation (57.10), that is $\dot{u} + au = f$, we have

$$\dot{e} - au = \dot{u} - au - \dot{U} + aU = f - \dot{U} + aU,$$

and thus we obtain the following representation of the error $|e(T)|^2$ in terms of the residual $R(U) = \dot{U} - aU - f$ and the dual solution φ,

$$|e(T)|^2 = \int_0^T (f + aU - \dot{U})\varphi \, dt + e(0)\varphi(0) = -\int_0^{t_N} R(U)\varphi \, dt + e(0)\varphi(0). \tag{57.12}$$

Next, we use the Galerkin orthogonality of cG(1),

$$\int_{t_{n-1}}^{t_n} R(U) \, dt = 0 \quad \text{for } n = 1, 2, \ldots,$$

to rewrite (57.12) as

$$e(T)^2 = -\int_0^T R(U)(\varphi - \bar{\varphi}) \, dt + e(0)\varphi(0), \tag{57.13}$$

where $\bar{\varphi}$ is the mean-value of φ over each time interval, that is

$$\bar{\varphi}(t) = \frac{1}{k_n} \int_{t_{n-1}}^{t_n} \varphi(s) \, ds \quad \text{for } t \in [t_{n-1}, t_n).$$

We shall now use

$$\int_{I_n} |\varphi - \bar{\varphi}| \, dt \leq k_n \int_{I_n} |\dot{\varphi}| \, dt,$$

which follows by integration from the facts that

$$\varphi(t) - \bar{\varphi}(t) = \frac{1}{k_n} \int_{t_{n-1}}^{t_n} (\varphi(t) - \varphi(s))\, ds,$$

and

$$|\varphi(t) - \varphi(s)| \le \int_s^t |\dot{\varphi}(\sigma)|\, d\sigma \le \int_{t_{n-1}}^{t_n} |\dot{\varphi}(\sigma)|\, d\sigma \quad \text{for } s, t \in [t_{n-1}, t_n].$$

Thus, (57.13) implies

$$
\begin{aligned}
|e(T)|^2 &\le \sum_{n=1}^N R_n \int_{I_n} |\varphi - \bar{\varphi}|\, dt + |e(0)||\varphi(0)| \\
&\le \sum_{n=1}^N k_n R_n \int_{I_n} |\dot{\varphi}|\, dt + |e(0)||\varphi(0)|,
\end{aligned}
\tag{57.14}
$$

where

$$R_n = \max_{t_{n-1} \le t \le t_n} |R(U(t))|.$$

Bringing out the max of $k_n R_n$ over n, we get

$$|e(T)|^2 \le \max_{1 \le n \le N} k_n R_n \int_0^{t_N} |\dot{\varphi}|\, dt + |e(0)||\varphi(0)|.$$

Recalling that $\varphi(T) = e(T)$ and using the definitions of $S_c(t_N)$ and $S_d(t_N)$, we get the following final estimate,

$$|e(T)| \le S_c(T) \max_{0 \le t \le T} |k(t) R(U(t))| + S_d(T)|e(0)|.$$

The stability factors $S_c(T)$ and $S_d(T)$ measure the effects of the accumulation of error in the approximation. To give the analysis a quantitative meaning, we have to give a quantitative bound of this factor. The following lemma gives an estimate for $S_c(T)$ and $S_d(T)$ in the cases $a \le 0$ and the case $a \ge 0$ with possibly vastly different stability factors. We notice that the solution $\varphi(t)$ of (57.11) is given by the explicit formula

$$\varphi(t) = e(T) \exp(a(T - t)).$$

We see that if $a \le 0$, then the solution $\varphi(t)$ decays as t decreases from T, and the case $a \le 0$ is thus the "stable case". If $a > 0$ then the exponential factor $\exp(aT)$ enters, and depending on the size of a this case is "unstable". More precisely, we conclude directly from the explicit solution formula that

Lemma 57.1 *The stability factors $S_c(T)$ and $S_d(T)$ satisfy if $a > 0$,*

$$S_d(T) \le \exp(aT), \quad S_c(T) \le \exp(aT), \tag{57.15}$$

and if $a \le 0$, then

$$S_d(T) \le 1, \quad S_c(T) \le 1. \tag{57.16}$$

57.5 Analysis of cG(1) for a General IVP

The extension of the a posteriori error analysis to a general IVP $\dot{u} = f(u)$ with $f : \mathbb{R}^d \to \mathbb{R}^d$ goes as follows. We recall that the linearized dual problem takes the form

$$-\dot{\varphi}(t) = A^\top(t)\varphi(t) \quad \text{for } 0 < t < T, \ \varphi(T) = e(T), \tag{57.17}$$

with

$$A(t) = \int_0^1 f'(su(t) + (1-s)U(t)) \, ds,$$

where $u(t)$ is the exact solution and $U(t)$ the approximate solution. We now use the fact that

$$A(t)e(t) = \int_0^1 f'(su(t) + (1-s)U(t))e(t) \, ds$$
$$= \int_0^1 \frac{d}{ds} f(su(t) + (1-s)U(t)) \, ds = f(u(t)) - f(U(t)), \tag{57.18}$$

where we used the Chain rule and the Fundamental Theorem of Calculus. We start from the identity

$$\|e(T)\|^2 = \|e(T)\|^2 + \int_0^T e \cdot (-\dot{\varphi} - A^\top \varphi) \, dt,$$

and integrate by parts to get the error representation,

$$\|e(T)\|^2 = \int_0^T (\dot{e} - Ae) \cdot \varphi \, dt + e(0) \cdot \varphi(0),$$

where we allow $U(0)$ to be different from $u(0)$, corresponding to an error in the initial value $u(0)$. Since u solves the differential equation $\dot{u} - f(u) = 0$, (57.18) implies

$$\dot{e} - Ae = \dot{u} - f(u) - \dot{U} + f(U) = -\dot{U} + f(U),$$

and thus we obtain the following representation of the error $\|e(T)\|^2$ in terms of the residual $R(U) = \dot{U} - f(U)$ and the dual solution φ,

$$\|e(T)\|^2 = -\int_0^{t_N} R(U)\varphi \, dt + e(0)\varphi(0). \tag{57.19}$$

From this point, the proof proceeds just as in the scalar case considered above and we thus obtain the following a posteriori error estimate

$$\|e(T)\| \leq S_c(T) \max_{0 \leq t \leq T} \|k(t)R(U(t))\| + S_d(T)\|e(0)\|,$$

which can be used a basis for adaptive time step control as described above. The stability factors $S_c(T)$ and $S_d(T)$ may be estimated by solving the dual problem with suitable initial data. The proof of the a posteriori error estimate shows that the stability factors may be defined by

$$S_d(T) = \frac{\|\varphi(t)\|}{\|e(T)\|},$$

$$S_c(T) = \frac{\int_0^t \|\dot{\varphi}(s)\|\, ds}{\|e(T)\|}, \tag{57.20}$$

where φ solves the linearized dual problem with initial data $\varphi(T) = e(T)$. As indicated, to compute the stability factors $S_d(T)$ and $S_c(T)$, we may solve the dual problem with some estimation of $e(T)$ obtained by solving the initial value problem with two tolerances and approximating $e(T)$ by the difference of the corresponding approximate solutions. Alternatively, choosing $\varphi(T) = e_i$, we obtain a posteriori error control for error component $e_i(T)$. If d is not large, we may this way control all components of the error, and if d is large, we may choose a couple different i at random.

The size of the stability factors indicate the degree of stability of the solution $u(t)$ being computed. If the stability factors are large, the residuals $R(U(t)$ and $e(0)$ have to be made correspondingly smaller by choosing smaller time steps and the computational problem is more demanding.

57.6   Analysis of Backward Euler for a General IVP

We now derive an a posteriori error estimate for the backward Euler method for the IVP (56.1):

$$U(t_n) = U(t_{n-1}) + k_n f(U(t_n)), \quad n = 1, 2, \ldots, N, \quad U(0) = u^0.$$

We associate a function $U(t)$ defined on $[0, T]$ to the function values $U(t_n)$, $n = 0, 1, \ldots, N$, as follows:

$$U(t) = U(t_n) \quad \text{for } t \in (t_{n-1}, t_n].$$

In other words, $U(t)$ is piecewise constant on $[0, T]$ and takes the value $U(t_n)$ on $(t_{n-1}, t_n]$, and thus takes a jump from the value $U(t_{n-1})$ to the value $U(t_n)$ at the time level t_{n-1}.

We can now write the backward Euler method as,

$$U(t_n) = U(t_{n-1}) + \int_{t_{n-1}}^{t_n} f(U(t))\, dt,$$

or equivalently

$$U(t_n) \cdot v = U(t_{n-1}) \cdot v + \int_{t_{n-1}}^{t_n} f(U(t)) \cdot v\, dt, \tag{57.21}$$

for all $v \in \mathbb{R}^d$. This method os also referred to as dG(0), that is the *discontinuous Galerkin method of order zero*, corresponding to *approximating the exact solution by a piecewise constant function $U(t)$ satisfying the orthogonality condition* (57.21).

We are now ready to derive an a posteriori error estimate following the same strategy as for the cG(1) method. We start from the identity

$$\|e(T)\|^2 = \|e(T)\|^2 + \sum_{n=1}^{N} \int_{t_{n-1}}^{t_n} e \cdot (-\dot\varphi - A^\top \varphi) \, dt,$$

and integrate by parts on each subinterval (t_{n-1}, t_n) to get the following error representation,

$$\|e(T)\|^2 = \sum_{n=1}^{N} \int_{t_{n-1}}^{t_n} (\dot e - Ae) \cdot \varphi \, dt$$
$$- \sum_{n=2}^{N-1} (U(t_n) - U(t_{n-1})) \varphi(t_{n-1}),$$

where the last term results from the jumps of $U(t)$ at the the nodes $t = t_{n-1}$ and we assume $U(0) = u(0)$ for simplicity. Since u solves the differential equation $\dot u - f(u) = 0$, (57.18) and the fact that $\dot U$ on (t_{n-1}, t_n) imply

$$\dot e - Ae = \dot u - f(u) - \dot U + f(U) = -\dot U + f(U) = f(U) \quad \text{on } (t_{n-1}, t_n),$$

and thus we obtain

$$\|e(T)\|^2 = -\sum_{n=2}^{N-1} (U(t_n) - U(t_{n-1})) \varphi(t_{n-1}) + \int_{0}^{t_N} f(U)\varphi \, dt.$$

Using (57.21) with $v = \bar\varphi$, the mean value of φ as above, we get

$$\|e(T)\|^2 = -\sum_{n=2}^{N-1} (U(t_n) - U(t_{n-1})) \cdot (\varphi(t_{n-1}) - \bar\varphi(t_{n-1}))$$
$$+ \sum_{n=1}^{n} \int_{t_{n-1}}^{t_n} f(U)(\varphi - \bar\varphi) \, dt.$$

We note that

$$\int_{t_{n-1}}^{t_n} f(U)(\varphi - \bar\varphi) \, dt = 0,$$

since $f(U(t))$ is constant on $(t_{n-1}, t_n]$, and $\bar\varphi$ is the mean value of φ, and thus the error representation takes the final form

$$\|e(T)\|^2 = -\sum_{n=2}^{N-1} (U(t_n) - U(t_{n-1})) \cdot (\varphi(t_{n-1}) - \bar\varphi(t_{n-1})).$$

Using

$$\|\varphi(t_{n-1}) - \bar{\varphi}(t_{n-1})\| \le \int_{t_{n-1}}^{t_n} \|\dot{\varphi}(t)\|\,dt,$$

we obtain the following a posteriori error estimate for the backward Euler method,

$$\|e(T)\| \le S_c(T) \max_{1 \le n \le N} \|U(t_n)) - U(t_{n-1}))\|. \qquad (57.22)$$

Note the very simple form of this estimate involving the jumps $\|U(t_n)) - U(t_{n-1}))\|$ playing the role the residual. The a posteriori error estimate (57.22) can be used as a basis for an algorithm for adaptive time step control of the following form: for $n = 1, 2, \ldots$, choose k_n so that

$$\|U(t_n)) - U(t_{n-1}))\| \approx \frac{TOL}{S_c(T)}.$$

57.7   Stiff Initial Value Problems

A *stiff* initial value problem $\dot{u} = f(u)$ may be characterized by the fact that the stability factors $S_d(T)$ and $S_c(T)$ are of moderate size even for large T, while the norm of the linearized operator $f'(u(t))$ is large, that is the Lipschitz constant L_f is very large. Such initial value problems are common for example in models of chemical reaction with reactions on a range of time scales from slow to fast. Typical solutions include so-called *transients* where the fast reactions make the solution change quickly over a short (initial) time interval, after which the fast reactions are "burned out" and the slow reactions make the solution change on a longer time scale.

The prototype of a stiff initial value problem has the form

$$\dot{u} = f(u) \equiv -Au \quad \text{for } t > 0, \ u(t) = u^0 = (u_i^0), \qquad (57.23)$$

where A is a constant symmetric positive semidefinite $d \times d$ matrix with non-negative eigenvalues λ_i ranging from zero to large positive values. Accordingly, the norm of the matrix A is large and L_f is large. By diagonalization, we may reduce to the case when A is a diagonal matrix with non-negative diagonal elements λ_i, in which case the solution $u(t) = (u_i(t))$ is given by

$$u_i(t) = \exp(-\lambda_i t) u_i^0 \quad \text{for } t > 0, \qquad (57.24)$$

with $u^0 = (u_i^0)$. This explicit solution formula shows that a component $u_i(t)$ corresponding to a large positive eigenvalue λ_i decays very quickly to zero, while a component with a small eigenvalue stays almost constant for a long time and eventually decays to zero. The sign of the eigenvalues is evidently crucial: if some λ_i was negative, then the corresponding solution component would explode exponentially more or less quickly depending on the size of λ_i. In particular, (57.24) with the λ_i non-negative implies

$$\|u(t)\| \leq \|u^0\| \quad \text{for } t > 0, \tag{57.25}$$

which indicates a form of stability with stability factor equal to 1 in the sense that the norm of the solution does not increase in time.

The dual problem corresponding to (57.23) takes the form

$$-\dot{\varphi} + A\varphi = 0 \quad \text{for } T > t > 0, \; \varphi(T) = \psi,$$

with ψ given data at time $t = T$. As a counterpart of (57.25), we conclude that $S_d(T) \leq 1$. We can similarly show that $S_c(T)$ grows very slowly with increasing T. We sum up: (57.23) represents a stiff problem; stability factors are of moderate size even for large T while the norm of the (linearized) operator A is large.

From numerical point of view, stiff problems may seem particularly friendly since the stability factors grow very slowly with time, but there is one hook that has attracted a lot of attention in the literature on numerical methods for initial value problems, namely the failure of an explicit method like the forward Euler method. We write this method for the equation $\dot{u} = -Au$ in the form

$$U^n = U^{n-1} - k_n A U^{n-1}$$

with U^n an approximation of $u(t_n)$ and $0 = t_0 < t_1 < \ldots$ an increasing sequence of time levels, and $k_n = t_n - t_{n-1}$. If A is diagonal with diagonal elements $\lambda_i \geq 0$, then

$$U_i^n = (1 - k_n \lambda_i) U_i^{n-1}$$

and if λ_i is large positive, then $|1 - k_n \lambda_i|$ may be much larger than 1 unless the time step k_n is sufficiently small ($k_n \leq 2/|\lambda_i|$ for all i) and the the numerical solution will then quickly explode to infinity, while the corresponding exact solution quickly decays to zero. The explicit Euler method will thus give completely wrong results unless sufficiently small time steps are used. This may lead to very inefficient time-stepping since after the transients have died out, the solution may vary only slowly and large time steps would be desirable. We note that the time step limit $k_n \leq 2/|\lambda_i|$ for all i, is set by the largest eigenvalue $\max \lambda_i$, while the time long-time scale is set by the smallest eigenvalue $\min \lambda_i$, so that if the quotient $\max \lambda_i / \min \lambda_i$ is large (which signifies a stiff problem), then explicit Euler would be inefficient outside transients.

On the other hand, the dG(0), or implicit Euler method,

$$U^n + k_n A U^n = U^{n-1}$$

with

$$U_i^n = (1 + k_n\lambda_i)^{-1}U_i^{n-1}$$

will be stable and work very well without step size limitation because $1 + k_n\lambda_i \geq 1$ for all $\lambda_i \leq 0$.

For the cG(1)-method, we will have

$$U_i^n = \frac{1 - k_n\lambda_i}{1 + k_n\lambda_i}U_i^{n-1}$$

and stability prevails because

$$\left|\frac{1 - k_n\lambda_i}{1 + k_n\lambda_i}\right| \leq 1$$

for all $\lambda_i \geq 0$.

We conclude that both dG(0) and cG(1) may be used for stiff problems, but both these methods are implicit and require the solution of system of equations at each time step. More precisely, dG(0) for a problem of the form $\dot{u} = f(u)$ takes the form

$$U^n - k_n f(U^n) = U^{n-1}.$$

At each time step we have to solve an equation of the form $v - k_n f(v) = U^{n-1}$ with U^{n-1} given. To this end we may try a damped fixed point iteration in the form

$$v^{(m)} = v^{(m-1)} - \alpha(v^{(m-1)} - k_n f(v^{(m-1)}) - U^{n-1}),$$

with α some suitable matrix (or constant in the simplest case). Choosing $\alpha = I$, and iterating once with $v^0 = 0$ corresponds to the explicit Euler method. Convergence of the fixed point iteration requires that

$$\|I + k_n\alpha f'(v)\| < 1$$

for relevant values of v, which could force α to be small (e.g. in the stiff case with $f'(v)$ having large negative eigenvalues) and result in slow convergence. A first try could be to choose α to be a diagonal matrix with $\alpha_i = (f'_{ii})(v^{m-1}))^{-1}$ (corresponding to *diagonal scaling*) and hope that the number of iterations would not be too large. In some cases more efficient iterative solvers would have to be used.

57.8   On Explicit Time-Stepping for Stiff Problems

We just learned that explicit time-stepping for stiff problems require small time steps outside transients and thus may be inefficient. We shall now indicate a way to get around this limitation through a process of stabilization, where a large time step is accompanied by a couple of small time steps. The resulting method has similarities with the control system of a modern (unstable) jet fighter like the Swedish JAS Gripen, the flight of which is controlled by quick small flaps of a pair of small extra wings ahead of the main wings, or balancing a stick vertically on the finger tips if we want a more domestic application.

We shall now explain the basic (simple) idea of the stabilization and present some examples, as illustrations of fundamental aspects of adaptive IVP-solvers and stiff problems. Thus to start with, suppose we apply explicit Euler to the scalar problem

$$\dot{u}(t) + \lambda u(t) = 0 \quad \text{for } t > 0.$$
$$u(0) = u^0, \tag{57.26}$$

with $\lambda > 0$ taking first a large time step K satisfying $K\lambda > 2$ and then m small time steps k satisfying $k\lambda < 2$, to get the method

$$U^n = (1 - k\lambda)^m (1 - K\lambda)U^{n-1}, \tag{57.27}$$

altogether corresponding to a time step of size $k_n = K + mk$. Here K gives a large unstable time step with $|1 - K\lambda| > 1$ and k is a small time step with $|1 - k\lambda| < 1$. Defining the polynomial function $p(x) = (1 - \theta x)^m(1 - x)$, where $\theta = \frac{k}{K}$, we can write the method (57.27) in the form

$$U^n = p(K\lambda)U^{n-1}.$$

For stability we need

$$|p(K\lambda)| \leq 1, \quad \text{that is } |1 - k\lambda|^m(K\lambda - 1) \leq 1,$$

or

$$m \geq \frac{\log(K\lambda - 1)}{-\log|1 - k\lambda|} \approx 2\log(K\lambda), \tag{57.28}$$

with $c = k\lambda \approx 1/2$ for definiteness.

We conclude that m may be quite small even if $K\lambda$ is large, since the logarithm grows so slowly, and then only a small fraction of the total time would be spent on stabilizing time-stepping with the small time steps k.

To measure the efficiency gain we introduce

$$\alpha = \frac{1 + m}{K + km} \in (1/K, 1/k),$$

which is the number of time steps per unit interval with stabilized explicit Euler method, and by (57.28) we have

$$\alpha \approx \frac{1 + 2\log(K\lambda)}{K + \log(K\lambda)/\lambda} \approx 2\lambda \frac{\log(K\lambda)}{K\lambda} \ll 2\lambda, \tag{57.29}$$

for $K\lambda \gg 1$. On the other hand, the number of time steps per unit interval for the usual explicit Euler is

$$\alpha_0 = 1/k = \lambda/2, \tag{57.30}$$

choosing a maximum time step $k = 2/\lambda$.

The cost reduction factor using the stabilized explicit Euler method would thus be

$$\frac{\alpha}{\alpha_0} \approx \frac{4\log(K\lambda)}{K\lambda}$$

which can be quite significant for large values of $K\lambda$.

We now present some examples using an adaptive cg(1) IVP-solver in stabilized explicit form with just a few iterations in each time step, which allows large time steps. In all problems we note the initial transient, where the solution components change quickly, and the oscillating nature of the time step sequence outside the transient with large time steps followed by some small stabilizing time steps.

Example 57.1. We apply the indicated method to the scalar problem equation (57.26) with $u^0 = 1$ and $\lambda = 1000$, and display the result in Figure 57.1. The cost reduction factor with comparison to a standard explicit method is large: $\alpha/\alpha_0 \approx 1/310$.

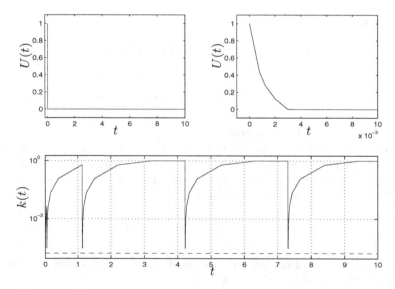

Fig. 57.1. Solution and time step sequence for (57.26), $\alpha/\alpha_0 \approx 1/310$

Example 57.2. We now consider the 2×2 diagonal system

$$\dot{u}(t) + \begin{pmatrix} 100 & 0 \\ 0 & 1000 \end{pmatrix} u(t) = 0 \quad \text{for } t > 0,$$
$$u(0) = u^0,$$

(57.31)

with $u^0 = (1, 1)$. There are now two eigenmodes with large eigenvalues that have to be stabilized. The cost reduction is $\alpha/\alpha_0 \approx 1/104$.

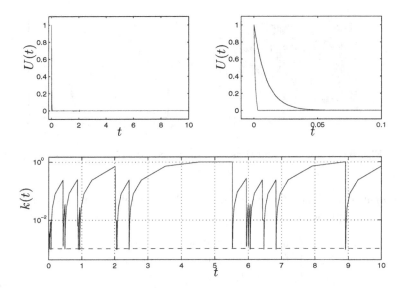

Fig. 57.2. Solution and time step sequence for (57.31), $\alpha/\alpha_0 \approx 1/104$

Example 57.3. The is the so-called HIRES problem ("High Irradiance RESponse") from plant physiology which consists of the following eight equations:

$$\begin{cases} \dot{u}_1 &= -1.71u_1 + 0.43u_2 + 8.32u_3 + 0.0007, \\ \dot{u}_2 &= 1.71u_1 - 8.75u_2, \\ \dot{u}_3 &= -10.03u_3 + 0.43u_4 + 0.035u_5, \\ \dot{u}_4 &= 8.32u_2 + 1.71u_3 - 1.12u_4, \\ \dot{u}_5 &= -1.745u_5 + 0.43u_6 + 0.43u_7, \\ \dot{u}_6 &= -280.0u_6u_8 + 0.69u_4 + 1.71u_5 - 0.43u_6 + 0.69u_7, \\ \dot{u}_7 &= 280.0u_6u_8 - 1.81u_7, \\ \dot{u}_8 &= -280.0u_6u_8 + 1.81u_7, \end{cases}$$

(57.32)

together with the initial condition $u^0 = (1.0, 0, 0, 0, 0, 0, 0, 0.0057)$. We present the solution and the time step sequence in Figure 57.3. The cost is now $\alpha \approx 8$ and the cost reduction factor is $\alpha/\alpha_0 \approx 1/33$.

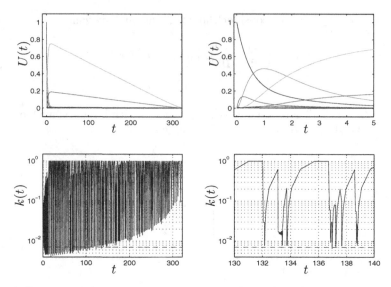

Fig. 57.3. Solution and time step sequence for (57.32), $\alpha/\alpha_0 \approx 1/33$

Example 57.4. The "Chemical Akzo-Nobel" problem consists of the following six equations:

$$\begin{cases} \dot{u}_1 &= -2r_1 + r_2 - r_3 - r_4, \\ \dot{u}_2 &= -0.5r_1 - r_4 - 0.5r_5 + F, \\ \dot{u}_3 &= r_1 - r_2 + r_3, \\ \dot{u}_4 &= -r_2 + r_3 - 2r_4, \\ \dot{u}_5 &= r_2 - r_3 + r_5, \\ \dot{u}_6 &= -r_5, \end{cases} \qquad (57.33)$$

where $F = 3.3 \cdot (0.9/737 - u_2)$ and the reaction rates are given by $r_1 = 18.7 \cdot u_1^4 \sqrt{u_2}$, $r_2 = 0.58 \cdot u_3 u_4$, $r_3 = 0.58/34.4 \cdot u_1 u_5$, $r_4 = 0.09 \cdot u_1 u_4^2$ and $r_5 = 0.42 \cdot u_6^2 \sqrt{u_2}$. We integrate over the interval $[0, 180]$ with initial condition $u^0 = (0.437, 0.00123, 0, 0, 0, 0.367)$. Allowing a maximum time step of $k_{max} = 1$ (chosen arbitrarily), the cost is only $\alpha \approx 2$ and the cost reduction factor is $\alpha/\alpha_0 \approx 1/9$. The actual gain in a specific situation is determined by the quotient between the large time steps and the small damping time steps, as well as the number of small damping steps that are needed. In this case the number of small damping steps is small, but the large time steps are not very large compared to the small damping steps. The gain is thus determined both by the stiff nature of the problem and the tolerance (or the size of the maximum allowed time step).

Example 57.5. We consider now Van der Pol's equation:

$$\ddot{u} + \mu(u^2 - 1)\dot{u} + u = 0,$$

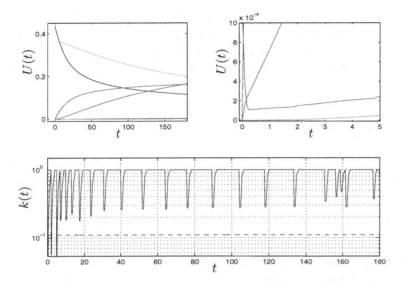

Fig. 57.4. Solution and time step sequence for (57.33), $\alpha/\alpha_0 \approx 1/9$

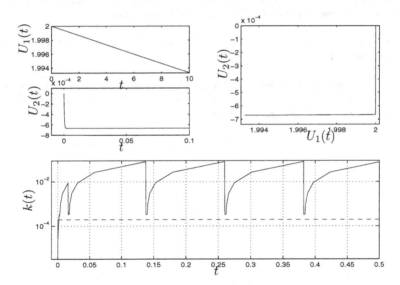

Fig. 57.5. Solution and time step sequence for (57.34), $\alpha/\alpha_0 \approx 1/75$

which we write as

$$\begin{cases} \dot{u}_1 = u_2, \\ \dot{u}_2 = -\mu(u_1^2 - 1)u_2 - u_1. \end{cases} \tag{57.34}$$

We take $\mu = 1000$ and solve on the interval $[0, 10]$ with initial condition $u^0 = (2, 0)$. The time step sequence behaves as desired with only a small portion of the time spent on taking small damping steps. The cost is now $\alpha \approx 140$ and the cost reduction factor is $\alpha/\alpha_0 \approx 1/75$.

Chapter 57   Problems

57.1. Compute the stability factors $S_d(T)$ and $S_c(T)$ for the linear scalar IVP $\dot{u}(t) = -\lambda(t)u(t)$ for $t > 0$, $u(0) = u^0$, where $\lambda(t)$ depends on time t and (a) $\lambda(t) \geq 0$, (b) $\lambda(t) < 0$.

57.2. Compute $S_d(T)$ and $S_c(T)$ for the linear 2×2 system $\dot{u}_1 = u_2$, $\dot{u}_2 = -u_1$ for $t > 0$, $u(0) = u^0$.

57.3. Implement adaptive IVP-solvers based on dG(0) and cG(1) and apply the solvers to different problems.

57.4. Show that the a posteriori error estimate for cG(1) may be written on the form $\|e(T)\| \leq S_c(T) \max_{0 \leq t \leq T} \|k(t)(f(U(t)) - \bar{f}(U(t)))\| + S_d(T)\|e(0)\|$, where $\bar{f}(U(t))$ is the mean-value of $f(U(t))$ over each time interval.

57.5. Show that choosing in the dual problem $\varphi(T) = e_i$ gives control of error component $e_i(T)$.

57.6. Develop explicit versions of dG(0) and cG(1) based on fixed point iteration at each time step. Show that with diagonal scaling such an explicit method may work very well for some stiff problems.

58

Lorenz and the Essence of Chaos*

I am convinced that chaos, along with its many associated concepts –
strange attractors, basin boundaries, period-doubling bifurcations
and the like – can readily be understood and relished by readers
who have no special mathematical or other scientific background...
(E. Lorenz, in Foreword to *The Essence of Chaos*)

58.1 Introduction

On December 29, 1972, the meteorologist Edward Lorenz presented in a session on the Global Atmospheric Research Program at the 139th meeting of the American Association for the Advancement of Science in Washington D.C., a talk with the title *Predictability: Does the Flap of a Butterfly's Wings in Brazil Set off a Tornado in Texas?* The talk by Lorenz with its "Butterfly effect" rocketed to fame a decade later during the development of "Chaos Theory" that became a fashion in mathematics and physics during the 80s, with the pretention of explaining a variety of phenomena from turbulent fluid flow to collapsing stock markets sharing qualities of *unpredictability*. A decade earlier, "Catastrophe Theory" played a similar role, while today very few remember this intriguing subject. Of course, unpredictability or "chaos" is a phenomenon that has long been familiar to mankind. The word "chaos" comes from early Greek cosmology and signifies the complete lack of order of the Universe before the creation of Gaea and Eros (Earth and Desire).

Lorenz' question is connected to the obvious difficulty of making reasonably reliable predictions of the daily weather over longer time than a week.
A weather forecast is made by numerically solving an IVP modeling the
evolution of the atmosphere, including variables such as temperature, wind
speed and pressure. There are many sources of errors in a weather forecast
made this way: errors in the initial value, modeling errors and numerical
errors, and it seems that these errors are magnified at a rate that limits the
predictions, depending on the scale from a few hours in very local models
to weeks in global circulation models.

Lorenz' Butterfly analogy indicates that in certain dynamical systems,
very small causes may have large effects after some time. We have already
met such a problem in the form of a pendulum being released starting from
the unstable top position: depending on the initial perturbation the position
of the pendulum will be vastly different after some time (one side or the
other). In meteorology, this corresponds to a situation where the weather-
man can't say if a certain low pressure will take this way or that way, and
thus can't be sure if it will rain in Göteborg tomorrow or not. In his book,
Lorenz gives other examples of unstable systems such as a pinball machine,
where very small changes in the action of the player can change the outcome
of the game completely. Of course there are many other examples from
real life of "small" causes having large effects, from soccer games to the
assassination of Archduke Francis Ferdinand by the Serb nationalist Gavrilo
Princip in Sarajevo on June 28, 1914, initiating the First World War.

58.2   The Lorenz System

Lorenz formulated an IVP of the form $\dot{u} = f(u)$ with $f : \mathbb{R}^3 \to \mathbb{R}^3$ given
by

$$f(u) = \left(-10u_1 + 10u_2, 28u_1 - u_2 - u_1u_3, -\frac{8}{3}u_3 + u_1u_2\right),$$

which is the famous *Lorenz system*. Lorenz found that the solution of this
system is very sensitive to perturbations. The system has some vague connection to a very simple model for fluid flow and has been given the role
of explaining properties of fluid motion, such as turbulence. This was not
Lorenz' original idea, who just wanted to make a connection to the apparent unpredictability and supposed sensitivity to perturbations of common
meteorological models. If the seemingly very harmless and innocent Lorenz
system could have unpredictable solutions, then there should be no surprise
that also the weather could be unpredictable.

More precisely, Lorenz found that two solutions of the Lorenz system
with very close initial data will stay close for some time but will eventually
move apart completely. The Lorenz system is therefore very difficult to solve

accurately using a numerical method over times longer than say 30 units. The numerical solution will stay close the the exact solution for some time, but will eventually move apart significantly. Of course there are many IVP:s sharing this property of instability. Even the simple pendulum has this property if the pendulum reaches the top position with small velocity. It is thus remarkable that the Lorenz system seemed to present some kind of surprise to the scientific world. But it did, and it has become quite popular to explain all sorts of phenomena, from turbulence to politics, by referring to the "strange attractor" supposedly being displayed in plots of solutions of the Lorenz system.

The Lorenz system in component form reads:

$$\begin{cases} \dot{u}_1 = -10u_1 + 10u_2, \\ \dot{u}_2 = 28u_1 - u_2 - u_1u_3, \\ \dot{u}_3 = -\frac{8}{3}u_3 + u_1u_2, \\ u_1(0) = u_{01},\ u_2(0) = u_{02},\ u_3(0) = u_{03}, \end{cases} \tag{58.1}$$

and u_0 is a given initial condition. The system (58.1) has three equilibrium points $\bar{u}$ with $f(\bar{u}) = 0$: $\bar{u} = (0,0,0)$ and $\bar{u} = (\pm 6\sqrt{2}, \pm 6\sqrt{2}, 27)$. The equilibrium point $\bar{u} = (0,0,0)$ is unstable with the corresponding Jacobian $f'(\bar{u})$ having one positive (unstable) eigenvalue and two negative (stable) eigenvalues. The equilibrium points $(\pm 6\sqrt{2}, \pm 6\sqrt{2}, 27)$ are slightly unstable with the corresponding Jacobians having one negative (stable) eigenvalue and two eigenvalues with very small positive real part (slightly unstable) and also an imaginary part. More precisely, the eigenvalues at the two non-zero equilibrium points are $\lambda_1 \approx -13.9$ and $\lambda_{2,3} \approx .0939 \pm 10.1i$.

In Fig. 58.1, we present two views of a solution $u(t)$ that starts at $u(0) = (1,0,0)$ computed to time 30 with an error tolerance of $TOL = 0.5$ using an adaptive IVP-solver of the form presented in Chapter *Adaptive IVP-solvers*. We can think of $u(t) = (x(t), y(t), z(t))$ as the position at time t of a particle that moves according to the equation $\dot{u} = f(u)$. In Fig. 58.1 we thus plot the trajectory or path followed by the particle as the particle moves with increasing time. The plotted trajectory is typical: the particle is kicked away from the unstable point $(0,0,0)$ and moves towards one of the non-zero equilibrium points. It then slowly orbits away from that point and at some time decides to cross over towards the other non-zero equilibrium point, again slowly orbiting away from that point and coming back again, orbiting out, crossing over, and so on. This pattern of some orbits around one non-zero equilibrium point followed by a transition to the other non-zero equilibrium point is repeated with a seemingly random number of revolutions around each equilibrium point.

As noted by Lorenz, a close inspection of the trajectory in Fig. 58.1 reveals quite a bit of structure in the behavior of the solution. From the path of the trajectory, it seems that, roughly speaking, there are two flat "lobes" in which the orbits around the non-zero equilibrium points are

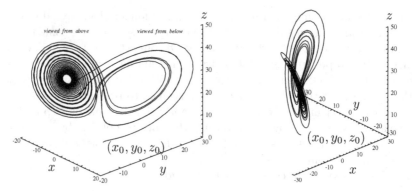

Fig. 58.1. Two views of a numerical trajectory of the Lorenz system over the time interval $[0, 30]$ starting at $(1, 0, 0)$ computed with absolute error tolerance 0.5

located. In each lobe, the spiraling segments of the trajectory seem to be grouped in "bands" that are made up of parts of the trajectory that are spiralling out from the equilibrium point and parts of the trajectory that have just crossed over from the other lobe. Only the trajectories in the outer band switch to the other equilibrium point. This causes a sharp separation between trajectories located in the outer band and those located in the next band inside as the trajectories approach the z-axis. We refer to this as *cutting* through the action of a "razor" separating the trajectories in the outer band. The trajectories in the outer band expand in width as they approach the other equilibrium point, with trajectories near the outside of the band ending up nearer to the fixed point. We refer to this as *expansion* and *flipping* respectively. The position of initial approach of the trajectory to an equilibrium point determines the number of orbits the trajectory makes in that lobe before returning to the other equilibrium point. Finally, we see that the orbits in one band come close to the next outer band after one revolution, this repeats with every band of the trajectory, until eventually they all end up in the outer band and leave towards the other equilibrium point. We refer to this as *interlacing*. In short, we can describe the dynamics of the Lorenz system as a never-ending process of cutting, expansion, flipping, and interlacing.

58.3   The Accuracy of the Computations

The first task is to measure the reliability of the computed error bound based on an a posteriori error estimate of the form presented in Chapter *Adaptive IVP-solvers*. Since we do not have the exact solution, we perform the following experiment: Using the initial data $(0, 1, 0)$, we compute twice using residual tolerances 10^{-5} and 10^{-9} and approximate the error in the less accurate computation by taking the difference between the values of the less accurate and more accurate computations. In Fig. 58.2, we plot the computed error bound and the approximate error. The error bound predicts the size of the error quite well in spite of the sensitivity of the solution to perturbations. Similar results are obtained for a variety of initial data.

To give some idea of the behavior of the error control, we plot the step sizes used in a computation with absolute error tolerance 0.75 in Fig. 58.3. The step sizes vary roughly by a factor of 6 over the interval of computation. In Fig. 58.3, we also plot the product of the time step and the residual for this computation. We note that these values are kept within

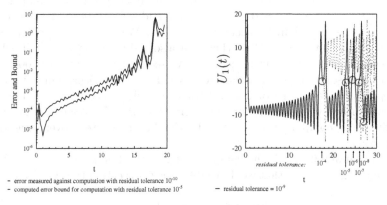

Fig. 58.2. On the *left*, we plot the results of the reliability test for the computed error bound with initial condition is $(0, 1, 0)$. On the *right*, we plot the effect of changing the residual tolerance on the accuracy in $U_1(t)$ component with initial condition is $(1, 0, 0)$

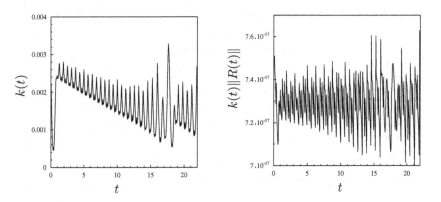

Fig. 58.3. Time steps and residual × time step versus time for a computation beginning at $(1, 0, 0)$ with absolute error tolerance 0.75

10% of a constant value. With more computational work, the size of the variations can be reduced, which produces a more smoothly-varying error bound.

58.4 Computability of the Lorenz System

Encouraged by these results we decrease the tolerance or, equivalently, the time step, and try to compute an accurate solution to the Lorenz system on an even longer time interval. Using the cG(1) method as described in Chapter *Adaptive IVP-solvers*, we compute solutions with smaller and smaller time steps, $k = 0.01$, $k = 0.001$ and $k = 0.0001$, and expect to produce more and more accurate solutions. We plot the U_1-component of the solution in Fig. 58.4 where we also indicate the points at which the solutions are no longer accurate. We see that even with $300,000$ time steps the solution is not accurate beyond $t = 26$. Decreasing the time step with a factor 10 or 100 will take us only a little further, but the computation will take 10 or 100 times longer. We conclude that it is difficult to compute the solution to the Lorenz system over long time intervals.

To examine in detail the computability of the Lorenz system we return to the error estimate that we derived for the error $e(t)$ of the cG(1) method:

$$\|e(t)\| \leq S_c(T) \max_{0 \leq t \leq T} \|k(t)R(t)\|. \tag{58.2}$$

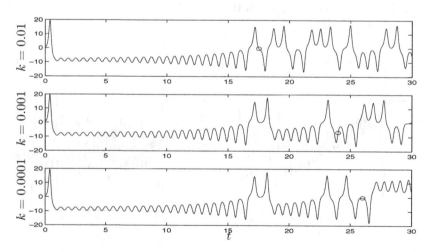

Fig. 58.4. The U_1-component of the cG(1) solution for different time steps. The *small circles* indicate the points at which the solutions are no longer accurate

Remember that the stability factor $S_c(T)$ for the Lorenz system is defined in terms of the solution to the linearized dual problem as

$$S_c(T) = \max_{\varphi_0 \in \mathbb{R}^3} \frac{\int_0^T \|\dot{\varphi}(t)\| \, dt}{\|\varphi_0\|}.$$

Judging by the error estimate we should be able to reach as far as we want if only the time step $k(t)$ and the residual $R(t)$ are small enough. However, a little more careful analysis reveals an additional error contribution, which is often ignored. Including also this term into our error estimate, we find:

$$\|e(t)\| \leq S_c(T) \max_{0 \leq t \leq T} \|k(t)R(t)\| + S_0(T) \max_{0 \leq t \leq T} \epsilon/k(t), \qquad (58.3)$$

where ϵ is the *machine precision* of the computer, i.e. the smallest number for which $1 + \epsilon \neq 1$ (in computer arithmetic) and $S_0(T)$ is a new stability factor. For a standard computer (in 2002) with so-called double-precision arithmetic, the machine precision is $\epsilon \approx 10^{-16}$. The stability factor $S_0(T)$ is defined in terms of the dual solution as

$$S_0(T) = \max_{\varphi_0 \in \mathbb{R}^3} \frac{\int_0^T \|\varphi(t)\| \, dt}{\|\varphi_0\|}.$$

The additional term in our refined error estimate (58.3) accounts for the round-off error made at every time step in the computation; when the new value $U(t_n)$ for the cG(1) solution is computed in every time step, it is unavoidable that we make a round-off error of size ϵ. As we shall see, it is the second term that sets the limit for the computability of the Lorenz system; the second term in (58.3) can be large even though the first term is small.

The difficulty of computing accurate solutions to the Lorenz system becomes obvious if we plot the size of the stability factors. In Fig. 58.5 we plot the size of the stability factor $S_0(T)$ associated with round-off errors as function of the final time T. Notice the logarithmic scale in this figure. A simple approximation of the growth of this stability factor is

$$S_0(T) \approx 10^{T/3},$$

and so the round-off error grows as $E_r = 10^{T/3} \cdot 10^{-16}/k = 10^{T/3-16}/k$. Notice how the error grows larger if we decrease the time step! This is natural (although unusual), since with a smaller time step we will have to take a larger number of time steps and thus make a larger number of round-off errors. To make the influence from round-off errors small we specify a large time step, say $k = 0.1$, for which the round-off error now grows as $10^{T/3-15}$. At time $T = 3 \cdot 15 = 45$ the accumulated round-off error is then $E_r = 1$, which means that we cannot expect to compute much beyond time $T = 45$, since then the round-off error will dominate anyway. Using the cG(1) method, we will not even reach $T = 45$, since we have to use a time step much smaller than $k = 0.1$ (as seen in Fig. 58.4) to make the first term in the error estimate small.

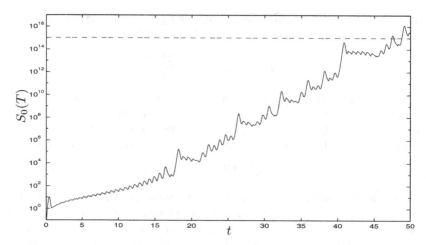

Fig. 58.5. The growth of the stability factor $S_0(T)$ for the Lorenz problem

58.5   The Lorenz Challenge

From the previous discussion it is now clear that the mysterious unpredictability and "chaotic" behavior of the Lorenz system only means that the stability factors grow quickly, making it difficult to compute accurate solutions over long time intervals. The obvious challenge is now, using the method of choice, *to compute an accurate solution to the Lorenz system over as long a time interval $[0, T]$ as possible.*

We saw in the previous section that brute force is not the way to go. It is not enough to use a very fast computer with very small and very many time steps. Using the cG(1) method we cannot reach much further than $T = 30$, no matter how small time steps we use since then the accumulated round-off error will grow large. A solution to this problem would be if we could design a method, similar to cG(1), which can be used with larger time steps than what is possible with cG(1). As one can expect, there exist corresponding methods cG(2), cG(3) and so on, which can be used with larger time steps. It can be proved that for these cG(q) methods, the error grows as k^{2q}, i.e. we have so called *a priori* error estimates of the type

$$\|e(T)\| \leq C(T)k^{2q},$$

where $C(T)$ is a constant (unknown!) depending on the exact solution $u(t)$. We say that the cG(q) method is of *order* $2q$. The standard cG(1) method is thus a second order method. (This is in agreement with (58.2) since one factor $k(t)$ is hidden inside $R(t)$.) With a higher order method, i.e. $q > 1$, we can thus obtain a smaller error with the same time step, which makes it possible to compute the solution with larger time steps. This in turn implies that with a higher order method, we can keep the round-off error smaller and thus reach further than what is possible with the cG(1) method. In Fig. 58.6 we plot the U_1-components of solutions to the Lorenz system, computed with time step $k = 0.1$, with a sequence of higher order methods. We see that with high enough order, the solutions agree to a point just beyond $T = 45$ as we predicted; the first term in our error estimate (58.3) has been reduced by increasing the order of the method and so the second term dominates. It is possible to reach beyond time $T = 50$, perhaps to $T = 100$, but to do this we have to go from double-precision arithmetic to quadruple-precision.

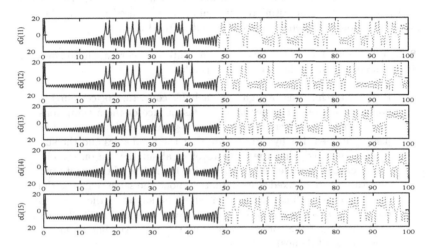

Fig. 58.6. The U_1-component of cG(q) solutions for $q = 11, 12, 13, 14, 15$ with time step $k = 0.1$. *Dashed lines* indicate where the solution is no longer accurate

Chapter 58 Problems

58.1. Verify that the three equilibrium points given in the text satisfy $f(u) = 0$. Linearize the system around these equilibrium points, i.e. compute the eigenvalues (and eigenvectors) for the Jacobian of f at the three equilibrium points.

58.2. Compute a solution to the Lorenz system and plot the orbit $(x(t), y(t), z(t))$ for $t \in [0, T]$. Do you agree with the description of the dynamics of the Lorenz system as never-ending process of cutting, expansion, flipping, and interlacing?

58.3. Repeat the experiment outlined in Sect. 58.4, i.e. compute solutions to the Lorenz system using the cG(1) method with a sequence of smaller and smaller time steps and examine the accuracy of the solutions (by comparing them to each other). Can you reach beyond $T = 25$?

58.4. Try the same experiment as in the previous problem but now with the lower order methods explicit Euler and implicit Euler. How far do you reach now?

58.5. Implement a simple version of the fourth-order cG(2) method given by

$$U(t_{n-1/2}) = U(t_{n-1}) + \int_{t_{n-1}}^{t^n} f(U(t), t) \cdot (5 - 6(t - t_{n-1})/k_n)/4 \, dt,$$
$$U(t_n) = U(t_{n-1}) + \int_{t_{n-1}}^{t^n} f(U(t), t) \, dt,$$

where $U(t)$ is the quadratic polynomial on $[t_{n-1}, t_n]$ determined by the three values $U(t_{n-1})$, $U(t_{n-1/2})$ and $U(t_n)$. How much further can you reach with this method?

58.6. Give a motivation for the additional term in the refined error estimate (58.3), starting from the estimate containing errors caused by using the wrong initial condition as presented in Chapter *Adaptive IVP-solvers*.

58.7. Take on the Lorenz Challenge, i.e. compute an accurate solution over $[0, T]$ with T as large as possible. No rules, all is allowed!

> But Aristarchus has brought out a book consisting of certain hypotheses, wherein it appears, as a consequence of the assumptions made, that the universe is many times greater than the 'universe' just mentioned. His hypotheses are that the fixed stars and the sun remain unmoved, that the earth revolves about the sun on the circumference of a circle, the sun lying in the middle of the orbit, and that the sphere of fixed stars, situated about the same centre as the sun, is so great that the circle in which he supposes the earth to revolve bears such a proportion to the distance of the fixed stars as the centre of the sphere bears to its surface. (Archimedes about Aristharcus of Samos)

59

The Solar System*

There is talk of a new astrologer who wants to prove that the earth moves and goes around instead of the sky, the sun, the moon, just as if somebody were moving in a carriage or ship might hold that he was sitting still and at rest while the earth and the trees walked and moved. But that is how things are nowadays: when a man wishes to be clever he must needs invent something special, and the way he does it must needs be the best! The fool wants to turn the whole art of astronomy upside-down. However, as Holy Scripture tells us, so did Joshua bid the sun to stand still and not the earth.
(Sixteenth century reformist M. Luther in his table book *Tischreden*, in response to Copernicus' pamphlet *Commentariolus*, 1514.)

59.1 Introduction

The problem of mathematical modeling of our Solar System including the Sun, the 9 planets Venus, Mercury, Tellus (the Earth), Mars, Jupiter, Saturn, Uranus, Neptune and Pluto together with a large number of moons and asteroids and occasional comets, has been of prime concern for humanity since the dawn of culture. The ultimate challenge concerns mathematical modeling of the Universe consisting of billions of galaxies each one consisting of billions of stars, one of them being our own Sun situated in the outskirts of the Milky Way galaxy.

According to the *geocentric* view presented by Aristotle (384–322 BC) in *The Heavens* and further developed by Ptolemy (87–150 AD) in *The Great*

System dominating the scene over 1800 years, the Earth is the center of the Universe with the Sun, the Moon, the other planets and the stars moving around the Earth in a complex pattern of circles upon circles (so-called epicycles). Copernicus (1473–1543) changed the view in *De Revolutionibus* and placed the Sun in the center in a new *heliocentric* theory, but kept the complex system of epicycles (now enlarged to a very complex system of 80 circles upon circles). Johannes Kepler (1572–1630) discovered, based on the extensive accurate observations made by the Swedish/Danish scientist Tycho Brahe (1546–1601), that the planets move in elliptic orbits with the Sun in one of the foci following *Kepler's laws*, which represented an enormous simplification and scientific rationalization as compared to the system of epicycles.

Fig. 59.1. The Earth with the Moon and some other planets in orbit. Jupiter and the Galilean satellites, Io, Europa, Ganymede, and Callisto

In fact, already Aristarchus (310–230 BC) of Samos understood that the Earth rotates around its axis and thus could explain the (apparent) motion of the stars, but these views were rejected by Aristotle arguing as follows: if the Earth is rotating, how is it that an object thrown upwards falls on the same place? How come this rotation does not generate a very strong wind? No one until Copernicus could question these arguments. Can you?

Newton (1642–1727) then cleaned up the theory by showing that the motion of the planets could be explained from one single hypothesis: the inverse square law of gravitation, see Chapter *Newton's nightmare* below. In particular, Newton derived Kepler's laws for the *two-body problem* with one (small) planet in an elliptic orbit around a (large) sun, see Chapter *Lagrange and the Principle of Least Action*. Leibniz criticized Newton for not giving any explanation of the inverse square law, which Leibniz believed

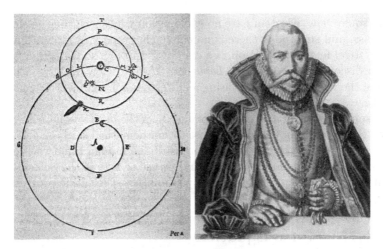

Fig. 59.2. Tycho Brahe: "I believe that the Sun and the Moon orbit around the Earth but that the other planets orbit around the Sun"

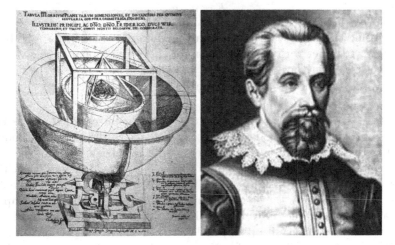

Fig. 59.3. Johannes Kepler: "I believe that the planets are separated by invisible regular polyhedra: tetrahedron, cube, octahedron, dodekahedron and ikosahedron, and further that the planets including the Earth move in elliptical orbits around the Sun"

could be derived from some basic fact, beyond one of "mutual love" which was quite popular. A sort of explanation was given by Einstein (1879–1955) in his theory of *General Relativity* with gravitation arising as a consequence of space-time being "curved" by the presence of mass. Einstein revolutionized *cosmology*, the theory of the Universe, but relativistic effects only add small corrections to Newton's model for our Solar System based on the inverse square law. Einstein gave no explanation why space-time gets curved by mass, and still today there is no convincing theory of gravitation with its mystical feature of "action at a distance" through some mechanism yet to be discovered. In Chapter *Laplacian Models* below we give a derivation of the inverse square law using a mathematical argument presented by Laplace.

Despite the lack of a physical explanation of the inverse square law, Newton's theory gave an enormous boost to mathematical sciences and a corresponding kick to the egos of scientists: if the human mind was capable of (so easily and definitely) understanding the secrets of the Solar System, then there could be no limits to the possibilities of scientific progress...

59.2 Newton's Equation

The basis of celestial mechanics is Newton's second law,

$$F = m \cdot a, \tag{59.1}$$

expressing that a *force* F results in an acceleration of size a for a body of mass m, together with the expression for the gravitational force given by the inverse square law:

$$F = G \frac{m m_a}{r^2}, \tag{59.2}$$

where $G \approx 6.67 \cdot 10^{-11} \mathrm{Nm}^2/\mathrm{kg}^2$ is the *gravitational constant*, m_a is the mass of the attracting body and r is the distance to the attracting body.

Together (59.1) and (59.2) give a set of differential equations for the evolution of the Solar System. If we know the initial positions and velocities for all bodies in the Solar System, we can solve the system of differential equations, using the same techniques as presented above in Chapter Adaptive IVP-Solvers. We discuss this in detail below in Sect. 59.4. As a preparation, we rewrite (59.1) and (59.2) in dimensionless form, which will be convenient. The three fundamental units appearing in the equations are those of *space*, *time* and *mass*, which are represented by the variables x (or r), t and m. We now introduce new dimensionless variables, $x' = x/\mathrm{AU}$, $t' = t/\mathrm{year}$ and $m' = m/M$, where 1 AU is the mean distance from the Sun to Earth and M is the mass of the Sun. We can use the chain rule to obtain the dimensionless acceleration, $a' = \frac{d}{dt'}\frac{d}{dt'}x' = \frac{dt}{dt'}\frac{d}{dt}\frac{dt}{dt'}\frac{d}{dt}x/\mathrm{AU} = \frac{\mathrm{year}^2}{\mathrm{AU}}a$. Combining (59.1) and (59.2) using our new dimensionless variables, we then obtain

$$m'M \frac{\mathrm{AU}}{\mathrm{year}^2} a' = G \frac{m'M \cdot m_a'M}{r'^2 \mathrm{AU}^2}, \tag{59.3}$$

or

$$a' = G' \frac{m_a'}{r'^2}, \tag{59.4}$$

where the new gravitational constant G' is given by

$$G' = \frac{G \cdot \text{year}^2 M}{\text{AU}^3}.$$

(59.5)

We leave it as an exercise to show that with suitable definitions of the units year and AU, the new dimensionless gravitational constant G' is given by

$$G' = 4\pi^2.$$

(59.6)

59.3   Einstein's Equation

In general relativity the basic concept is not *force*, as in Newtonian theory, but instead the *curvature* of space-time. Einstein explains the motion of the planets in our Solar System in the following way: the planets move through space-time along straight lines, *geodesics*, which appear as circular (or elliptical) orbits only because space-time is curved by the large mass of the Sun. We shall now try to give an idea of how this works.

The curvature of space-time is given by its *metric*. A metric defines the distance between two nearby points in space-time. In Euclidean geometry that we have studied extensively in this book, the distance between to points $x = (x_1, x_2, x_3)$ and $y = (y_1, y_2, y_3)$ is given by the square root of the scalar product $dx \cdot dx$, where dx is the difference $dx = x - y$. With the notation $ds = |x - y|$ we thus have

$$ds = \sqrt{dx \cdot dx} = \left(\sum_{i=1}^{3} dx_i^2 \right)^{1/2},$$

(59.7)

or

$$ds^2 = \sum_{i=1}^{3} dx_i^2.$$

(59.8)

In the notation of general relativity, the Euclidean metric is then given by the matrix (tensor)

$$g = \begin{bmatrix} 1 & 0 & 0 \\ 0 & 1 & 0 \\ 0 & 0 & 1 \end{bmatrix},$$

(59.9)

as

$$ds^2 = dx^T g \, dx.$$

(59.10)

In space-time we include time t as a fourth coordinate and every event in space-time is given by a vector (t, x_1, x_2, x_3). In flat or *Minkowski* space-time in the absence of masses, the curvature is zero and the metric is given by

$$g = \begin{bmatrix} -1 & 0 & 0 & 0 \\ 0 & 1 & 0 & 0 \\ 0 & 0 & 1 & 0 \\ 0 & 0 & 0 & 1 \end{bmatrix}, \tag{59.11}$$

which gives

$$ds^2 = -dt^2 + dx_1^2 + dx_2^2 + dx_3^2. \tag{59.12}$$

In the presence of masses, we obtain a different metric which does not even have to be diagonal.

From the metric g one can find the straight lines of space-time, which give the orbits of the planets. The metric itself is determined by the distribution of mass in space-time, and is given by the solution of Einstein's equation,

$$R_{ij} - \frac{1}{2} R g_{ij} = 8\pi T_{ij}, \tag{59.13}$$

where (R_{ij}) is the so-called *Ricci-tensor*, R is the so-called *scalar curvature* and (T_{ij}) is the so-called *stress-energy tensor*. Now (R_{ij}) and R depend on derivatives of the metric $g = (g_{ij})$ so (59.13) is a partial differential equation for the metric g.

The solution for the orbits of the planets obtained from Einstein's equation are a little different than the solution obtained from (59.4) given by Newton. Although the difference is small, it has been verified in observations of the orbit of the planet Mercury which is the planet closest to the Sun. We will not include these "relativistic effects" in the next section where we move on to the computation of the evolution of the Solar System.

59.4   The Solar System as a System of ODEs

To use the techniques developed in Chapter Adaptive IVP-Solvers to compute the evolution of the Solar System, we need to rewrite the second-order system of ODEs given by (59.4) in the standard form $\dot{u} = f$. We start by introducing coordinates $x^i(t) = (x_1^i(t), x_2^i(t), x_3^i(t))$ for all bodies in the Solar System, including the nine planets, then Sun and the Moon. This gives a total of $n = 9 + 2 = 11$ bodies and a total of $3n = 33$ coordinates. To rewrite the equations as the first-order system $\dot{u} = f$ we need to include also the velocities of all bodies, $\dot{x}^i(t) = (\dot{x}_1^i(t), \dot{x}_2^i(t), \dot{x}_3^i(t))$, giving a total of $N = 6n = 66$ coordinates. We collect all these coordinates in the vector $u(t)$ of length N in the following order:

$$\begin{aligned} u(t) = (&x_1^1(t), x_2^1(t), x_3^1(t), \ldots, x_1^n(t), x_2^n(t), x_3^n(t), \\ &\dot{x}_1^1(t), \dot{x}_2^1(t), \dot{x}_3^1(t), \ldots, \dot{x}_1^n(t), \dot{x}_2^n(t), \dot{x}_3^n(t)), \end{aligned} \tag{59.14}$$

so that the first half of the vector $u(t)$ contains the positions of all bodies and the second half contains the corresponding velocities.

To obtain the differential equation for $u(t)$, we take the time-derivative and notice that the derivative of the first half of $u(t)$ is equal to the second half of $u(t)$:

$$\dot{u}_i(t) = u_{3n+i}(t), \quad i = 1, \ldots, 3n, \tag{59.15}$$

i.e. for $n = 11$ we have $\dot{u}_1(t) = \dot{x}_1^1(t) = u_{34}(t)$ and so on.

The derivative of the second half of $u(t)$ will contain the second derivatives of the positions, i.e. the accelerations, and these are given by (59.4). Now (59.4) is written as a scalar equation and we have to rewrite it in vector form. For every body in the Solar System, we need to compute the contribution to the total force on the body by summing the contributions from all other bodies. Assuming that we work in dimensionless variables (but writing x instead of x', m_i instead of m_i' and so on for convenience) we then need to compute the sum:

$$\ddot{x}_i(t) = \sum_{j \neq i} \frac{G' m_j}{|x^j - x^i|^2} \frac{x^j - x^i}{|x^j - x^i|}, \tag{59.16}$$

where the unit vector $\frac{x^j - x^i}{|x^j - x^i|}$ gives the direction of the force, see Fig. 59.4.

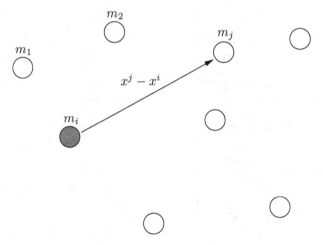

Fig. 59.4. The total force on body i is the sum of the contributions from all other bodies

Our final differential equation for the evolution of the Solar System in the form $\dot{u} = f$ is then given by

$$\dot{u}(t) = f(u(t)) = \begin{bmatrix} u_{3n+1}(t) \\ \vdots \\ u_{6n}(t) \\ \displaystyle\sum_{j\neq 1} \frac{G'm_j}{|x^j - x^1|^2} \frac{x_1^j - x_1^1}{|x^j - x^i|} \\ \vdots \\ \displaystyle\sum_{j\neq n} \frac{G'm_j}{|x^j - x^n|^2} \frac{x_3^j - x_3^n}{|x^j - x^n|} \end{bmatrix}, \qquad (59.17)$$

Table 59.1. Initial data for the Solar System at 00.00 Universal Time (UT1, approximately GMT) January 1, 2000 for dimensionless positions and velocities scaled with units 1 AU $= 1.49597870 \cdot 10^{11}$ m (one astronomical unit), 1 year $= 365.24$ days and $M = 1.989 \cdot 10^{30}$ kg (one solar mass)

	Position	Velocity	Mass
Mercury	$x^1(0) =$ -0.147853935 -0.400627944 -0.198916163	$\dot{x}^1(0) =$ 7.733816715 -2.014137426 -1.877564183	$1.0/6023600$
Venus	$x^2(0) =$ -0.725771746 -0.039677000 0.027897127	$\dot{x}^2(0) =$ 0.189682646 -6.762413869 -3.054194695	$1.0/408523.5$
Earth	$x^3(0) =$ -0.175679599 0.886201933 0.384435698	$\dot{x}^3(0) =$ -6.292645274 -1.010423954 -0.438086386	$1.0/328900.5$
Mars	$x^4(0) =$ 1.383219717 -0.008134314 -0.041033184	$\dot{x}^4(0) =$ 0.275092348 5.042903370 2.305658434	$1.0/3098710$
Jupiter	$x^5(0) =$ 3.996313003 2.731004338 1.073280866	$\dot{x}^5(0) =$ -1.664796930 2.146870503 0.960782651	$1.0/1047.355$
Saturn	$x^6(0) =$ 6.401404019 6.170259699 2.273032684	$\dot{x}^6(0) =$ -1.565320566 1.286649577 0.598747577	$1.0/3498.5$
Uranus	$x^7(0) =$ 14.423408013 -12.510136707 -5.683124574	$\dot{x}^7(0) =$ 0.980209400 0.896663122 0.378850106	$1.0/22869$
Neptune	$x^8(0) =$ 16.803677095 -22.983473914 -9.825609566	$\dot{x}^8(0) =$ 0.944045755 0.606863295 0.224889959	$1.0/19314$
Pluto	$x^9(0) =$ -9.884656563 -27.981265594 -5.753969974	$\dot{x}^9(0) =$ 1.108139341 -0.414389073 -0.463196118	$1.0/150000000$
Sun	$x^{10}(0) =$ -0.007141917 -0.002638933 -0.000919462	$\dot{x}^{10}(0) =$ 0.001962209 -0.002469700 -0.001108260	1
Moon	$x^{11}(0) =$ -0.177802714 0.884620944 0.384016593	$\dot{x}^{11}(0) =$ -6.164023246 -1.164502534 -0.506131880	$1.0/2.674 \cdot 10^7$

where we have kept the notation $x^1 = (x_1^1, x_2^1, x_3^1)$ rather than (u_1, u_2, u_3) and so on in the right-hand side for simplicity. The evolution of our Solar System can now be computed by the standard techniques developed in Chapter Adaptive IVP-Solvers, using the initial data supplied in Table 59.1.

59.5 Predictability and Computability

Two important questions that arise naturally when we study numerical solutions of the evolution of our Solar System, such as the one in Fig. 59.5, are the questions of *predictability* and *computability*.

The predictability of the Solar System is the question of the accuracy of a computation given the accuracy in initial data. If initial data is known with an accuracy of say five digits, and the numerical computation is exact, how long does it take until the solution is no longer accurate even to one digit?

The computability of the Solar System is the question of the accuracy in a numerical solution given exact initial data, i.e. how far we can compute an accurate solution with available resources such as method, computational power and time.

Both the predictability and the computability are determined by the growth rate of errors. Luckily, the error does not grow exponentially as we saw for the Lorenz system. If we imagine that we displace Earth slightly from its orbit and start a computation, the orbit and velocity of Earth will

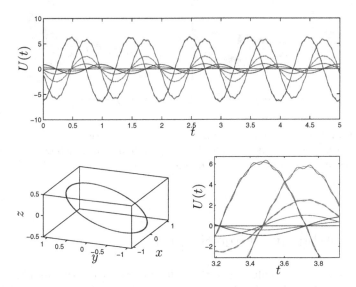

Fig. 59.5. A numerical computation of the evolution of the Solar System, including Earth, the Sun and the Moon

be slightly different, resulting in an error that grows *linearly* with time. This means that the predictability of the Solar System is quite good, since every extra digit of accuracy in initial data means that the limit of predictability is increased by a factor ten. If now the solution is computed using a numerical method, such as the adaptive cG(1) method, this will result in additional errors. We can think of the error caused by a numerical method as a small perturbation introduced with every new time step. Adding the contributions from all time steps we find that the numerical error grows *quadratically*, see Problem 59.2.

As it turns out however, the error does not grow quadratically but only linearly for the cG(1) method as shown in Fig. 59.6. This pleasant surprise is the result of an important property of the cG(1) method: it conserves energy. As a result, the cG(1) method performs better on a long time interval than the higher-order (more accurate) dG(2) method.

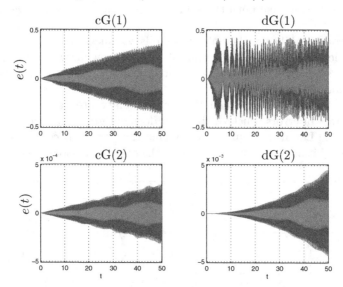

Fig. 59.6. The growth of the numerical error in simulations of the Solar System using different numerical methods. The two methods on the *left* conserve energy, which results in linear rather than quadratic error growth

59.6 Adaptive Time-Stepping

If we compute the evolution of the Solar System using the adaptive cG(1) method, we find that the time steps need to be small enough to follow the orbit of the Moon (or Mercury if we do not include the Moon). This is inefficient since the time scales for the other bodies are much larger: the period of the Moon is one month and the period of Pluto is 250 years, and so the time steps for Pluto should be roughly a factor 3,000 larger that the time steps for the Moon. It has been shown recently that the standard methods cG(q), including cG(1), and dG(q) can be extended to individual, *multi-adaptive*, time-stepping for different components. In Figure 59.7 we show a computation made with individual time steps for the different planets. Notice how the error grows quadratically, indicating that the method does not conserve energy. (It is possible to construct also multi-adaptive methods which conserve energy.)

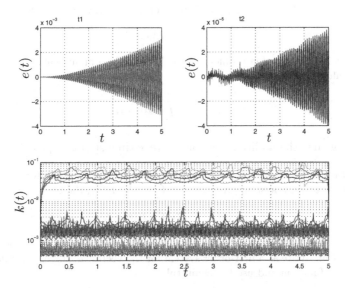

Fig. 59.7. A computation of the evolution of the Solar System with individual, multi-adaptive, time steps for the different planets

59.7   Limits of Computability and Predictability

Using the multiadaptive cG(2) it appears that the limit of computability of the Solar System (with the Moon and the nine planets) using double precision, is of the order 10^6 years. Concerning the predictability of the same system it appears that for every digit beyond 5 in the precision of data we gain a factor of ten in time, so that for example predicting the position of the Moon 1000 years ahead would require about 8 correct digits in e.g. the initial positions and velocities, masses and gravitational constant. We conclude that it appears that normally the precision in data would set the limit for accurate simulations of the evolution of the Solar System, if we use a high order multiadaptive solver.

Chapter 59   Problems

59.1. Prove that with suitable definitions of the units year and AU the gravitational constant is $G' = 4\pi^2$. *Hint:* assume that Earth is in a circular orbit where the centripetal force mv^2/r is balanced by the gravitational force GmM/r^2.

59.2. Motivate the quadratic growth of the numerical error for the Solar System. *Hint:* Assume that an error of size ϵ is added to the velocity of a planet in every time step.

59.3. (*Hard!*) Prove that in general if an error in initial data grows as

$$|e(T)| \le S(T)|e(0)|,$$

for a specific initial value problem, then the error in a numerical solution of the initial value problem grows as

$$|e(T)| \lesssim \epsilon \int_0^T S(t)\, dt,$$

assuming that the additional error in every time step is kept below $k_n \epsilon$.

59.4. Prove that the cG(1) method conserves energy for a Hamiltonian system, i.e. prove that for a system given by $\ddot{x} = F = -\nabla_x P(x)$, the total energy

$$E(t) = K(\dot{x}(t)) + P(x(t)),$$

is conserved. Here $P(x)$ is a given potential field, and $K(\dot{x}) = \frac{\dot{x}^2}{2}$ is the kinetic energy. *Hint:* Write as a first-order system for the vector $[u, v] = [x, \dot{x}]$, take $[\dot{v}, \dot{u}]$ as the test function and use the chain rule.

59.5. Investigate numerically the predictability and computability of the Solar System. Can you verify the linear error growth for the cG(1) method?

60

Optimization

1. All living beings are driven by passion to seek maximal Pleasure. 2. There is Pleasure of the Body and Pleasure of the Soul. In the Pleasure of the Soul, the Body cannot take part, whereas the Pleasure of the Body is equally shared by the Soul. (the 2 first of the 14 basic principles of *Anthropologica physica*, by King Karl XII of Sweden, 1717)

60.1 Introduction

In this chapter, we expand on some basic aspects of optimization touched upon in the previous chapter in connection to minimization. Optimization is very rich subject and we shall return to other aspects below. The issues we discuss here are connected to the very basics of Calculus and are considered as "deep" and understandable only by the very best math majors. You may test your own reaction to the discussions presented, and if you get the expected feeling of confusion, don't worry, just proceed to the next chapter. If on the other hand, against all odds, you get the feeling of grasping the main ideas, then you may congratulate yourself for being more gifted for mathematics than you thought!

In our modern world, *optimization* is a code word. To *optimize* is to use available resources as efficiently as possible, or to find the best of available alternatives. In our private lives, we may want our car to use as little fuel as possible, to buy an item at lowest possible price, to use as little effort as possible to clean the house, or to get maximal enjoyment out of the vacation trip.

In automatized production, the leading principle is always to optimize and seek to use as little energy, material and human resources as possible to produce a certain amount of goods. A basic idea in our capitalistic system is that in the long run the most efficient mode of production will win the market.

A basic problem of optimization is to find the maximum or minimum value of a given function $f : \Omega \to \mathbb{R}$ defined on some set of numbers Ω. Typically, Ω may be a domain in $\mathbb{R}^d$ with $d = 1, 2, 3, \ldots$, that may be bounded or unbounded, or Ω may be a finite set such as the set of natural numbers $1, 2 \ldots, 100$. More precisely, finding a *minimum point* $\bar{x}$ in Ω amounts to finding a point $\bar{x} \in \Omega$ such that

$$f(x) \geq f(\bar{x}) \quad \text{for all } x \in \Omega, \tag{60.1}$$

and we then say that $f(\bar{x})$ is the *minimum value* of $f : \Omega \to \mathbb{R}$. Note that there may be several minimum points, but of course there may be only one minimum value. If in an Olympic 100 meter race, three runners share the best time of 9.99 seconds, then all the three runners may share the gold medal. However, there cannot be two runners with different final times who both get a gold medal.

We now consider the problem of finding the minimum value and corresponding minimum point(s) of a given function $f : \Omega \to \mathbb{R}$. We may distinguish between the following two cases: (a) Ω is a domain of $\mathbb{R}^d$ with infinitely many points, as when Ω is the unit disc $\{x \in \mathbb{R}^d : \|x\| \leq 1\}$; (b) Ω contains finitely many points, as for example $\Omega = \{1, 2, 3, \ldots, 10\}$. The case (a) is "continuous" and (b) is "discrete". The two cases are not fully disjoint; there may be a gradual passage from "discrete" to "continuous" as the number of elements in Ω increases. In the case Ω is discrete with finitely many points, we may find the minimum value and corresponding minimum point(s) of $f : \Omega \to \mathbb{R}$ by different algorithms for *sorting*. If Ω is "continuous" with infinitely many points, sorting may be impossible and different algorithms that use information from the derivative of $f(x)$ in variations of steepest descent are often used.

60.2   Sorting if Ω Is Finite

If Ω is a finite set of numbers, for example if $\Omega = \{1, 2, \ldots, 9, 10\}$, then we just make a list of the corresponding 10 function values $f(1), f(2), \ldots, f(10)$ and by sorting these values according to magnitude in increasing order, we can find the minimum value $f(\bar{x})$ and the corresponding argument $\bar{x}$. Of course, we don't have to sort all the numbers according to magnitude to find the smallest one. We just have to find the first element in the list sorted according to magnitude in increasing order. Repeating this process we can sort all the numbers in the given list of numbers.

Example 60.1. For example, suppose that $\Omega = \{1, 2, \ldots, 9, 10\}$ and $f(1) = 143$, $f(2) = 538$, $f(3) = 67$, $f(4) = 964$, $f(5) = 287$, $f(6) = 64$, $f(7) = 123$, $f(8) = 333$, $f(9) = 63$, $f(10) = 88$. By direct inspection, we see that the minimum point is $\bar{x} = 9$ and the minimum value is $f(\bar{x}) = 63$.

While sorting sounds simple, it turns out to be an interesting problem to do sorting efficiently when there are a large number of values to be sorted. So there are different algorithms for sorting and sorting algorithms hold a prominent place in computer science. A simple algorithm for finding the minimum m of N numbers $f(1), \cdots, f(N)$ goes as follows:

1. Set $m = f(1)$ and $\bar{x} = 1$

2. For $x = 2, \cdots, N$, if $f(x) < m$ set $m = f(\bar{x})$ and $\bar{x} = x$.

The minimum value is then $m = f(\bar{x})$ and the minimum point is $x = \bar{x}$. The algorithm is based on successive comparison of pairs of numbers (if $f(x) < m$ then we update m and $\bar{x}$ and set $m = f(\bar{x})$ and $\bar{x} = x$). The number of comparisons in the indicated algorithm is apparently $N - 1$.

Repeating the algorithm with $f(\bar{x})$ eliminated, we can get a complete sorting according to magnitude using $(N - 1) + (N - 2) + \cdots + 1 \approx \frac{1}{2}N^2$ comparisons.

60.3   What if Ω Is Not Finite?

If Ω is an interval of real numbers, for example $\Omega = [0, 1]$, then Ω contains infinitely many points and sorting the values $f(x)$ with $x \in \Omega$ of a given function $f : \Omega \to \mathbb{R}$ by pairwise comparison appears impossible in practice because we cannot perform infinitely many comparisons. Of course, in practice we replace Ω by a finite set of numbers, for example by using a single precision floating point representation of the points in Ω. So in principle, we can then apply the above sorting strategy. But, the procedure will be computationally intensive. With seven digits we would have 10^7 values $f(x)$ to compare, which using the above algorithm requires on the order of 10^7 comparisons to find the minimum. If the interval Ω is larger and the desired precision higher, then the number of comparisons would be correspondingly larger. The total computational cost would also involve as a multiplicative factor the cost of evaluating the function value $f(x)$ for a given x, which itself could require many arithmetic operations. The total cost in direct comparison thus may be prohibitively large.

We now seek efficient algorithms to handle the case that Ω is a domain of $\mathbb{R}^d$ interval and the function $f(x)$ is Lipschitz continuous with a Lipschitz constant L. In this case, the function values $f(x)$ cannot change more than the argument of x changes times L. If we want to find the minimum value up to a certain tolerance TOL, then we need to do approximately $(L/TOL)^d$ comparisons if the diameter of Ω is of order one. Depending on the choice of the tolerance TOL and L this may be acceptable or not.

If the function $f(x)$ is differentiable then we may restrict the search further using information from the derivative, as we shall see below.

60.4 Existence of a Minimum Point

How can we be sure that there in fact is a minimum point? We discuss the proof of the following basic theorem addressing this question below.

Theorem 60.1 *If $f : \Omega \to \mathbb{R}$ is Lipschitz continuous, where Ω is a closed and bounded subset of $\mathbb{R}^d$, then there is a minimum point $\bar{x} \in \Omega$ such that $f(\bar{x}) \leq f(x)$ for all $x \in \Omega$.*

The assumption that Ω is closed and bounded is essential to guarantee existence of the minimum, as the following examples show.

Example 60.2. The function $f : (0, 1) \to \mathbb{R}$ with $f(x) = x$ does not have a minimum point in $(0, 1)$. In this case $\Omega = (0, 1)$ is not closed.

Example 60.3. The function $f : [1, \infty) \to \mathbb{R}$ with $f(x) = 1/x$ does not have a minimum point in $[1, \infty)$. In this case $\Omega = [1, \infty)$ is not bounded.

Note however that a function $f : \Omega \to \mathbb{R}$ may have a minimum even if Ω is unbounded. In particular, if $f(x)$ increases to infinity as $\|x\|$ increases, then we can effectively reduce the search for a minimum to a bounded set.

Example 60.4. The function $f : [0, \infty)$ given by $f(x) = x^2 - 2x$ attains a minimum value $f(1) = -1$; since $f(x) \geq 0$ for $x \geq 2$, we may restrict the search for a minimum to $[0, 2]$.

60.5 The Derivative Is Zero
at an Interior Minimum Point

We assume that $f : \Omega \to \mathbb{R}$ is a given Lipschitz continuous differentiable function, where Ω is a domain in $\mathbb{R}^d$. We shall now prove that if $\bar{x}$ is an *interior minimum point* of $f : \Omega \to \mathbb{R}$, that is $\bar{x}$ is a minimum point and the ball $\{x \in \mathbb{R}^d : \|x - \bar{x}\| < \delta\}$ is included in Ω for some positive number δ, then $f'(\bar{x}) = \nabla f(\bar{x}) = 0$, where $f' = \nabla f$ is the gradient of f. This follows by writing

$$f(x) = f(\bar{x}) + f'(\bar{x}) \cdot (x - \bar{x}) + E_f(x, \bar{x})$$

with $|E_f(x, \bar{x})| \leq K_f(\bar{x})\|x - \bar{x}\|^2$. If now $f'(\bar{x}) \neq 0$, we may choose $x = \bar{x} - \epsilon f'(\bar{x}) \in \Omega$ with $\epsilon > 0$ and estimate to get

$$f(x) \leq f(\bar{x}) - \epsilon\|f'(\bar{x})\|^2 + \epsilon^2 K_f(\bar{x})\|f'(\bar{x})\|^2$$
$$= f(\bar{x}) - \epsilon\|f'(\bar{x})\|^2(1 - \epsilon K_f(\bar{x})) < f(\bar{x}).$$

For ϵ sufficiently small, we get a contradiction to the assumption that $\bar{x}$ is a minimum point. We have proved the following basic result, see Fig. 60.1 and Fig. 60.2.

Theorem 60.2 *Suppose* $f : \Omega \to \mathbb{R}$ *has a minimum point at an interior point* $\bar{x}$ *in* Ω, *and suppose that* $f : \Omega \to \mathbb{R}$ *is differentiable at* $\bar{x}$. *Then* $f'(\bar{x}) = 0$.

Using this result, we may search for interior minimum points among the zeros of the derivative $f'(x)$ in Ω. To find these zeros we may use some algorithm for computing roots, like Fixed Point Iteration or Newton or the Bisection algorithm. There is thus a strong connection between algorithms for finding interior minimum points of $f : \Omega \to \mathbb{R}$ and algorithms for computing roots of $f'(x) = 0$.

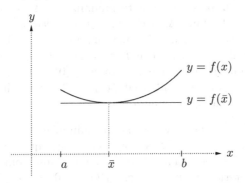

Fig. 60.1. $f'(\bar{x}) = 0$ at an interior minimum point $\bar{x}$

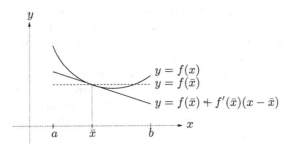

Fig. 60.2. $f'(\bar{x}) < 0$ implies that $f(x) < f(\bar{x})$ for x close to $\bar{x}$ with $\bar{x} > x$, that is, $\bar{x}$ cannot be a minimum point

Note that if the minimum point $\bar{x}$ of $f : \Omega \to \mathbb{R}$ is not interior to Ω, i.e. $\bar{x}$ lies on the boundary of Ω, then the derivative $f'(\bar{x})$ may be non-zero, see Fig. 60.3.

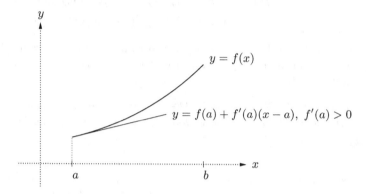

Fig. 60.3. $f'(\bar{x})$ may be non-zero at a minimum $\bar{x}$ on the boundary

Example 60.5. Suppose we want to minimize $f : \Omega \to \mathbb{R}$ with $\Omega = [0, 2]$ and $f(x) = x^2 - 2x$. Since Ω is closed and bounded and $f(x)$ is Lipschitz continuous, we know that there is a minimum point $\bar{x} \in [0, 2]$. If $\bar{x}$ is interior to $[0, 2]$, that is if $0 < \bar{x} < 2$, then $f'(\bar{x}) = 2\bar{x} - 2 = 0$ and thus $\bar{x} = 1$. We compare the value $f(1) = -1$ to the values $f(0) = 0$ and $f(2) = 0$ on the boundary of $[0, 2]$ and conclude that $f(1) = -1$ is the minimum value and $\bar{x} = 1$ the corresponding minimum point.

Example 60.6. Suppose we want to minimize $f : Q \to \mathbb{R}$ with $f(x) = f(x_1, x_2) = x_1^2 + x_2^2 - 2x_1 - x_2$ on a closed square $Q = [0, 2] \times [0, 2]$, see Fig. 60.4. We know that there is a minimum point in Q. We first compute the interior points $\hat{x}$ where $f'(\hat{x}) = 0$. Since $f'(x) = (2x_1 - 2, 2x_2 - 1)$, $\hat{x} = (1, 0.5)$ with the function value $f(1, 0.5) = -1.25$. It remains to study the variation of $f(x)$ on the boundary of Q to see if we find a value smaller than -1.25. We do this by considering each piece of the boundary separately. On the part $x_2 = 0$, we have $f(x) = x_1^2 - 2x_1$ with $x_1 \in [0, 2]$, and we see arguing as in the previous example that the minimum value is $f(1, 0) = -1$. On the part $x_2 = 2$, we have $f(x_1, 2) = x_1^2 - 2x_1 + 3$ with minimum $f(1, 2) = 2$. On the part $x_1 = 0$, we have $f(0, x_2) = x_2^2 - x_2$ with minimum $f(0, 0.5) = -0.25$, and on the part $x_1 = 2$, we have $f(2, x_2) = x_2^2 - x_2$ with minimum $f(2, 0.5) = -0.25$. We conclude that the minimum point is the interior point $\bar{x} = (1, 0.5)$ and that the minimum value is $f(1, 0.5) = -1.25$.

Example 60.7. You are asked to design a box (without top) of a given volume using as little material as possible. Letting the sides of the box be

$$x_3 = f(x_1, x_2) = x_1^2 + x_2^2 - 2x_1 - x_2$$

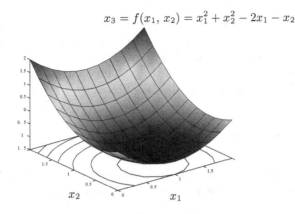

Fig. 60.4. Minmizing $f(x) = x_1^2 + x_2^2 - 2x_1 - x_2$ on $Q = [0,2] \times [0,2]$

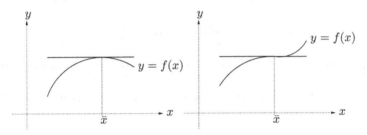

Fig. 60.5. $f'(\bar{x}) = 0$ may correspond also to a maximum point or an inflection point

x_1, x_2 and x_3, the volume is $x_1 x_2 x_3 = V$ and the surface to be minimized is $x_1 x_2 + 2x_1 x_3 + 2x_2 x_3$. Eliminating x_3 gives

$$f(x_1, x_2) = x_1 x_2 + 2V \left(\frac{1}{x_1} + \frac{1}{x_2} \right),$$

which is to be minimized over $\Omega = [0, \infty) \times [0, \infty)$. Seeking points $\hat{x}$ with $f'(\hat{x}) = (0,0)$, we find $\hat{x}_1 = \hat{x}_2 = (2V)^{1/3}$ with the corresponding height $\hat{x}_3 = \frac{1}{2}(2V)^{1/3}$, and the area

$$f(\hat{x}) = (2V)^{2/3} + 2(2V)^{2/3}.$$

Comparing with (x_1, x_2) with x_1 or x_2 very large or small give large values to $f(x_1, x_2)$ and thus the minimum point is $\hat{x}$. The solution is a box with square bottom and height half of the width.

We also remark that a minimum value may be attained at an interior point where the given function is *nondifferentiable*. For example, the minimum value of the function $f(x) = |x - 1|$ on $[0, 2]$ is attained at $\bar{x} = 1$

with minimum value $f(\bar{x}) = f(1) = 0$. This type of interior minimum points must be considered separately. Thus, to find all possible minimum points we have to consider the points $\bar{x}$ for which $f'(\bar{x}) = 0$, and in addition to these the end points of the domain of definition and interior points where $f(x)$ is not differentiable, see Fig. 60.5.

60.6 The Role of the Hessian

We know that if $\bar{x}$ is an interior minimum point of a function $f : \Omega \to \mathbb{R}$, then $f'(\bar{x}) = 0$. But it is not true in general that if $f'(\bar{x}) = 0$, then $\bar{x}$ is a minimum point. A point $\bar{x}$ with $f'(\bar{x}) = 0$ may e.g. be a *maximum point*, or an *inflection point*, see Fig. 60.5. If the Hessian H of $f : \Omega \to$ is positive definite close to $\bar{x}$ and $f'(\bar{x}) = 0$, then we have by Taylor's theorem

$$f(x) = f(\bar{x}) + \frac{1}{2}(x - \bar{x})^\top H(y) \cdot (x - \bar{x}) > f(\bar{x})$$

for x close to $\bar{x}$ and some y between x and $\bar{x}$, and thus $\bar{x}$ is a *local minimum point*.

We recall that an $n \times n$ matrix A is said to be positive definite if

$$v^\top A v > 0$$

for all non-zero $v \in \mathbb{R}^n$. The Spectral Theorem implies that A is positive definite if and only if the eigenvalues of A are positive.

Example 60.8. If $A = (a_{ij})$ is a symmetric 2×2 matrix, then A is positive definite if

$$a_{11}a_{22} - a_{12}^2 > 0 \quad \text{and} \ a_{11} > 0.$$

This follows by completing squares in

$$v^\top A v = a_{11}v_1^2 + a_{22}v_2^2 + 2a_{12}v_1v_2.$$

60.7 Minimization Algorithms: Steepest Descent

We discuss briefly how to find candidates for minimum points of a given function $f : \Omega \to \mathbb{R}$, where Ω is a domain in $\mathbb{R}^d$. We assume that $f : \Omega \to \mathbb{R}$ is Lipschitz continuous and differentiable on Ω.

In the *steepest descent method*, we construct a sequence $\{x_i\}$ in $\mathbb{R}^d$ that hopefully converges to a (local) minimum point by means of the iteration

$$x_{i+1} = x_i - \alpha_i f'(x_i), \tag{60.2}$$

where α_i is a positive parameter. Since $\alpha_i > 0$, if $f'(x_i) > 0$ then $x_{i+1} < x_i$, and if $f'(x_i) < 0$ then $x_{i+1} > x_i$. This means that if $f'(x_i) > 0$, so that $f(x)$ is increasing at $x = x_i$, then taking $x_{i+1} < x_i$ should result in $f(x_{i+1}) < f(x_i)$, and thus x_{i+1} should be closer to a minimum point than x_i. A similar argument applies in the case $f'(x_i) < 0$.

It is clear that the choice of the parameter α_i is important. If α_i is too small, then the convergence will be slow, and if α_i is too large, the sequence x_i may start to oscillate.

Note that we may view the gradient method (60.2) for minimization of $f(x)$ as Fixed Point Iteration for computing a root of $f'(x) = 0$.

If steepest descent leads to the boundary Γ of Ω, then we may replace the steepest descent iteration by the *projected gradient method* defined

$$x_{i+1} = x_i - \alpha_i P f'(x_i),$$

where $Pf'(x_i)$ is the projection of $f'(x_i)$ onto the tangent plane to Γ at $x_i \in \Gamma$.

The general idea is thus to find roots of $f'(x) = 0$ using steepest descent for the minimization of $f(x)$ or equivalently fixed point iteration for $f'(x) = 0$. Once the roots of $f'(x) = 0$ have been determined, the minimization is reduced to a search on the boundary of Ω and the interior zeros of $f'(x)$.

60.8   Existence of a Minimum Value and Point

We return to proof of the fundamental result which says that if $f : \Omega \rightarrow \mathbb{R}$ is Lipschitz continuous and Ω is a closed and bounded domain of $\mathbb{R}^d$, then there is a minimum point $\bar{x} \in \Omega$ with corresponding minimum value $f(\bar{x})$. We carry out the proof in for $d = 1$ so that $\Omega = [a, b]$ is a bounded closed interval. The proof in the case $d > 1$ is similar.

We shall prove that a Lipschitz continuous function $f : [a, b] \rightarrow \mathbb{R}$ on a closed and bounded interval $[a, b]$ has a minimum point by "constructing" a minimum point using the Bisection algorithm. We shall see that the "construction" is controversial at one step. Trying to resolve this issue yields added insight into the nature of minimization algorithms.

Normally, the proof we present here is considered so "difficult" that it is given only in "advanced" senior undergraduate or beginning graduate courses. With our good preparation on the Bisection algorithm and the nature of real numbers, we can plunge into the proof, and we will see that it is "easy" up to the non-constructive aspects.

We first recall that the Lipschitz continuity of $f(x)$ and the fact that $[a, b]$ is bounded implies that $f : [a, b] \rightarrow \mathbb{R}$ is bounded from above and below. In particular, there is some $m \in \mathbb{R}$ such that

$$f(x) \geq m \quad \text{for all } x \in [a, b]. \tag{60.3}$$

We say that m is a *lower bound* of $f : [a, b] \to \mathbb{R}$ if (60.3) holds. Clearly, there are many lower bounds since if m is a lower bound, any number $\underline{m} < m$ is also a lower bound.

In the proof, we shall use the concept of *greatest lower bound* defined as follows: we say that $\overline{m}$ is a greatest lower bound of $f : [a, b] \to \mathbb{R}$ if

$$f(x) \geq \overline{m} \quad \text{for all } x \in [a, b] \tag{60.4}$$

$$\text{for all } M > \overline{m} \quad \text{there is some } x \in [a, b] \quad \text{such that } f(x) < M. \tag{60.5}$$

In words, $\overline{m}$ is a greatest lower bound of $f : [a, b] \to \mathbb{R}$ if $\overline{m}$ is a lower bound of $f : [a, b] \to \mathbb{R}$ and any number bigger than $\overline{m}$ is not a lower bound for $f : [a, b] \to \mathbb{R}$. The concept of greatest lower bound has played an important role in the development of Calculus during the 20th century.

The proof now proceeds in two steps:

Step 1: Existence of a Greatest Lower Bound $\overline{m}$ of $f : [a, b] \to \mathbb{R}$

We shall prove the existence of a greatest lower bound $\overline{m}$ by using the Bisection method. Let m be a lower bound of $f : [a, b] \to \mathbb{R}$, whose existence was established above. Set $y_0 = m$ and $Y_0 = f(b)$ and define $\hat{y}_1 = \frac{1}{2}(y_0 + Y_0) = \frac{1}{2}(m + f(b))$. Note that $y_0 \leq \hat{y}_1 \leq Y_0$. If $f(x) \geq \hat{y}_1$ for all $x \in [a, b]$, then set $y_1 = \hat{y}_1$ and $Y_1 = Y_0$. If not, then there is an $x \in [a, b]$ such that $f(x) < \hat{y}_1$, and we set $y_1 = m$ and $Y_1 = \hat{y}_1$. We have now passed from the pair (y_0, Y_0), or interval (y_0, Y_0), to the interval (y_1, Y_1). By construction, $f(x) \geq y_i$ for all $x \in [a, b]$ and $i = 0, 1$ and there is some $x \in [a, b]$ such that $f(x) < Y_i$ unless Y_0 or Y_1 is already a greatest lower bound.

Repeating this process, we get two sequences $\{y_i\}$ and $\{Y_i\}$ such that for $i = 0, 1, 2, \ldots,$

$$y_i < Y_i, \quad y_{i+1} \geq y_i \quad Y_{i+1} \leq Y_i,$$

$$0 < Y_i - y_i = 2^{-i}(Y_0 - m),$$

$$f(x) \geq y_i \quad \text{for all } x \in [a, b],$$

$$\text{there is an } x \in [a, b] \quad \text{such that } f(x) < Y_i,$$

or some Y_i is a greatest lower bound. As in Chapter $\sqrt{2}$, we see that the sequences $\{y_i\}$ and $\{Y_i\}$ are Cauchy sequences and both converge to one real number, which we denote by $\overline{m}$. The number $\overline{m}$ is the greatest lower bound of $f : [a, b] \to \mathbb{R}$ since $\overline{m}$ satisfies the following two conditions:

$$(f(x) \geq \overline{m} \quad \text{for all } x \in [a, b],$$

$$\text{for any } M > \overline{m} \text{ there is an } x \in [a, b] \quad \text{such that } f(x) < M.$$

We have now proved the existence of a greatest lower bound to the Lipschitz continuous function $f : [a, b] \to \mathbb{R}$ on the closed and bounded interval $[a, b]$. Note that this result also holds if (a, b) is a bounded open interval. We have thus not yet used the fact that $[a, b]$ is closed.

Step 2: Existence of a Minimum Point

We now construct a convergent sequence $\{x_i\}$ with $x_i \in [a, b]$ and

$$\lim_{i \to \infty} f(x_i) = \overline{m}.$$

Setting $\bar{x} = \lim_{i \to \infty} x_i$, we have $f(\bar{x}) = \overline{m}$ and thus $\bar{x}$ is a minimum point and we are done.

To construct $\{x_i\}$ we again use the Bisection algorithm as follows: set $x_0 = a$, and $X_0 = b$, and define $\hat{x}_1 = \frac{1}{2}(x_0 + X_0)$. If $f(x) > \overline{m}$ for all x such that $\hat{x}_1 < x \le X_1$, then we set $x_1 = x_0$ and $X_1 = \hat{x}_1$. If not, we set $x_1 = \hat{x}_1$ and $X_1 = X_0$. Repeating the process, we obtain a convergent sequence $\{x_i\}$ with limit $\bar{x}$ and by construction we have $f(\bar{x}) = \overline{m}$. Note that to guarantee that $\bar{x} \in [a, b]$, we need $[a, b]$ to be closed. We note that the minimum value (of course) is equal to the greatest lower bound.

We summarize in the following theorem:

Theorem 60.3 (Existence of minimum point) *Suppose $f : I \to \mathbb{R}$ is Lipschitz continuous and $I = [a, b]$ is a closed and bounded interval. Then there is a point $\bar{x} \in [a, b]$, where $f : I \to \mathbb{R}$ assumes a minimum value $\bar{m}$, that is, $f(x) \ge \bar{m}$ for all $x \in [a, b]$, and $f(\bar{x}) = \bar{m}$.*

In the proof of this theorem we used the Bisection algorithm twice. Setting $y = f(x)$, we may say that we first used the Bisection algorithm in the variable y to prove existence of a greatest lower bound $\overline{m}$ and then in the variable x to prove existence of a minimum point $\bar{x}$ satisfying $f(\bar{x}) = \overline{m}$.

60.9   Existence of Greatest Lower Bound

If we examine the proof of existence of a greatest lower bound to the Lipschitz continuous function $f : I \to \mathbb{R}$, we see that the crucial fact behind the proof is that $f : [a, b] \to \mathbb{R}$ is bounded below, that is there is a real number m such that $f(x) \ge m$ for all $x \in [a, b]$. We can interpret this in terms of a property of the range $R(f) = \{y : y = f(x) \text{ for some } x \in D(f) = [a, b]\}$, namely

$$y \ge m \quad \text{for all } y \in R(f).$$

This says that the set $R(f)$ is *bounded below*.

More generally, we say that a set A of real numbers is bounded from below if there is a real number m such that $y \ge m$ for all $y \in A$. Using the same argument as just used in the case $A = R(f)$, we obtain the following fundamental property of real numbers.

Theorem 60.4 (Existence of greatest lower bound) *Suppose A is a set of real numbers which is bounded from below, that is, there is a real number m such that $x \ge m$ for $x \in A$. Then the set A has a greatest lower* *bound* $\overline{m} \in \mathbb{R}$ *satisfying* $x \geq \overline{m}$ *for all* $x \in A$ *and for all* $M > \overline{m}$ *there is an* $x \in A$ *such that* $x < M$.

60.10 Constructibility of a Minimum Value and Point

We now discuss to what extent the above existence proof is constructive. There are two issues: (i) construction of the greatest lower bound, which is the same as the minimum value, and (ii) construction of a minimum point.

In the application of the Bisection algorithm in (i), we have to check if

$$f(x) \geq \hat{y}_i \quad \text{for all } x \in [a, b],$$

while in the application in (ii), we have to check if

$$f(x) > \overline{m} \text{ for all } x \text{ such that } \hat{x}_1 < x \leq X_1.$$

Both checks appear to involve *infinitely many* values of x. In the worst case this would require infinitely many comparisons. The number may be reduced if $f(x)$ is differentiable by using information concerning $f'(x)$. For example, the sign of $f'(x)$ indicates if $f(x)$ is increasing or decreasing which may be used to reduce the amount of comparison.

Thus, depending on the nature of the given function $f : I \to \mathbb{R}$, the given proof of existence of a minimum value and minimum point may be more or less constructive in nature.

Is it possible to make the proof fully constructive? We expect this to be possible if we accept to determine the minimum value up to a tolerance $TOL > 0$. Suppose then that the function $f(x)$ is Lipschitz continuous with Lipschitz constant L. We can then reduce all comparisons to a discrete grid of points of mesh size $\frac{1}{L}TOL$ between neighboring points.

To sum up, if $f : I \to \mathbb{R}$ is Lipschitz continuous and $[a, b]$ is bounded, then it is possible to determine the minimum value $f : I \to \mathbb{R}$ up to a given tolerance with a finite number of operations.

To determine an interior minimum point amounts to finding a root of $f'(x) = 0$ and thus the constructibility of a minimum point can be reduced to the constructibility of a root of $f'(x) = 0$. We discussed the cost of computing roots in Chapters *Fixed Point Iteration* and *Newton's method*.

60.11 A Decreasing Bounded Sequence Converges!

Suppose $\{x_i\}$ is a bounded decreasing sequence, that is $x_1 \geq x_2 \geq \cdots \geq x_n \geq x_{n+1} \geq \ldots$, and $x_n \geq m$ for all n for some number m. Then the set of all numbers x_n is bounded below, and thus has a greatest lower bound $\bar{m}$. We shall prove that $\lim_{n\to\infty} x_n = \bar{m}$. By the definition of greatest lower bound, for all $\epsilon > 0$ there is an x_N such that $\bar{m} \le x_N \le \bar{m} + \epsilon$. Since $x_n \le x_N$ for $n \ge N$, and $x_n \ge \bar{x}$, it follows that $\bar{m} \le x_n \le \bar{m} + \epsilon$ for all $n \ge N$, which proves the desired result. We summarize in the following theorem which is a cornerstone of the analysis of functions of a real variables.

Theorem 60.5 *Suppose $\{x_i\}_{i=1}^{\infty}$ is a decreasing sequence that is bounded below or an increasing sequence that is bounded above. Then $\{x_i\}_{i=1}^{\infty}$ is convergent.*

Chapter 60 Problems

60.1. Find the maximum and minimum values of the function $f(x_1, x_2) = x_1^2 + 2x_2^2 - x_1$ on the unit disc $x_1^2 + x_2^2 \le 1$.

60.2. Find the point of the plane $3x_1 + 4x_2 - x_3 = 26$ which is closest to the origin.

60.3. Find the shape of a box (with top included) which for given surface area has maximal volume.

60.4. Seek minimum and maximum values of the following functions:

(a) $f(x_1, x_2) = (1 + x_1^2 + x_2^2)^{-1}$ for $(x_1, x_2) \in \mathbb{R}^2$, (b) $f(x_1, x_2) = x_1 x_2$ for $x_1^2 + x_2^2 \le 1$, (c) $f(x_1, x_2, x_3) = x_1 + x_2 + x_3$ for $x_1^2 + x_2^2 + x_3^2 \le 1$.

60.5. Show that the function $x_1^4 + x_2^4 + x_3^4 - 4x_1 x_2 x_3$ has a minimum point at $(x_1, x_2.x_3) = (1, 1, 1)$.

60.6. Find the triangle of largest area that can be inscribed in a given circle.

60.7. Find the point on the curve $x_2 = x_1^2$ which is closest to the point $(0, 1)$.

60.8. Determine the constants a_0 and a_1 which minimize for a given function $f : [0, 1] \to \mathbb{R}$, the integral

$$\int_0^1 (f(x) - a_0 - a_1 x)^2 \, dx.$$

60.9. Find the maximum value of $x_1 + x_2 + \ldots + x_n$ subject to the condition $x_1^2 + x_2^2 + \ldots + x_n^2 \le 1$.

60.10. A *stationary point* of a function $f : \mathbb{R}^n \to \mathbb{R}$ is a point $x \in \mathbb{R}^n$ such that $f'(x) = 0$. Determine if any of the stationary points of the following functions is a maximum or minimum point: (a) $f(x_1, x_2, x_3) = x_1^2 + x_2^2 + x_3^2 - x_1 - x_2 + x_3 + 1$, (b) $f(x_1, x_2, x_3) = x_1^2 + x_2^2 + 2x_3^2 + 4x_1 - x_2 + x_3 + 5$, (c) $f(x_1, x_2, x_3) = \cos(x_1) + \cos(x_2) + \cos(x_3)$.

61

The Divergence, Rotation and Laplacian

...Stokes was a very important formative influence on subsequent generations of Cambridge men, including Maxwell. With Green, who in turn had influenced him, Stokes followed the work of the French, especially Lagrange, Laplace, Fourier, Poisson and Cauchy. This is seen most clearly in his theoretical studies in optics and hydrodynamics; but it should also be noted that Stokes, even as an undergraduate, experimented incessantly. Yet his interests and investigations extended beyond physics, for his knowledge of chemistry and botany was extensive, and often his work in optics drew him into those fields. (Parkinson)

Appointed professor of mathematics at the Ecole Polytechnique in 1809 Ampère held posts there until 1828. Ampère and Cauchy shared the teaching of analysis and mechanics and there was a great contrast between the two with Cauchy's rigorous analysis teaching leading to great mathematical progress but found extremely difficult by students who greatly preferred Ampère's more conventional approach to analysis and mechanics. (O'Connor and Robertson)

61.1 Introduction

We saw previously that the gradient of a function of several variables is a practically useful *differential operator*. In this chapter, we introduce some other useful operators, including the *divergence*, *rotation* and the *Laplacian*, together with the gradient play a fundamental role in mathematical modeling in science and engineering. We first define the operators in $\mathbb{R}^2$ and then in $\mathbb{R}^3$, noting that the rotation takes somewhat different forms in $\mathbb{R}^2$ and $\mathbb{R}^3$.

Fig. 61.1. Napoleon to Laplace (1749–1827): "You have written this huge book on the system of the world without once mentioning the Author of the Universe". Laplace to Napoleon: "Sire, I had no need of this hypothesis"

61.2   The Case of $\mathbb{R}^2$

We recall that the *gradient* of a function $u : \mathbb{R}^2 \to \mathbb{R}$, denoted grad u or ∇u, is the vector-valued function formed by the first order partial derivatives of u, i.e.

$$\text{grad } u = \nabla u = \left(\frac{\partial u}{\partial x_1}, \frac{\partial u}{\partial x_2} \right).$$

The *divergence* of a vector function $u = (u_1, u_2) : \mathbb{R}^2 \to \mathbb{R}^2$, denoted div u or $\nabla \cdot u$, is the scalar function defined by

$$\text{div } u = \nabla \cdot u = \frac{\partial u_1}{\partial x_1} + \frac{\partial u_2}{\partial x_2}.$$

Formally, we have

$$\nabla \cdot u = \left(\frac{\partial}{\partial x_1}, \frac{\partial}{\partial x_2} \right) \cdot (u_1, u_2)$$

where we may think of $(\frac{\partial}{\partial x_1}, \frac{\partial}{\partial x_2})$ "as a vector" and let the dot indicate a "scalar product". This idea applies to all the formulas below involving ∇ combined with the operators $\cdot$ and $\times$.

The *rotation* of a vector function $u : \mathbb{R}^2 \to \mathbb{R}^2$, denoted by rot u or $\nabla \times u$, is the scalar function

$$\text{rot } u = \nabla \times u = \frac{\partial u_2}{\partial x_1} - \frac{\partial u_1}{\partial x_2} = \left(\frac{\partial}{\partial x_1}, \frac{\partial}{\partial x_2} \right) \times (u_1, u_2).$$

If $u : \mathbb{R}^2 \to \mathbb{R}$ is a scalar function, then rot $u = \nabla \times u$ is defined as the vector function

$$\text{rot } u = \nabla \times u = \left(\frac{\partial u}{\partial x_2}, -\frac{\partial u}{\partial x_1} \right).$$

The different appearances of rot $u = \nabla \times u$, with u a scalar or $u = (u_1, u_2)$ a vector function will be explained when we pass to $\mathbb{R}^3$ below. For now, it may be helpful to recall the different appearances of $a \times b$ with $a, b \in \mathbb{R}^2$ or $a, b \in \mathbb{R}^3$.

The following identities follow directly from the definitions for any function u:

$$\begin{aligned} \nabla \cdot (\nabla \times u) = \text{div (rot } u) = 0, & \quad (u : \mathbb{R}^2 \to \mathbb{R}^2) \\ \nabla \times (\nabla u) = \text{rot (grad } u) = 0, & \quad (u : \mathbb{R}^2 \to \mathbb{R}). \end{aligned} \tag{61.1}$$

Finally, the *Laplacian* Δu of a function $u : \mathbb{R}^2 \to \mathbb{R}$ is defined by

$$\Delta u = \nabla \cdot (\nabla u) = \text{div (grad } u) = \frac{\partial^2 u}{\partial x_1^2} + \frac{\partial^2 u}{\partial x_2^2},$$

where $\frac{\partial^2 u}{\partial x_i^2} = \frac{\partial}{\partial x_i} \left(\frac{\partial u}{\partial x_i} \right)$.

61.3 The Laplacian in Polar Coordinates

In polar coordinates $x = (x_1, x_2) = (r \cos(\theta), r \sin(\theta))$ with $r \geq 0$ and $0 \leq \theta < 2\pi$, the Laplacian takes the form

$$\Delta u = \frac{1}{r} \frac{\partial}{\partial r} \left(r \frac{\partial u}{\partial r} \right) + \frac{1}{r^2} \frac{\partial^2 u}{\partial \theta^2}. \tag{61.2}$$

This follows by a routine computation using that the Jacobian of the mapping $x = (r \cos(\theta), r \sin(\theta))$, in the notation (54.9) is given by

$$\frac{d(x_1, x_2)}{d(r, \theta)} = \begin{pmatrix} \cos(\theta) & -r \sin(\theta) \\ \sin(\theta) & r \cos(\theta) \end{pmatrix},$$

so that

$$\frac{d(r, \theta)}{d(x_1, x_2)} = \begin{pmatrix} \cos(\theta) & \sin(\theta) \\ -\sin(\theta)/r & \cos(\theta)/r \end{pmatrix},$$

and thus by the Chain rule

$$\frac{\partial}{\partial x_1} = \cos(\theta) \frac{\partial}{\partial r} - \frac{\sin(\theta)}{r} \frac{\partial}{\partial \theta} \quad \text{and} \quad \frac{\partial}{\partial x_2} = \sin(\theta) \frac{\partial}{\partial r} + \frac{\cos(\theta)}{r} \frac{\partial}{\partial \theta}.$$

61.4   Some Basic Examples

The function $u : \mathbb{R}^2 \to \mathbb{R}^2$ given by $u(x) = \frac{1}{2}(x_1, x_2)$, satisfies

$$\nabla \cdot u(x) = 1.$$

The function $v : \mathbb{R}^2 \to \mathbb{R}^2$ given by $v(x) = \frac{1}{2}(-x_2, x_1)$, satisfies

$$\nabla \times v(x) = 1.$$

The function $w : \mathbb{R}^2 \to \mathbb{R}$ given by $w(x) = \frac{1}{4}(x_1^2 + x_2^2)$, satisfies

$$\Delta w = 1.$$

We plot these basic examples in Fig. 61.2

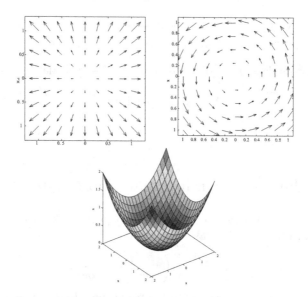

Fig. 61.2. Basic examples satisfying $\nabla \cdot u = 1$, $\nabla \times v = 1$ and $\Delta w = 1$, respectively

We see that $u(x)$ "explodes", $v(x)$ "rotates" and $w(x)$ is a "hump".

61.5   The Laplacian
Under Rigid Coordinate Transformations

It follows from the form of the Laplacian in polar coordinates, that the
Laplacian is invariant under rotations and translations in $\mathbb{R}^2$, i.e. so-called

rigid transformations of the form

$$\tilde{x}_1 = \cos(\alpha)x_1 + \sin(\alpha)x_2 + a_1$$
$$\tilde{x}_2 = -\sin(\alpha)x_1 + \cos(\alpha)x_2 + a_2,$$

where (x_1, x_2) are the old coordinates and $(\tilde{x}_1, \tilde{x}_2)$ the new ones. In other words, the Laplacian takes exactly the same form in the two coordinate systems:

$$\frac{\partial^2 u}{\partial x_1^2} + \frac{\partial^2 u}{\partial x_2^2} = \frac{\partial^2 u}{\partial \tilde{x}_1^2} + \frac{\partial^2 u}{\partial \tilde{x}_2^2}.$$

This fact is reflected in the observation that the Laplace operator typically occurs in *isotropic* models that have the same properties in all directions.

61.6 The Case of $\mathbb{R}^3$

The *gradient* of a function $u : \mathbb{R}^3 \to \mathbb{R}$, denoted grad u or ∇u, is the vector-valued function formed by the set of first order partial derivatives of u, i.e.

$$\text{grad } u = \nabla u = \left(\frac{\partial u}{\partial x_1}, \frac{\partial u}{\partial x_2}, \frac{\partial u}{\partial x_3} \right).$$

For a vector function $u : \mathbb{R}^3 \to \mathbb{R}^3$, the divergence div u is a scalar function defined by

$$\text{div } u = \sum_{i=1}^{3} \frac{\partial u_i}{\partial x_i},$$

and rot u is the vector function

$$\text{rot } u = \nabla \times u = \left(\frac{\partial u_3}{\partial x_2} - \frac{\partial u_2}{\partial x_3}, \frac{\partial u_1}{\partial x_3} - \frac{\partial u_3}{\partial x_1}, \frac{\partial u_2}{\partial x_1} - \frac{\partial u_1}{\partial x_2} \right).$$

We now explain the relation of the operator of rotation $\nabla\times$ in $\mathbb{R}^3$ to the operator of rotation $\nabla\times$ in $\mathbb{R}^2$ introduced above. Consider first a function $u : \mathbb{R}^3 \to \mathbb{R}^3$ of the form $u = (u_1, u_2, 0)$ with u_1 and u_2 being independent of x_3 so that effectively $u_i : \mathbb{R}^2 \to \mathbb{R}$ with $u_i = u_i(x_1, x_2)$ for $i = 1, 2$. We have

$$\nabla \times u = \left(0, 0, \frac{\partial u_2}{\partial x_1} - \frac{\partial u_1}{\partial x_2} \right) = (0, 0, \nabla \times (u_1, u_2)).$$

Secondly, if $u : \mathbb{R}^3 \to \mathbb{R}^3$ has the form $u = (0, 0, u_3)$ with u_3 independent of x_3, so that effectively $u_3 : \mathbb{R}^2 \to \mathbb{R}$, then

$$\nabla \times u = \left(\frac{\partial u_3}{\partial x_2}, -\frac{\partial u_3}{\partial x_1}, 0 \right) = (\nabla \times u_3, 0).$$

We conclude that $\nabla \times u$ for $u : \mathbb{R}^2 \to \mathbb{R}$ and $\nabla \times u$ for $u : \mathbb{R}^2 \to \mathbb{R}^2$, may be viewed as special cases of $\nabla \times u$ for $u : \mathbb{R}^3 \to \mathbb{R}^3$.

The *Laplacian* Δu of a function $u : \mathbb{R}^2 \to \mathbb{R}$ is defined by

$$\Delta u = \nabla \cdot (\nabla u) = \operatorname{div} (\operatorname{grad} u) = \sum_{i=1}^{3} \frac{\partial^2 u}{\partial x_i^2}.$$

By direct computation we verify the following identities:

$$\nabla \cdot (\nabla \times u) = 0,$$
$$\nabla \times (\nabla u) = 0, \qquad (61.3)$$
$$\nabla \times (\nabla \times u) = -\Delta u + \nabla(\nabla \cdot u).$$

61.7   Basic Examples, Again

The function $u : \mathbb{R}^3 \to \mathbb{R}^3$ given by $u(x) = \frac{1}{3}x$, satisfies

$$\nabla \cdot u(x) = 1.$$

The function $v : \mathbb{R}^3 \to \mathbb{R}^3$ given by $v(x) = \frac{1}{2}(-x_2, x_1, 0)$, satisfies

$$\nabla \times v(x) = (0, 0, 1).$$

The function $w : \mathbb{R}^3 \to \mathbb{R}$ given by $w(x) = \frac{1}{6}\|x\|^2$, satisfies

$$\Delta w = 1.$$

We plot these basic examples in Fig. 61.3

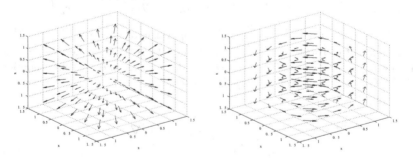

Fig. 61.3. Basic examples in $\mathbb{R}^3$ satisfying $\nabla \cdot u = 1$, $\nabla \times v = 1$

We see again that $u(x)$ "explodes", $v(x)$ "rotates" along the x_3 axis while the "hump" $w(x)$ is difficult to visualize.

61.8 The Laplacian in Spherical Coordinates

In *spherical coordinates.*

$$x = (x_1, x_2, x_3) = (r \sin(\varphi) \cos(\theta), r \sin(\varphi) \sin(\theta), r \cos(\varphi)),$$

where $r \geq 0$, $0 \leq \theta < 2\pi$ and $0 \leq \varphi < \pi$, the Laplacian is given by

$$\Delta u = \frac{1}{r^2} \frac{\partial}{\partial r} \left(r^2 \frac{\partial u}{\partial r} \right) + \frac{1}{r^2 \sin(\theta)} \frac{\partial}{\partial \theta} \left(\sin(\theta) \frac{\partial u}{\partial \theta} \right) + \frac{1}{r^2 \sin^2(\theta)} \frac{\partial^2 u}{\partial \varphi^2}. \quad (61.4)$$

The Laplacian is invariant under orthogonal coordinate transformations in $\mathbb{R}^3$.

Example 61.1. Consider the velocity field generated by rotation around a vector $\omega \in \mathbb{R}^3$ with angular speed $\|\omega\|$, that is the vector field

$$v(x) = \omega \times x.$$

We compute

$$\nabla \times v(x) = \nabla \times (\omega_2 x_3 - \omega_3 x_2, \omega_3 x_1 - \omega_1 x_3, \omega_1 x_2 - \omega_2 x_1)$$
$$= (2\omega_1, 2\omega_2, 2\omega_3) = 2\omega.$$

We conclude that the rotation $\nabla \times v(x)$ of a velocity field $v(x)$ generated by a rotation according to a given vector ω is equal to 2ω. This motivates the name of the differential operator $\nabla \times$ as the "rotation".

Example 61.2. A basic formula of electromagnetics expressing Ampère's law states that the *magnetic field H* generated by a unit electrical current flowing through the x_3-axis in the positive direction, is given by

$$H(x) = H(x_1, x_2, x_3) = \frac{1}{2\pi} \frac{(-x_2, x_1, 0)}{x_1^2 + x_2^2} \quad \text{for } x_1^2 + x_2^2 > 0. \quad (61.5)$$

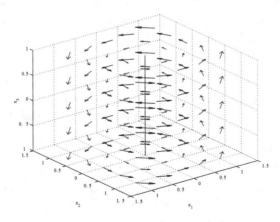

Fig. 61.4. The magnetic field around a current through the x_3-axis

We compute

$$\nabla \times H(x) = \frac{1}{2\pi}\left(0, 0, \frac{\partial}{\partial x_1}\frac{x_1}{x_1^2 + x_2^2} - \frac{\partial}{\partial x_2}\frac{-x_2}{x_1^2 + x_2^2}\right) = 0 \text{ for } x_1^2 + x_2^2 > 0.$$

Thus $\nabla \times H(x) = 0$ for $x_1^2 + x_2^2 > 0$, which is just *Amperes's Law* $\nabla \times H = J$, where J is the current density, noting that $J(x) =$ for $x_1^2 + x_2^2 > 0$, i.e. outside the x_3-axis. Amperes's Law is one of *Maxwell's equations*. Below we shall   show   how   to   interpret   the   equation $\nabla \times H(x) = J(x)$ for $x_1^2 + x_2^2 = 0$ and motivate the factor $\frac{1}{2\pi}$ in (61.5).

Chapter 61 Problems

61.1. Let $F = (5x_1 - 3x_1x_2 + x_3^2, \sin(x_1)\cos(x_1) + x_1, \sin(x_1)\exp(x_1x_2))$. With $x = (1, 2, 3)$, compute (a) $\nabla \cdot F$, (b) $\nabla \times F$, (c) $\nabla(\nabla \cdot F)$, (d) $\nabla \times (\nabla \times F)$.

61.2. Interpret the expression $(\nabla \times \nabla)u$ in a reasonable way and show that $(\nabla \times \nabla)u = 0$ for any u. Compare with $\nabla \times (\nabla \times u)$.

61.3. Show that for appropriate function u and v

1. $\nabla(uv) = (\nabla u)v + u(\nabla v)$,
2. $\nabla \cdot (uv) = (\nabla u) \cdot v + u(\nabla \cdot v)$,
3. $\nabla \times (uv) = (\nabla u) \times v + u(\nabla \times v)$,
4. $\nabla \cdot (u \times v) = v \cdot (\nabla \times u) - u \cdot (\nabla \times v)$,
5. $\nabla \times (u \times v) = (v \cdot \nabla)u - (\nabla \cdot u)v - (u \cdot \nabla)v + (\nabla \cdot v)u$,
6. $\nabla(u \cdot v) = (u \cdot \nabla)v + (v \cdot \nabla)u + u \times (\nabla \times v) + v \times (\nabla \times u)$.

61.4. Compute $\nabla(r \cdot F(r))$ where $r = \|x\|$.

61.5. Prove that the velocity field $v(x) = w \times x$, where $w \in \mathbb{R}^3$ is a given vector, satisfies $\nabla \cdot v(x) = 0$. Interpret the result in fluid mechanical terms.

61.6. Prove directly using the Chain rule that the Laplacian in $\mathbb{R}^2$ and $\mathbb{R}^3$ is invariant under rigid coordinate transformations.

61.7. Prove (61.3), (61.2) and (61.4).

61.8. Show that if $u : \mathbb{R}^2 \to \mathbb{R}$, then $\nabla \times (\nabla \times u) = \text{rot}(\text{rot } u) = -\Delta u$.

61.9. Show that the function $u : \mathbb{R}^2 \to \mathbb{R}$ given by $u(x) = c_1 \log(\|x\|) + c_2$ with c_1 and c_2 constants, is a solution of the Laplace equation $\Delta u(x) = 0$ in $\mathbb{R}^2$ for $x \neq 0$.

61.10. Prove that the function $u : \mathbb{R}^3 \to \mathbb{R}$ given by $u(x) = c_1\|x\|^{-1} + c_2$, with c_1 and c_2 constants, is a solution of Laplace's equation $\Delta u(x) = 0$ in $\mathbb{R}^3$ for $x \neq 0$.

61.11. Show that the divergence is invariant under rigid coordinate transformations. Does the rotation have the same property?

> All the effects of Nature are only the mathematical consequences of a small number of immutable laws. (Laplace)

62

Meteorology and Coriolis Forces*

Any teacher who stands up in front of a class and says that Coriolis force determines which way the water flows from a sink or bathtub, should not only read Fraser's Bad Coriolis Web page (www.ems.psu.edu/ fraser/Bad/BadCoriolis.html), but be required to copy it on the blackboard 100 times.
(Jack Williams, USA TODAY)

62.1 Introduction

A common weather map shows the level curves of the air pressure p, the so-called *isobars*. Intuition might suggest that the wind will blow from high pressure to low pressure, i.e. in the opposite direction to the pressure gradient ∇p and orthogonal to the isobars. However, this turns out to be completely false. In fact, the wind circles around a center of low pressure in a counter-clockwise direction on the North hemisphere and in a clockwise direction on the Southern hemisphere, and in the opposite directions around centers of high pressure. Thus the wind blows along the isobars, instead of orthogonal to the isobars. This fact is well-known to sailors, making it possible to easily and accurately predict the wind direction if the centers of the low and high pressures are known. The reason is that the Earth is rotating, which creates a force of acceleration called the *Coriolis force*. This causes the wind to deviate to the right on the Northern hemisphere and to the left on the Southern hemisphere (away from the equator). The effect is that the wind circles around a center of low pressure in a counter-clockwise direction on the Northern hemisphere, as any weather map in a newspaper indicates. The Coriolis force is felt on a turn-around when seeking to change position in the radial direction, which causes an (unexpected) force in the tangential direction.

62.2   A Basic Meteorological Model

We shall now derive a simple model for the motion of the atmosphere, which predicts that the wind should revolve around centers of low and high pressure. The model takes the form

$$\nabla p = \rho 2\omega \times v, \tag{62.1}$$

where p is the pressure, v is the wind velocity, $\omega \in \mathbb{R}^3$ is the angular velocity of the Earth, and ρ is the density of the atmosphere. The quantity $2\omega \times v$ is an approximation of the Coriolis acceleration and the equation $\nabla p = \rho 2\omega \times v$ gives a balance of the pressure force ∇p and the Coriolis force $\rho 2\omega \times v$. Here ∇p represents the gradient in the plane of the surface of the Earth and the model applies to "caps" on the Northern or Southern away from the Equator, say above or below the 60 degree latitude, where we can approximate the surface of the Earth by a flat disc, see Fig. 62.1, that is, the "world" of the sailor and the wind is a big flat turn-around.

We see that (62.1) states that ∇p is orthogonal to the direction of the wind. If we know p, we can determine the wind direction and speed from (62.1).

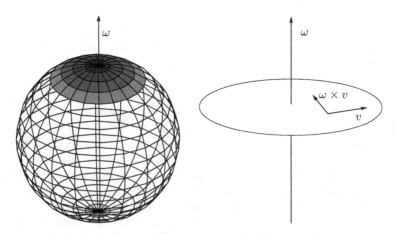

Fig. 62.1. Northern Hemisphere **Change w to ω**

62.3   Rotating Coordinate Systems and Coriolis Acceleration

To derive the expression $2\omega \times v$ of the Coriolis force, we need to study coordinate transformations from one fixed coordinate system to a rotating coordinate system. We thus let $\{e_1, e_2, e_3\}$ be a fixed orthonormal reference coordinate system for $\mathbb{R}^3$, and we let $\{\bar{e}_1, \bar{e}_2, \bar{e}_3\}$ be another orthonormal coordinate system with the same origin, which rotates around the fixed vector $\omega \in \mathbb{R}^3$ with the angular speed $\|\omega\|$. More precisely, if $x(t) = x_1(t)e_1 + x_2(t)e_2 + x_3(t)e_3$ are the reference coordinates of a fixed point in the rotating coordinate system, then according to Fig. 62.2 we have

$$\frac{dx}{dt} = \omega \times x, \tag{62.2}$$

since $\frac{dx}{dt}$ is perpendicular to both ω and x, and $\|\frac{dx}{dt}\| = \|\omega\|\|x\| \sin(\theta)$, where $\theta \in [0, \pi]$ is the angle between ω and x. In particular we have for the basis vectors of the moving coordinate system

$$\frac{d\bar{e}_i}{dt} = \omega \times \bar{e}_i, \quad i = 1, 2, 3. \tag{62.3}$$

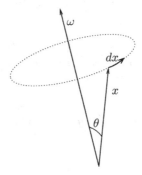

Fig. 62.2. A vector x rotating with angular velocity ω.**Change w to ω and q to θ**

Consider now a moving point with coordinates $x(t)$ in the fixed reference system and coordinates $\bar{x}(t)$ in the rotating system, so that

$$x(t) = x_1(t)e_1 + x_2(t)e_2 + x_3(t)e_3,$$
$$\bar{x}(t) = \bar{x}_1(t)\bar{e}_1(t) + \bar{x}_2(t)\bar{e}_2(t) + \bar{x}_3(t)\bar{e}_3(t),$$

and of course $x(t) = \bar{x}(t)$. In particular, we may this way seek the coordinates of the basis vectors $\bar{e}_i(t)$ in the fixed system $\{e_1, e_2, e_3\}$. We now compute the velocity $\frac{dx}{dt}$ by differentiating $x(t) = \bar{x}(t)$ with respect to t to get

$$\frac{dx}{dt} = \frac{d}{dt}\bar{x}(t) = \frac{d\bar{x}_1}{dt}\bar{e}_1 + \frac{d\bar{x}_2}{dt}\bar{e}_2 + \frac{d\bar{x}_3}{dt}\bar{e}_3 + \bar{x}_1\frac{d\bar{e}_1}{dt} + \bar{x}_2\frac{d\bar{e}_2}{dt} + \bar{x}_3\frac{d\bar{e}_3}{dt}$$
$$= \frac{d\bar{x}_1}{dt}\bar{e}_1 + \frac{d\bar{x}_2}{dt}\bar{e}_2 + \frac{d\bar{x}_3}{dt}\bar{e}_3 + \bar{x}_1(\omega \times \bar{e}_1) + \bar{x}_2(\omega \times \bar{e}_2) + \bar{x}_3(\omega \times \bar{e}_3),$$

where we used (62.3). We can write this expression as

$$\frac{dx}{dt} = \frac{\bar{d}\bar{x}}{dt} + \omega \times x, \tag{62.4}$$

if we agree to write

$$\frac{\bar{d}\bar{x}}{dt} = \frac{d\bar{x}_1}{dt}\bar{e}_1 + \frac{d\bar{x}_2}{dt}\bar{e}_2 + \frac{d\bar{x}_3}{dt}\bar{e}_3.$$

The velocity of $x(t) = \bar{x}(t)$ in the fixed reference system is $\frac{dx}{dt}$, while $\frac{\bar{d}\bar{x}}{dt}$ is the velocity vs the rotating system involving the derivatives $\frac{\bar{d}}{dt}\bar{x}_i(t)$. In particular, if the point is fixed in the rotating system so that $\frac{\bar{d}\bar{x}}{dt} = 0$, then we retrieve (62.2) and (62.3).

We now seek a corresponding formula for the accelerations. We differentiate with respect to t once more, and using (62.4) with x replaced by $\frac{\bar{d}\bar{x}}{dt}$, we get

$$\frac{d^2x}{dt^2} = \frac{d}{dt}\left(\frac{\bar{d}\bar{x}}{dt} + \omega \times x\right) = \frac{d}{dt}\left(\frac{\bar{d}\bar{x}}{dt}\right) + \omega \times \frac{dx}{dt}$$

$$= \frac{\bar{d}}{dt}\left(\frac{\bar{d}\bar{x}}{dt}\right) + \omega \times \frac{\bar{d}\bar{x}}{dt} + \omega \times \left(\frac{\bar{d}\bar{x}}{dt} + \omega \times x\right).$$

We can write this as

$$\frac{d^2x}{dt^2} = \frac{\bar{d}^2\bar{x}}{dt^2} + 2\omega \times \frac{\bar{d}\bar{x}}{dt} + \omega \times (\omega \times x). \tag{62.5}$$

Here, $\omega \times (\omega \times x)$ represents the *centripetal acceleration* and $2\omega \times \frac{\bar{d}\bar{x}}{dt}$ the *Coriolis acceleration*, and $\frac{d^2x}{dt^2}$ is the acceleration vs the reference system and $\frac{\bar{d}^2\bar{x}}{dt^2}$ the acceleration vs the rotating system.

By Newton's Law $F = ma$, *acceleration* is directly coupled to *force*, and thus both the centripetal and the Coriolis acceleration show up as forces in the fixed reference system. Both these forces in fact have a somewhat mysterious character; we have through massive daily experience become quite familiar with the centripetal acceleration, while the Coriolis force still presents surprises to most of us.

If the rotation speed $\|\omega\|$ is relatively small, then we can neglect the centripetal acceleration and we get

$$\frac{d^2x}{dt^2} \approx \frac{d^2\bar{x}}{dt^2} + 2\omega \times \frac{d\bar{x}}{dt}, \tag{62.6}$$

which leads to the model (62.1). Note that we use the rotating coordinate system in our "world", and thus $\frac{d\bar{x}}{dt}$ is the relevant velocity.

Chapter 62   Problems

62.1. Motivate (62.1) using (62.6).

62.2. Inspect the isobars of a weather map and compute wind direction from (62.1) and compare with the wind direction of the map.

62.3. Study the effect of the Coriolis acceleration at the Equator.

62.4. Show that the centripetal acceleration of a body moving in a circle with radius r with speed v is equal to $\frac{v^2}{r}$.

62.5. The Gulf Stream is the reason Scandinavia is not deep frozen like Alaska. Explain why the Gulf Stream bends over from North America to North Europe.

62.6. Consider a car driving East-West along a certain latitude. At what speed is the Coriolis force on the car of the same size as the centripetal force? Determine this speed as a function of the latitude and find out at which latitudes the minimum and maximum is attained.

62.7. A bucket of water is spinning around its center with angular velocity ω. What is the shape of the water surface?

62.8. A pendulum of length l swings back and forth once every period of length $t = \sqrt{l/g}$, where g is the acceleration of gravity. Compute the Coriolis force on the pendulum at latitude θ (i.e. at an angle θ from the equator). This Coriolis force makes the plane in which the pendulum swings rotate, i.e. if the pendulum swings north–south at one instant, it will later swing west–east. Find the time T after which the pendulum swings in the initial direction once again as function of the latitude. What is the period on you latitude?

63
Curve Integrals

We can scarcely believe that Ampère really discovered the law of action by means of the experiments which he describes. We are led to suspect, what, indeed, he tells us himself, that he discovered the law by some process which he has not shown us, and that when he had afterwards built up a perfect demonstration he removed all traces of the scaffolding by which he had raised it. (Maxwell about Ampères *Memoir on the Mathematical Theory of Electrodynamic Phenomena, Uniquely Deduced from Experience*)

63.1 Introduction

In this chapter we introduce the concept of an *integral over a curve* or *curve integral*, and develop some applications including *arc length*, *work* and *line integrals*. We start with plane curves parameterized by functions $s : I \to \mathbb{R}^2$, where $I = [a, b]$ is an interval of the real line $\mathbb{R}$. We then generalize to curves in $\mathbb{R}^n$ parameterized by functions $s : I \to \mathbb{R}^n$ with $n \geq 2$.

63.2 The Length of a Curve in $\mathbb{R}^2$

Let Γ be a curve in $\mathbb{R}^2$ given by the function $s : I \to \mathbb{R}^2$, where $I = [a, b]$ is an interval of $\mathbb{R}$, that is, $\Gamma = \{s(t) \in \mathbb{R}^2 : t \in I\}$, or $\Gamma = s(I)$, see Fig. 63.1. We now try to determine the *length* of Γ. We shall see that this leads to the introduction of the notion of an integral over a curve or a curve integral.

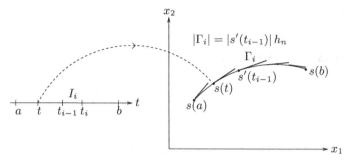

Fig. 63.1. The total length of curve is the sum of the lengths of little pieces of the curve

To define the length of a curve, we view the curve Γ as the being made up of little pieces of Γ. If the little pieces are sufficiently small, we can get away with approximating them by straight segments, and the length of a straight piece of curve is easy to compute. To find the total length of Γ, we will sum the lengths of all the little pieces forming Γ. We will find the integral is useful for this purpose.

Let $a = t_0 < t_1 < \ldots < t_n = b$ be a subdivison of I into intervals $I_i = (t_{i-1}, t_i]$. Consider the following linear approximation of the mapping $s(t)$ restricted to the subinterval I_i, see Fig. 63.1,

$$\bar{s}(t) = s(t_{i-1}) + (t - t_{i-1})s'(t_{i-1}).$$

The mapping $\bar{s}$ maps I_i onto the line segment Γ_i of length

$$\|s'(t_{i-1})\|(t_i - t_{i-1}),$$

and it is thus natural to use

$$L_n(\Gamma) = \sum_{i=1}^{n} \|s'(t_{i-1})\|(t_i - t_{i-1})$$

as an approximation of the length of Γ. Assuming that $\|s'(t)\|$ is Lipschitz continuous on I and assuming that $\max_i(t_i - t_{i-1})$ tends to zero as n tends to infinity, we can use the usual arguments to show that $\{L_n(\Gamma)\}_{n=1}^{\infty}$ is a Cauchy sequence and thus converges to a limit, which we denote by $L(\Gamma)$. We define this limit to be the *length* of Γ:

$$L(\Gamma) = \int_I \|s'(t)\| \, dt. \tag{63.1}$$

This formula expresses the length of a curve $\Gamma = s(I)$ as an integral over the parameter domain I of Γ with the modulus $\|s'(t)\|$ of the derivative of the representing function $s : I \to \mathbb{R}^2$ as a weight. Formally, we have

$ds = \|s'(t)\|dt$, where ds represents the increase of the length of the curve corresponding to an increase dt of the parameter t; the function $\|s'(t)\|$ gives the local "change of scale" between the "element of curve length" ds; and the "parameter element" dt, see Fig. 63.1. We are thus led to write

$$L(\Gamma) = \int_\Gamma ds = \int_I \|s'(t)\| \, dt.$$

We will return to this notation in the next section.

Example 63.1. We compute the length of the circumference Γ of a circle of radius 1 centered at the origin. The curve Γ is given by the function $s : [0, 2\pi) \to \mathbb{R}^2$ with $s(t) = (\cos(t), \sin(t))$ and $0 \le t < 2\pi$. We have $s'(t) = (-\sin(t), \cos(t))$ and $\|s'(t)\| = 1$, and thus

$$L(\Gamma) = \int_0^{2\pi} \|s'(t)\| \, dt = \int_0^{2\pi} dt = 2\pi.$$

We conclude that the the length of the circumference of a circle of radius 1 is equal to 2π (no big surprise). We check the result using a different parametrization. The upper semi-circle Γ_+ of Γ can be parameterized by $s : [-1, 1] \to \mathbb{R}^2$ given by $s(t) = (t, \sqrt{1 - t^2})$ with $-1 \le t \le 1$. We have

$$s'(t) = \left(1, -\frac{t}{\sqrt{1 - t^2}}\right), \quad \|s'(t)\| = \frac{1}{\sqrt{1 - t^2}},$$

and thus

$$L(\Gamma) = 2L(\Gamma_+)$$
$$= \int_{-1}^1 \frac{1}{\sqrt{1 - t^2}} \, dt = 2\left[\arcsin(t)\right]_{-1}^1 = 2\left(\frac{\pi}{2} - \left(-\frac{\pi}{2}\right)\right) = 2\pi.$$

63.3   Curve Integral

Let $\Gamma = s(I)$ be a curve in $\mathbb{R}^2$ given by the function $s : I \to \mathbb{R}^2$, where $I = [a, b]$ is an interval of $\mathbb{R}$, and let $u : \Gamma \to \mathbb{R}$ be a function defined on Γ. We assume that the tangent $s' : I \to \mathbb{R}^2$ and the function $u : \Gamma \to \mathbb{R}$ are both Lipschitz continuous, which guarantees that $\|s'(t)\|$ and $u(s(t))$ are both Lipschitz continuous on I. We define the *integral of u over* Γ by

$$\int_\Gamma u \, ds \equiv \int_\Gamma u(x) \, ds(x) \equiv \int_a^b u(s(t)) \|s'(t)\| \, dt.$$

Formally, we have $ds = ds(x) = \|s'(t)\| \, dt$, where $x = s(t)$.

Example 63.2. If Γ is an interval $[a, b]$ on the x_1-axis given by $s(t) = (t, 0)$, $a \le t \le b$, then $s'(t) = (1, 0)$, $\|s'(t)\| = 1$, and

$$\int_\Gamma u \, ds = \int_a^b u(x_1, 0) \, dx_1 = \int_a^b u(t, 0) \, dt.$$

Example 63.3.   Let $\Gamma = s(I)$ be the semicircle given by $s(t) = (\cos(t), \sin(t))$, $0 \le t \le \pi$, and $u(x) = u(x_1, x_2) = x_1^2$. Using $\|s'(t)\| = 1$, we get

$$\int_\Gamma u \, ds = \int_0^\pi \cos^2(t) \, dt = \frac{1}{2} \int_0^\pi (1 - \cos(2t)) \, dt = \frac{\pi}{2}.$$

63.4   Reparameterization

An important observation is that the value of a curve integral is independent of the parameterization of the curve. To see this, consider two different parameterizations $s : [a, b] \to \Gamma$ and $\sigma : [c, d] \to \Gamma$ of a curve Γ in $\mathbb{R}^2$. Associate to each $\tau \in [c, d]$ the unique value $t \in [a, b]$ such that $s(t) = \sigma(\tau)$, which defines $t = t(\tau)$ as a function of τ (assuming that the curve does not cross itself), so that $\sigma(\tau) = s(t(\tau))$, see Fig. 63.2.

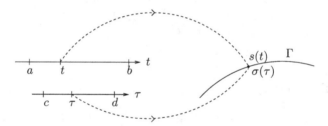

Fig. 63.2. Reparametrization of a curve

We now use the formula for change of integration variables and the fact that by the Chain rule

$$\sigma'(\tau) = \frac{d\sigma}{d\tau} = \frac{ds}{dt} \frac{dt}{d\tau} = s'(t) \frac{dt}{d\tau},$$

to see that, assuming $\frac{dt}{d\tau} \ge 0$,

$$\int_a^b u(s(t)) \|s'(t)\| \, dt = \int_c^d u(s(t(\tau))) \|s'(t(\tau))\| \frac{dt}{d\tau} \, d\tau$$

$$= \int_c^d u(\sigma(\tau)) \|\sigma'(\tau)\| \, d\tau.$$

This shows that the curve integral

$$\int_\Gamma u \, ds = \int_\Gamma u \, d\sigma$$

is independent of the parametrization $s : [a, b] \to \Gamma$ or $\sigma : [c, d] \to \Gamma$ of Γ.

Example 63.4. We reparameterize the semicircle Γ in the previous example by $s(t) = (t, \sqrt{1 - t^2})$ with $-1 \le t \le 1$ and get with $u(x) = x_1^2$, integrating by parts

$$\int_\Gamma u \, ds = \int_{-1}^{1} t \frac{t}{\sqrt{1 - t^2}} \, dt = \left[-t\sqrt{1 - t^2} \right]_{-1}^{1} + \int_{-1}^{1} \sqrt{1 - t^2} \, dt$$

$$= \int_{-\pi}^{0} \sqrt{1 - \cos^2(\theta)} (-\sin(\theta)) \, d\theta = \int_0^{\pi} \sin^2(\theta) \, d\theta = \frac{\pi}{2}.$$

63.5   Work and Line Integrals

Let $F : \mathbb{R}^2 \to \mathbb{R}^2$ be a vector function representing a variable force, or a *force field*, defined in $\mathbb{R}^2$, and let Γ be a curve in $\mathbb{R}^2$ given by $s : [a, b] \to \mathbb{R}^2$ starting at $A = s(a)$ and ending at $B = s(b)$. Consider a particle acted upon by the force F moving along Γ from A to B, see Fig. 63.3. The projection $F_s(s(t))$ of the force $F(s(t))$ on the direction $s'(t)$ of the tangent to $s(t)$ is equal to

$$F_s(s(t)) = F(s(t)) \cdot s'(t) \frac{1}{\|s'(t)\|}. \tag{63.2}$$

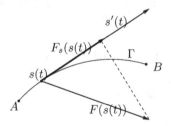

Fig. 63.3. Force field F and curve Γ, and projection of F onto $s'(t)$

Using the idea that "the work is equal to the projection of the force in the direction of the displacement × displacement", the *work* performed by the force $F(s(t))$ as the particle moves from $s(t_{i-1})$ to $s(t_i)$ is

$$F(s(t_i)) \cdot s'(t_i) \frac{1}{\|s'(t_i)\|} \|s(t_i) - s(t_{i-1})\|$$

$$\approx F(s(t_i)) \cdot s'(t_i) \frac{1}{\|s'(t_i)\|} \|s'(t_i)\|(t_i) - t_{i-1}) = F(s(t_i)) \cdot s'(t)(t_i - t_{i-1}).$$

As above $a = t_0 < t_1 < \ldots t_{i-1} < t_i < \ldots < t_n = b$ is an increasing sequence of discrete time levels, where we think of the time steps $t_i - t_{i-1}$ as tending to zero. We are now led to define the *total work* $W(F, \Gamma)$ as the particle moves from $A = s(a)$ to $B = s(b)$ along Γ, as

$$W(F, \Gamma) = \int_a^b F(s(t)) \cdot s'(t)\, dt.$$

Setting $ds = s'(t)\, dt$, we also write

$$\int_\Gamma F \cdot ds = \int_a^b F(t) \cdot s'(t)\, dt,$$

which we call a *line integral*. To sum up, we have

$$W(F, \Gamma) = \int_\Gamma F \cdot ds = \int_a^b F(t) \cdot s'(t)\, dt$$

$$= \int_a^b (F_1(t)s_1'(t) + F_2(t)s_2'(t))\, dt. \qquad (63.3)$$

Alternatively, we can write

$$W(F, \Gamma) = \int_\Gamma F_s\, ds = \int_a^b F_s \|s'(t)\|\, dt = \int_\Gamma F \cdot ds, \qquad (63.4)$$

with F_s being the projection of F onto $s'(t)$ according to (63.2).

Example 63.5. Assume that $F(x) = (x_2, -x_1)$ and let Γ be given by $s(t) = (\cos(t), \sin(t))$, $0 \le t < 2\pi$. We have

$$W(F, \Gamma) = \int_\Gamma F \cdot ds = \int_0^{2\pi} (\sin(t), -\cos(t)) \cdot (-\sin(t), \cos(t))\, dt$$

$$= -\int_0^{2\pi} dt = -2\pi.$$

63.6   Work and Gradient Fields

There is an important special case. If $F = \nabla \varphi$, that is the force field F is the *gradient field* of a *potential* $\varphi(x)$, then the Chain rule implies

$$W(F, \Gamma) = \int_\Gamma F \cdot ds = \int_a^b \nabla \varphi(s(t)) \cdot s'(t)\, dt$$

$$= \int_a^b \frac{d}{dt} \varphi(s(t))\, dt = \varphi(B) - \varphi(A).$$

We conclude that if the force field F is the gradient field $F = \nabla\varphi$ of a potential $\varphi(x)$, then the work performed by F along a curve Γ from A to B is equal to the difference $\varphi(B) - \varphi(A)$ of the values of the potential φ at the end point B and the starting point A. In other words, the work is independent of the curve from A to B. In particular, if the curve is *closed* so that $B = s(b) = s(a) = A$, then the work is zero.

Below we consider the problem of finding conditions guaranteeing that a given force $F(x)$ is the gradient of a potential so that $F(x) = \nabla\varphi(x)$ for some scalar function $\varphi(x)$.

Example 63.6. As a basic application, we consider the attractive *gravitational force* $F(x) = \nabla\varphi(x)$ with $\varphi(x) = 1/\|x\|$ being the *Newtonian potential*, corresponding to a unit mass at the the origin, that is

$$F(x) = -\frac{1}{\|x\|^2}\frac{x}{\|x\|},$$

with normalization of the gravitational constant to one. We note that $F(x)$ is directed towards the origin and obeys the inverse square law: $\|F(x)\| = \|x\|^{-2}$. We have

$$W(F,\Gamma) = \frac{1}{\|B\|} - \frac{1}{\|A\|},$$

which corresponds to the work performed as a unit mass moves in the gravitational field from a distance $\|A\|$ to the distance $\|B\|$ from the origin. In particular, if $\|A\| = \infty$, then $W(F,\Gamma) = 1/\|B\|$. We conclude that the work required to "lift" a particle of unit mass from a distance r of an attracting gravitational field of unit strength at the origin to an infinite distance is equal to $1/r$.

63.7 Using the Arclength as a Parameter

Note that if $u(x) = 1$ for all $x \in \Gamma$, then

$$\int_\Gamma ds = \int_\Gamma 1\,ds = \int_\Gamma u(x)\,ds(x) = \int_a^b \|s'(t)\|\,dt$$

is the length of the curve $\Gamma = s(I)$ with $I = [a, b]$. In particular,

$$\sigma(\bar{t}) = \int_a^{\bar{t}} \|s'(t)\|\,dt$$

is the *arclength* of the part of the curve from $s(a)$ to $s(\bar{t})$. The Fundamental Theorem of Calculus implies

$$\sigma'(\bar{t}) = \|s'(\bar{t})\|. \tag{63.5}$$

We may now choose the arclength $\sigma = \sigma(t)$ as the parameter instead of t since to each t, there is a unique arclength $\sigma(t)$ and vice versa. This gives a reparameterization of $s(t) = \bar{s}(\sigma)$ with

$$\|\bar{s}'(\sigma)\| = \|\frac{ds}{dt}\|\,|\frac{dt}{d\sigma}| = \|s'(t)\|\frac{1}{|\sigma'(t)|} = \frac{\|s'(t)\|}{\|s'(t)\|} = 1.$$

We conclude that if the arclength σ to used to parameterize the curve $s : I \to \mathbb{R}^2$, then $\|s'(\sigma)\| = 1$ and, see Fig. 63.4,

$$L(\Gamma) = \int_\Gamma ds = \int_0^{L(\Gamma)} d\sigma.$$

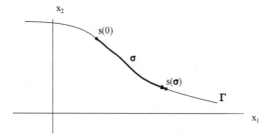

Fig. 63.4. A curve Γ parameterized by arclength σ

63.8   The Curvature of a Plane Curve

The *curvature* of a curve $s : [a, b] \to \mathbb{R}^2$ measures of how quickly the curve bends as we move along the curve. It is defined by

$$\kappa = \frac{d\theta}{d\sigma},$$

where θ is the polar angle of the tangent vector $s' = (s_1', s_2')$ defined by $\theta(t)) = \tan^{-1}(s_2'/s_1')$ and σ is arclength. In the case of a straight line, the polar angle $\theta(t)$ is constant and the curvature is zero, see Fig. 63.5.

The arc length $\sigma(t)$ satisfies, recalling (63.5), $\frac{d\sigma}{dt} = |s'|$, and thus $\frac{dt}{d\sigma} = |s'|^{-1}$. The chain rule implies

$$\kappa(t) = \frac{d\theta}{dt}\frac{dt}{d\sigma} = \frac{\theta'(t)}{\|s'(t)\|}.$$

Computing $\theta'(t)$, we find that

$$\kappa(t) = \frac{s_1'(t)s_2''(t) - s_1''(t)s_2'(t)}{\left(s_1'(t)^2 + s_2'(t)^2\right)^{3/2}}.$$

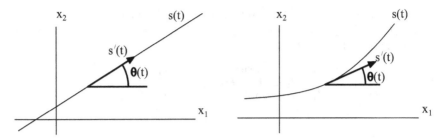

Fig. 63.5. The polar angle θ of the tangent vector of a straight line is constant as shown on the right. The tangent vector of a curve that bends, like the example on the left, has a different polar angle at each point

In particular if the curve is parameterized by $s(x_1) = (x_1, f(x_1))$, where $f : \mathbb{R} \to \mathbb{R}$ has two continuous derivatives, then the curvature at the point $(x_1, f(x_1))$, is given by

$$\kappa(x_1) = \frac{f''(x_1)}{\left(1 + (f'(x_1))^2\right)^{3/2}}.$$

We define the *circle of curvature* at a point $P = s(t)$ on a curve $s : [a, b] \to \mathbb{R}^2$, as the circle of radius $|\kappa|^{-1}(t)$ (assuming $\kappa \neq 0$) that shares the same tangent line as Γ at P and points to the left of T if $\kappa > 0$ and to the right if $\kappa < 0$, see Fig. 63.6. The *radius of curvature* at P is $|\kappa|^{-1}(t)$.

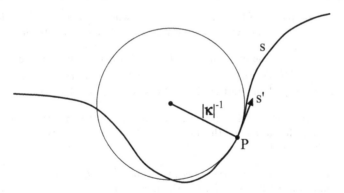

Fig. 63.6. The circle of curvature of Γ at P

63.9   Extension to Curves in $\mathbb{R}^n$

The definitions of integrals over curves and line integrals directly extend to curves in $\mathbb{R}^n$ represented by $s : [a, b] \to \mathbb{R}^n$ with $n \geq 2$.

Example 63.7. Consider the circular helix Γ in $\mathbb{R}^3$ given by $s(t) = (\cos(t), \sin(t), t)$, $0 \le t \le 20\pi$, and let $u(x) = x_3^2$. We have since $s'(t) = (-\sin(t), \cos(t), 1)$ and thus $\|s'(t)\| = \sqrt{2}$,

$$\int_\Gamma u \, ds = \int_0^{20\pi} t^2 \sqrt{2} \, dt = \frac{\sqrt{2}}{3}(20\pi)^3.$$

Chapter 63 Problems

63.1. (a) Compute the length of (a) the *catenary* (hanging chain curve) given by $s(t) = (t, \cosh(t))$ with $-1 \le t \le 1$, (b) the *circular helix* $s(t) = (\cos(t), \sin(t), t)$ with $0 \le t \le 4\pi$, (c) the *cycloid* $s(t) = (t - \sin(t), 1 - \cos(t))$ with $0 \le t \le 2\pi$, (d) the *semi-cubical parabola* $s(t) = (t^3, t^2)$ with $0 \le t \le 2$, (e) the *four-cusped hypocycloid or astroid* $s(t) = (\cos^3(t), \sin^3(t))$.

63.2. Let Γ be the circular helix $s(t) = (\cos(t), \sin(t), t)$ with $t \in [0, 2\pi)$. Compute the value of the curve integral $\int_\Gamma u \, ds$ for (a) $u(x) = 1$, (b) $u(x) = x_3$, (c) $u(x) = x_1 x_2 x_3$.

63.3. Compute the curve integral $\int_\Gamma x_1 x_2 \, ds$, where (a) Γ is the part of the unit circle in the $x_1 x_2$-plane from $(1, 0, 0)$ to $(0, 1, 0)$, (b) Γ is the part of the unit square in the $x_1 x_2$-plane from $(1, 0, 0)$ to $(0, 1, 0)$. (c) Γ is the shortest path from $(1, 0, 0)$ to $(0, 1, 0)$.

63.4. (a) Compute the line integral $\int_\Gamma x \cdot ds$ where Γ is the unit circle in the $x_1 x_2$-plane. (b) Try other choices of closed curves Γ and evaluate the integral.

63.5. Compute the line integral $\int_\Gamma F \cdot ds$ with Γ the unit circle in the $x_1 x_2$ plane and (a) $F(x) = \frac{(x_1, x_2)}{|x|^2}$, (b) $F(x) = \frac{(-x_2, x_1)}{|x|^2}$. Does the result depend on whether you integrate around the unit circle clockwise or counter-clockwise?

63.6. A particle is moved counter-clockwise around the square $0 \le x_1, x_2 \le 1$, $x_3 = 0$ under the action of the force field $f(x) = ((x_1 - x_2)^2, 2x_2 + x_1^2, x_1)$. Compute the work done.

63.7. Let $f(x) = (2x_1 + x_2, 3x_1 - 2x_2)$. Compute $\int_\Gamma f \cdot ds$ with Γ given by (a) the straight line from $(0, 0)$ to $(1, 1)$, (b) the parabola $x_2 = x_1^2$ from $(0, 0)$ to $(1, 1)$, (c) the curve $x_2 = \sin(\pi x_1/2)$ from $(0, 0)$ to $(1, 1)$, (d) the curve $x_2 = x_1^n$ with $n > 0$ from $(0, 0)$ to $(1, 1)$.

63.8. Compute the integral of $u = x_1 x_2$ over the boundary of the unit square $[0, 1] \times [0, 1]$.

63.9. Find the circle of curvature of $x_2 = x_1^2$ at $x_1 = 0$.

63.10. Find the curvature of the plane curve $(R\cos(\theta), R\sin(\theta))$ where R is constant. Conclude that the curvature of a circle of radius R is R^{-1}.

63.11. Verify the two formulas for the curvature.

63.12. (a) Compute the curvature of the curve (x_1, x_1^2). (b) Do the same for (x_1, x_1^3), and then discuss what happens at the inflection point.

63.13. Consider a hanging chain described by a function $y(x)$ with $-1 \leq x \leq 1$ and $y(-1) = y(1)$. Let for $0 \leq x \leq 1$, $T(x)$ be the modulus of the chain force at x, and let $s(x)$ be the length of the chain from 0 to x. Derive the vertical equilibrium equation

$$y'(x) = cs(x) = c \int_0^1 \sqrt{1 + (y'(x))^2} \, dx,$$

with c a constant. Show that this equation is satisfied with $y'(x) = \sinh(\frac{x}{c})$, and conclude that $y(x) = c \cosh(\frac{x}{c})$.

63.14. Find the direction of the tangent at the point $(1, 1, 1)$ of the curve cut out on the surface $x_1^2 + x_1^2 x_2 + x_2^2 x_3 + x_3^2 = 0$. Hint: Use implicit differentiation.

63.15. Show that if a plane curve Γ is represented in polar coordinates $(\rho(\theta), \theta)$ with $\rho(theta)$ a function of θ and $a \leq \theta \leq b$, then $ds^2 = \rho^2 \, d\theta^2 + d\rho^2$ and thus

$$L(\Gamma) = \int_a^b (\rho^2 + (\rho')^2)^{1/2} \, d\theta.$$

Compute the the length of the *cardioid* $\rho = (1 - \cos(\theta))$ with $0 \leq \theta \leq 2\pi$.

63.16. Compute the length of a string which is wound around a circular cylinder with a uniform pitch.

64
Double Integrals

To understand this for sense it is not required that a man should be a geometrician or a logician, but that he should be mad. ["This" is that the volume generated by revolving the region under $1/x$ from 1 to infinity has finite volume.] (Hobbes 1588–1679)

He was 40 years old before he looked on geometry; which happened accidentally. Being in a gentleman's library, Euclid's Elements lay open, and "twas the 47 El. libri I" [Pythagoras' Theorem]. He read the proposition. "By God", sayd he, "this is impossible:" So he reads the demonstration of it, which referred him back to such a proposition; which proposition he read. That referred him back to another, which he also read. Et sic deinceps, that at last he was demonstratively convinced of that trueth. This made him in love with geometry. (About Thomas Hobbes by John Aubrey 1626–1697)

64.1 Introduction

We have studied the integral

$$\int_0^1 f(x)\, dx,$$

where $f : [0, 1] \to \mathbb{R}$ is a Lipschitz continuous function of one variable. We call this a *one-dimensional integral*. We generalize this idea to the *double integral*

$$\int_0^1 \int_0^1 f(x_1, x_2)\, dx_1\, dx_2, \tag{64.1}$$

which has *two integration variables* x_1 and x_2 that run from 0 to 1. Here $f : Q \to \mathbb{R}$ is a Lipschitz continuous function defined on the unit square $Q = [0,1] \times [0,1] = \{x = (x_1, x_2) : 0 \le x_1 \le 1, 0 \le x_2 \le 1\}$, satisfying

$$|f(x) - f(y)| \le L_f \|x - y\| \quad \text{for } x, y \in Q. \tag{64.2}$$

64.2 Double Integrals over the Unit Square

Recall that we define the one dimensional integral as

$$\int_0^1 f(x)\, dx = \lim_{n \to \infty} \sum_{i=1}^N f(x_i^n) h_n, \tag{64.3}$$

where $0 = x_0^n < x_1^n < \ldots < x_N^n = 1$ is a subdivision of the interval $[0,1]$ with $x_i^n = ih_n$, $i = 1, \ldots, N$ and $h_n = 2^{-n}$ and $N = 2^n$.

To define the double integral, we let $0 = x_{1,0}^n < x_{1,1}^n < \ldots < x_{1,N}^n = 1$ and $0 = x_{2,0}^n < x_{2,1}^n < \ldots < x_{2,N}^n = 1$ be a subdivisions of the interval $[0,1]$ with $x_{1,i}^n = ih_n$, $i = 0, \ldots, N$, and $x_{2,j}^n = jh_n$, $j = 0, \ldots, N$, where $h_n = 2^{-n}$ and $N = 2^n$. This corresponds to a subdivision of the unit square $Q = [0,1] \times [0,1]$ into sub-squares $Q_{i,j}^n = I_i^n \times J_j^n$ of area $h_n h_n$, where $I_i^n = (x_{1,i-1}^n, x_{1,i}^n]$ $J_j^n = (x_{2,j-1}^n, x_{2,j}^n]$, where $i, j = 1, \ldots, N$, see Fig. 64.1.

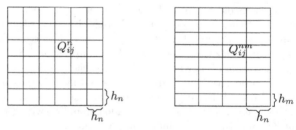

Fig. 64.1. Partition of the unit square Q into quadratic or rectangular sub-domains Q_{ij}^n or Q_{ij}^{nm}

We shall prove that the limit $\lim_{n \to \infty} S_n$ exists, where

$$S_n = \sum_{i=1}^N \sum_{j=1}^N f(x_{1,i}^n, x_{2,j}^n) h_n h_n \tag{64.4}$$

is a *Riemann sum* over all the sub-squares $Q_{i,j}^n$. We define

$$\int_0^1 \int_0^1 f(x_1, x_2)\, dx_1\, dx_2 = \lim_{n \to \infty} \sum_{i=1}^N \sum_{j=1}^N f(x_{1,i}^n, x_{2,j}^n) h_n h_n. \tag{64.5}$$

We begin by estimating the difference $S_n - S_{n+1}$ with the goal of proving that $\{S_n\}$ is a Cauchy sequence. Each sub-square $Q_{i,j}^n$ consists of the four sub-squares $Q_{2i,2j}^{n+1}$, $Q_{2i-1,2j}^{n+1}$ $Q_{2i,2j-1}^{n+1}$, and $Q_{2i-1,2j-1}^{n+1}$, see Fig. 64.2. We have

$$S_n - S_{n+1} = \sum_{i=1}^{N} \sum_{j=1}^{N} a_{ij} h_n h_n,$$

where, see Fig. 64.2,

$$a_{ij} = f(x_{1,i}^n, x_{2,j}^n) - \frac{1}{4}\Big(f(x_{1,2i}^{n+1}, x_{2,2j}^{n+1}) + f(x_{1,2i-1}^{n+1}, x_{2,2j}^{n+1})$$
$$+ f(x_{1,2i}^{n+1}, x_{2,2j-1}^{n+1}) + f(x_{1,2i-1}^{n+1}, x_{2,2j-1}^{n+1})\Big).$$

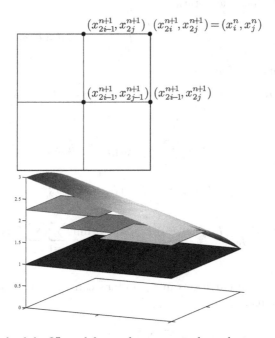

Fig. 64.2. On the *left*: Q_{ij}^n and four sub-squares and quadrature point. On the *right*: A function $f(x_1, x_2)$ and its piecewise constant approximation on Q_{ij}^n and on the four sub-squares

The Lipschitz continuity condition (64.2) implies

$$|a_{ij}| \leq \frac{1}{4} L_f \left(h_{n+1} + h_{n+1} + \sqrt{2} h_{n+1} \right) \leq L_f h_{n+1},$$

and thus

$$|S_n - S_{n+1}| \le \sum_{i=1}^{N}\sum_{j=1}^{N} |a_{ij}|h_n h_n \le L_f h_{n+1} \sum_{i=1}^{N}\sum_{j=1}^{N} h_n h_n = L_f h_{n+1}.$$

The usual arguments show that for $m > n$,

$$|S_n - S_m| \le 2L_f h_{n+1} = L_f h_n,$$

which proves that $\{S_n\}$ is a Cauchy sequence and thus converges to a real number. We decide, following our dear friends Leibniz and Cauchy as usual, to denote this real number by

$$\int_0^1 \int_0^1 f(x_1, x_2)\, dx_1\, dx_2 = \lim_{n\to\infty} S_n = \lim_{n\to\infty} \sum_{i=1}^{N}\sum_{j=1}^{N} f(x_{1,i}^n, x_{2,j}^n) h_n h_n.$$

We shall also use the notation

$$\int_Q f(x)\, dx = \int_0^1 \int_0^1 f(x_1, x_2)\, dx_1\, dx_2.$$

We summarize as follows:

Theorem 64.1 *If $f : [0,1] \times [0,1] \to \mathbb{R}$ is Lipschitz continuous, then the limit*

$$\lim_{n\to\infty} \sum_{i=1}^{N}\sum_{j=1}^{N} f(x_{1,i}^n, x_{2,j}^n) h_n h_n,$$

exists, where $h_n = 2^{-n}$ and $N = 2^n$, $x_{1,i}^n = ih_n$, $x_{2,j}^n = jh_n$, and we define

$$\int_Q f(x)\, dx = \int_0^1 \int_0^1 f(x_1, x_2)\, dx_1\, dx_2 = \lim_{n\to\infty} \sum_{i=1}^{N}\sum_{j=1}^{N} f(x_{1,i}^n, x_{2,j}^n) h_n h_n.$$

$$(64.6)$$

In general, the partitions in x_1 and x_2 can be independent, leading to Riemann sums of the form

$$S_{nm} = \sum_{i=1}^{N}\sum_{j=1}^{M} f(x_{1,i}^n, x_{2,j}^m) h_n h_m, \qquad (64.7)$$

where $h_n = 2^{-n}$ and $N = 2^n$, $h_m = 2^{-m}$ and $M = 2^m$. This corresponds to a subdivision of Q into sub-squares $Q_{ij}^{nm} = I_i^n \times J_j^m$. The proof above directly generalizes to prove that if $\bar{n} \ge n$ and $\bar{m} \ge m$ then

$$|S_{nm} - S_{\bar{n}\bar{m}}| \le L_f \max(h_n, h_m).$$

This proves the following generalization of the previous theorem.

Theorem 64.2 *Suppose* $f : [0,1] \times [0,1] \to \mathbb{R}$ *is Lipschitz continuous. Then the following limit exists*

$$\lim_{n,m\to\infty} \sum_{i=1}^{N} \sum_{j=1}^{M} f(x_{1,i}^n, x_{2,j}^m) h_n h_m,$$

where $h_n = 2^{-n}$, $N = 2^n$, $h_m = 2^{-m}$, $M = 2^m$, $x_{1,i}^n = ih_n$, $x_{2,j}^m = jh_m$, *and*

$$\int_Q f(x)\, dx = \int_0^1 \int_0^1 f(x_1, x_2)\, dx_1\, dx_2 = \lim_{n,m\to\infty} \sum_{i=1}^{N} \sum_{j=1}^{M} f(x_{1,i}^n, x_{2,j}^m) h_n h_m.$$

$$(64.8)$$

64.3 Double Integrals via One-Dimensional Integration

To compute the Riemann sum S_{nm}, we have to perform a summation over all the sub-squares Q_{ij}^{nm} covering Q. The summation may be performed in different orders, row by row, column by column, or in some other order. We thus obtain the following alternative expressions for the double integral of $f(x_1, x_2)$ over Q:

$$\int_0^1 \int_0^1 f(x_1, x_2)\, dx_1\, dx_2 = \lim_{n,m\to\infty} \sum_{i=1}^{N} \sum_{j=1}^{M} f(x_{1,i}^n, x_{2,j}^m) h_n h_m$$

$$= \lim_{n,m\to\infty} \sum_{i=1}^{N} \left(\sum_{j=1}^{M} f(x_{1,i}^n, x_{2,j}^m) h_m \right) h_n$$

$$= \lim_{n,m\to\infty} \sum_{j=1}^{M} \left(\sum_{i=1}^{N} f(x_{1,i}^n, x_{2,j}^m) h_n \right) h_m,$$

where $\sum_{i=1}^{N} \sum_{j=1}^{M}$ indicates an arbitrary order of summation, $\sum_{i=1}^{N} \left(\sum_{j=1}^{M} \right)$ summation column by column, and $\sum_{j=1}^{M} \left(\sum_{i=1}^{N} \right)$ summation row by row over the subdomains Q_{ij}^{nm} of Q in the $x_1 x_2$-plane, see Fig. 64.3

We can also perform the limits with respect to n and m independently, and we then arrive at the formula

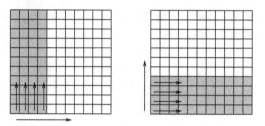

Fig. 64.3. Different orders of summation

$$\int_0^1 \int_0^1 f(x_1, x_2) \, dx_1 \, dx_2 = \lim_{n,m \to \infty} \sum_{i=1}^{N} \sum_{j=1}^{M} f(x_{1,i}^n, x_{2,j}^m) h_n h_m$$

$$= \lim_{n \to \infty} \sum_{i=1}^{N} \left(\lim_{m \to \infty} \sum_{j=1}^{M} f(x_{1,i}^n, x_{2,j}^m h_m) \right) h_n$$

$$= \lim_{m \to \infty} \sum_{j=1}^{M} \left(\lim_{n \to \infty} \sum_{i=1}^{N} f(x_{1,i}^n, x_{2,j}^m h_n) \right) h_m.$$

This corresponds to the following formula:

$$\int_0^1 \int_0^1 f(x_1, x_2) \, dx_1 \, dx_2 = \int_0^1 \left(\int_0^1 f(x_1, x_2) \, dx_2 \right) dx_1$$

$$= \int_0^1 \left(\int_0^1 f(x_1, x_2) \, dx_1 \right) dx_2,$$

or

$$\int_0^1 \int_0^1 f(x_1, x_2) \, dx_1 \, dx_2 = \int_0^1 g_1(x_1) \, dx_1 = \int_0^1 g_2(x_2) \, dx_2,$$

where

$$g_1(x_1) = \int_0^1 f(x_1, x_2) \, dx_2 = \lim_{m \to \infty} \sum_{j=1}^{M} f(x_1, x_{2,j}^m) h_m$$

and

$$g_2(x_2) = \int_0^1 f(x_1, x_2) \, dx_1 = \lim_{n \to \infty} \sum_{i=1}^{N} f(x_{1,i}^n, x_2) h_n$$

define functions $g_1(x_1)$ and $g_2(x_2)$ of x_1 and x_2 respectively. In other words, the double integral of $f(x_1, x_2)$ over $[0, 1] \times [0, 1]$ equals the integral of $g_2(x_2)$ over $[0, 1]$,

$$\int_0^1 \int_0^1 f(x_1, x_2) \, dx_1 \, dx_2 = \int_0^1 g_2(x_2) dx_2 = \lim_{n \to \infty} \sum_{j=1}^{M} g_2(x_{2,j}^m) h_m,$$

and equals the integral of $g_1(x_1)$ over $[0, 1]$,

$$\int_0^1 \int_0^1 f(x_1, x_2) \, dx_1 \, dx_2 = \int_0^1 g_1(x_1) dx_1 = \lim_{n \to \infty} \sum_{i=1}^N g_1(x_{1,i}^n) h_n.$$

We conclude that a double integral can be computed by repeated, or iterated, integration in one dimension. We may summarize this experience as follows:

Theorem 64.3 *If $f : [0, 1] \times [0, 1] \to \mathbb{R}$ is Lipschitz continuous, then*

$$\int_Q f(x) \, dx = \int_0^1 \int_0^1 f(x_1, x_2) \, dx_1 \, dx_2 =$$

$$= \int_0^1 \left(\int_0^1 f(x_1, x_2) \, dx_2 \right) dx_1 = \int_0^1 \left(\int_0^1 f(x_1, x_2) \, dx_1 \right) dx_2.$$

We can interpret the statement of this theorem as a *change of order of integration* in the sense that integrating with respect to x_1 and then with respect to x_2 gives the same result as integrating first with respect to x_2 and then with respect to x_1. The usual way to evaluate a double integral is to use iterated one-dimensional integration in some order.

Example 64.1.

With $Q = [0, 1] \times [0, 1]$,

$$\int_Q x_1 x_2^3 \, dx = \int_0^1 \int_0^1 x_1 x_2^3 \, dx_1 dx_2 = \int_0^1 x_1 \left(\int_0^1 x_2^3 \, dx_2 \right) dx_1$$

$$= \int_0^1 x_1 \left[\frac{x_2^4}{4} \right]_0^1 dx_1 = \frac{1}{4} \int_0^1 x_1 dx_1 = \frac{1}{4} \left[\frac{x_1^2}{2} \right]_0^1 = \frac{1}{8}.$$

$$\int_Q x_1 x_2^3 \, dx = \int_0^1 \int_0^1 x_1 x_2^3 \, dx_1 dx_2 = \int_0^1 x_2^3 \left(\int_0^1 x_1 \, dx_1 \right) dx_2$$

$$= \int_0^1 x_2^3 \left[\frac{x_1^2}{2} \right]_0^1 dx_2 = \frac{1}{2} \int_0^1 x_2^3 dx_2 = \frac{1}{2} \left[\frac{x_2^4}{4} \right]_0^1 = \frac{1}{8}.$$

Alternatively, we may first integrate with respect to x_1 and then with respect to x_2 to get,

$$\int_Q x_1 x_2^3 \, dx = \int_0^1 \int_0^1 x_1 x_2^3 \, dx_1 dx_2 = \int_0^1 x_2^3 \left(\int_0^1 x_1 \, dx_1 \right) dx_2$$

$$= \int_0^1 x_2^3 \left[\frac{x_1^2}{2} \right]_0^1 dx_2 = \frac{1}{2} \int_0^1 x_2^3 dx_2 = \frac{1}{2} \left[\frac{x_2^4}{4} \right]_0^1 = \frac{1}{8}.$$

64.4 Generalization to an Arbitrary Rectangle

The double integral defined on the unit square generalizes directly to integrals over arbitrary rectangles $Q = [a_1, b_1] \times [a_2, b_2]$ with sides parallel to the axis. If $f : Q \to \mathbb{R}$ is Lipschitz continuous, then

$$\int_Q f(x)\, dx = \int_Q f(x_1, x_2)\, dx_1 dx_2 = \int_{a_1}^{b_1} \left(\int_{a_2}^{b_2} f(x_1, x_2)\, dx \right) dx_1$$

$$= \int_{a_2}^{b_2} \left(\int_{a_1}^{b_1} f(x_1, x_2)\, dx_1 \right) dx_2.$$

64.5 Interpreting the Double Integral as a Volume

The sum

$$\sum_{i=1}^{N} \sum_{j=1}^{N} f(x_{1,i}^n, x_{2,j}^n) h_n h_n \tag{64.9}$$

represents the sum of the volumes

$$f(x_{1,i}^n, x_{2,j}^n) h_n h_n \tag{64.10}$$

of thin boxes with cross-section of area $h_n h_n$ and height $f(x_{1,i}^n, x_{2,j}^n)$. Intuitively, this is an approximation of the volume under the graph of $f(x_1, x_2)$ with (x_1, x_2) varying over Q. It is thus natural to define the volume $V(f, Q)$ under the graph of $f(x_1, x_2)$ over Q to be

$$V(f, Q) = \int_0^1 \int_0^1 f(x_1, x_2)\, dx_1\, dx_2 \tag{64.11}$$

Example 64.2. We compute the volume of a pyramid of height 1 with base $[0, 2] \times [0, 2]$, see Fig. 64.4. One quarter of the volume is equal to the integral $\int_Q f(x)\, dx$, where $Q = [0, 1] \times [0, 1]$, $f(x) = x_2$ for $x \in Q$ such that $x_2 \le x_1$ and $f(x) = x_1$ for $x \in Q$ such that $x_1 \le x_2$. We have

$$V(f, Q) = \int_Q f(x)\, dx = \int_0^1 \left(\int_0^{x_1} x_2\, dx_2 + \int_{x_1}^1 x_1\, dx_2 \right) dx_1$$

$$= \int_0^1 \left(\frac{x_1^2}{2} + x_1(1 - x_1)\, dx_1 \right) = \left[\frac{x_1^2}{2} - \frac{x_1^3}{6} \right]_0^1 = \frac{1}{2} - \frac{1}{6} = \frac{1}{3}.$$

We conclude that the volume of the pyramid is equal to $\frac{4}{3}$. This agrees with the standard formula stating that the volume of a pyramid is equal to $\frac{1}{3} Bh$, where B is the area of the base and h is the height.

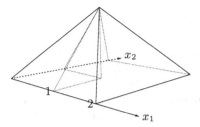

Fig. 64.4. Volume of pyramid

64.6 Extension to General Domains

We next define the double integral of a function $f(x)$ over a more general domain Ω in the plane. We start by assuming that the boundary Γ of Ω is described by two curves $x_2 = \gamma_1(x_1)$ and $x_2 = \gamma_2(x_1)$ for $0 \leq x_1 \leq 1$, as shown in Fig. 64.5, so that $\Omega = \{x \in [0,1] \times \mathbb{R} : \gamma_1(x_1) \leq x_2 \leq \gamma_2(x_1)\}$ We assume that the functions $\gamma_i : [0,1] \to \mathbb{R}$ are Lipschitz continuous with Lipschitz constant L_γ. We further assume that $f : \Omega \to \mathbb{R}$ is Lipschitz continuous with Lipschitz constant L_f.

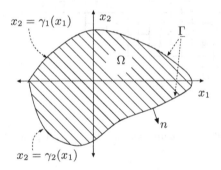

Fig. 64.5. The domain Ω in the plane

We assume that Ω is contained in the unit square Q. We partition Q as above into squares $I_i^n \times J_j^n$ of area $h_n h_n$, where $I_i^n = (x_{1,i-1}^n, x_{1,i}^n]$ $J_j^n = (x_{2,j-1}^n, x_{2,j}^n]$. We denote by ω_n the set of indices (i,j) such that the square $I_i^n \times J_j^n$ intersects Ω, and we let Ω_n be the union of the squares $I_i^n \times J_j^n$ with indices $(i,j) \in \omega_n$. In other words, Ω_n is an approximation of Ω consisting of all the squares $I_i^n \times J_j^n$ in Q that intersect Ω. We consider the Riemann sum

$$S_n = \sum_{(i,j) \in \omega_n} f(x_{1,i}^n, x_{2,j}^n) h_n h_n. \tag{64.12}$$

We shall prove that $\lim_{n\to\infty} S_n$ exists and then naturally define

$$\int_\Omega f(x)\,dx = \lim_{n\to\infty} \sum_{(i,j)\in\omega_n} f(x_{1,i}^n, x_{2,j}^n)h_n h_n. \tag{64.13}$$

To this end, we estimate the difference $S_n - S_{n+1}$, which now has contributions from two sources; (i) from the variation of $f(x)$ over each sub-square $I_i^n \times J_j^n$, and (ii) from the difference between Ω_n and Ω_{n+1}.

The first contribution can be shown to be bounded by $L_f h_n$ by arguing just as for integration over a square. The second contribution is bounded by $2A(1 + L_\gamma)h_n$, where A is a bound for $|f(x)|$, that is $|f(x)| \le A$ for $x \in \Omega$. This follows from the observation that if a square $I_i^n \times J_j^n$ of Ω_n, is entirely outside or inside Ω, then so are all the four squares of Ω_{n+1} within $I_i^n \times J_j^n$. The difference between Ω_n and Ω_{n+1} arises from the squares $I_i^n \times J_j^n$ which are partly inside and partly outside Ω. The area of these squares is bounded by $2L_\gamma h_n$, where the factor 2 arises from the fact that there are two curves γ_1 and γ_2, see Fig. 64.6. The difference in area between Ω_n and Ω_{n+1} is thus bounded by $2L_\gamma h_n$.

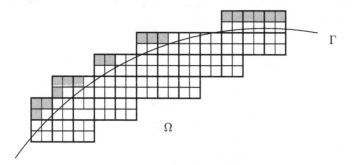

Fig. 64.6. Approximation of integral over general domain

Together, this shows that

$$|S_n - S_{n+1}| \le (L_f + 2AL_\gamma)h_n, \tag{64.14}$$

which as proves that $\lim_{n\to\infty} S_n$ exists. We summarize as follows:

Theorem 64.4 *Let $\Omega = \{x \in [0,1] \times \mathbb{R} : \gamma_2(x_1) \le x_2 \le \gamma_1(x_1)\}$, where $\gamma_i : [0,1] \to \mathbb{R}$ are Lipschitz continuous, and let $f : \Omega \to \mathbb{R}$ be Lipschitz continuous. Then $\lim_{n\to\infty} S_n$ exists, where S_n is the Riemann sum defined by (64.12), and we define*

$$\int_\Omega f(x)\,dx = \int_\Omega f(x_1, x_2)\,dx_1 dx_2 = \lim_{n\to\infty} S_n. \tag{64.15}$$

64.7 Iterated Integrals over General Domains

The integral of a function $f(x)$ over a domain $\Omega = \{x \in [0,1] \times \mathbb{R} : \gamma_2(x_1) \leq x_2 \leq \gamma_1(x_1)\}$ may be computed by iterated integration in one dimension as follows

$$\int_\Omega f(x)\, dx = \int_\Omega f(x_1, x_2)\, dx_1\, dx_2 = \int_0^1 \left(\int_{\gamma_2(x_1)}^{\gamma_1(x_1)} f(x_1, x_2)\, dx_2 \right) dx_1.$$

(64.16)

This is another way of expressing the fact that,

$$\int_\Omega f(x)\, dx = \lim_{n \to \infty} \sum_{(i,j) \in \omega_n} f(x_{1,i}^n, x_{2,j}^n) h_n h_n$$

$$= \lim_{n \to \infty} \sum_{i=1}^N \left(\sum_{j : (i,j) \in \omega_n} f(x_{1,i}^n, x_{2,j}^n) h_n \right) h_n$$

The role of x_1 and x_2 may be interchanged and the integral is independent of the particular representation of Γ. To handle a more general domain Ω, we split Ω into appropriate sub-domains Ω_j and define $\int_\Omega f\, dx = \sum_j \int_{\Omega_j} f\, dx$. Again the integral of f over Ω represents the volume of the domain under the graph of f over Ω.

Evaluation of an integral over a two-dimensional domain by repeated integration was used by Euler in 1738, when he computed the gravitational attraction of an elliptic lamina.

Example 64.3. We compute the double integral

$$I = \int_\Omega (x_1^2 + x_2)\, dx,$$

over the domain $\Omega = \{x \in \mathbb{R}^2 : x_1^2 \leq x_2 \leq x_1, \, 0 \leq x_1 \leq 1\}$. We have

$$I = \int_0^1 \left(\int_{x_1^2}^{x_1} (x_1^2 + x_2)\, dx_2 \right) dx_1 = \int_0^1 \left[x_1^2 x_2 + \frac{x_2^2}{2} \right]_{x_1^2}^{x_1} dx_1$$

$$= \int_0^1 \left(x_1^3 + \frac{x_1^2}{2} - x_1^4 - \frac{x_1^4}{2} \right) dx_1 = \frac{1}{4} + \frac{1}{6} - \frac{1}{5} - \frac{1}{10} = \frac{7}{60}.$$

Example 64.4. We compute the double integral

$$I = \int_\Omega \frac{1}{x_2}\, dx$$

over the domain $\Omega = \{x \in \mathbb{R}^2 : 1 \leq x_2 \leq \exp(x_1), 0 \leq x_1 \leq 1\}$. We have

$$I = \int_0^1 \left(\int_1^{\exp(x_1)} \frac{1}{x_2} \, dx_2 \right) dx_1 = \int_0^1 [\log(x_2)]_1^{\exp(x_1)} \, dx_1$$

$$= \int_0^1 x_1 \, dx_1 = \frac{1}{2}.$$

64.8   The Area of a Two-Dimensional Domain

We define the *area* $A(\Omega)$ of a domain Ω in $\mathbb{R}^2$ by

$$A(\Omega) = \int_\Omega dx, \qquad (64.17)$$

i.e. by integration of the constant function $f(x) = 1$ over Ω. If $\Omega = \{x \in [0,1] \times \mathbb{R} : \gamma_1(x_1) \leq x_2 \leq \gamma_2(x_1)\}$, then

$$A(\Omega) = \int_0^1 \left(\int_{\gamma_2(x_1)}^{\gamma_1(x_1)} dx_2 \right) dx_1 = \int_0^1 (\gamma_2(x_1) - \gamma_1(x_1)) \, dx_1,$$

which conforms with the previous formula of the area between the curves $\gamma_1(x_1)$ and $\gamma_1(x_1)$ as the integral of the difference $\gamma_2(x_1) - \gamma_1(x_1)$.

Example 64.5. The area of the triangle Ω with corners at $(0,0)$, $(1,0)$ and $(1,1)$, can be computed as follows

$$A(\Omega) = \int_\Omega dx = \int_0^1 \left(\int_0^{x_1} dx_2 \right) dx_1 = \int_0^1 \frac{1}{2} \, dx_1 = \frac{1}{2}.$$

64.9   The Integral as the Limit of a General Riemann Sum

We defined the integral using uniform subdivisions in x_1 and x_2, resulting in approximate subdivisions of a given domain Ω in $\mathbb{R}^2$ into squares or rectangles. We can however use more general subdivisions of Ω. Suppose that $f : \Omega \to \mathbb{R}$ is a Lipschitz continuous function and that the boundary of a domain Ω can be made up of pieces of Lipschitz curves $x_2 = \gamma(x_1)$ or $x_1 = \gamma(x_2)$. For $N = 1, 2, \ldots$, we divide Ω into a collection $\{\Omega_i\}_{i=1}^N$ of pairwise disjoint sets Ω_i such that the union of the Ω_i is equal to Ω. Let $d\Omega_i$ be the area of Ω_i and let d_N be the maximal diameter of Ω_i for $i = 1, \ldots, N$, see Fig. 64.7. We assume that d_N tends to zero as N tends to infinity.

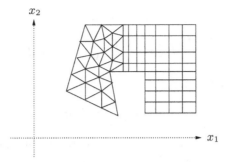

Fig. 64.7. Subdivison of general domain

The arguments used above show that

$$\int_\Omega f(x)\,dx = \lim_{N\to\infty} \sum_{i=1}^N f(x_i)d\Omega_i, \qquad (64.18)$$

where x_i is a point in Ω_i for $i = 1,\ldots,N$. The first step in proving this result is to use the estimate

$$|f(x) - f(y)| \le L_f d_N \quad \text{if } x,y \in \Omega_i, \qquad (64.19)$$

which implies that the variation of $f(x)$ with x ranging over Ω_i is small if the diameter of Ω_i is small. The second step involves the Lipschitz continuity of the boundary of Ω and the boundedness of $f(x)$. By the way, a byproduct of the proof of this result is the estimate

$$\left| \int_\Omega f(x)\,dx - \sum_{i=1}^N f(x_i)d\Omega_i \right| \le L_f d_N A(\Omega), \qquad (64.20)$$

where $A(\Omega)$ is the area of Ω.

64.10   Change of Variables in a Double Integral

We next extend the idea of changing variables in a one-dimensional integral to a two dimensional integral. More precisely, we want to make a change of variables in an integral

$$\int_\Omega f(x)\,dx = \int_\Omega f(x_1,x_2)\,dx_1 dx_2, \qquad (64.21)$$

where Ω is a given domain in $\mathbb{R}^2$ and the integration variable x runs over Ω. We assume that $g : \tilde\Omega \to \Omega$ is a one-to-one mapping of $y \in \tilde\Omega$ onto $x = g(y)$ in Ω that represents the change of variables. We shall prove that (64.21) with respect to x can be rewritten as an integral with respect to y in the form

$$\int_\Omega f(x)\,dx = \int_{\tilde{\Omega}} f(g(y))\,G(y)\,dy, \tag{64.22}$$

where $G(y)$ is defined as

$$G(y) = |\det g'(y)|.$$

That is $G(y)$ is the absolute value of the determinant of the Jacobian $g'(y)$ of $g(y)$. Formally, this gives $dx = |\det g'(y)|\,dy$ or $|\det g'(y)| = |\det \frac{dx}{dy}|$, and $|\det g'(y)|$ is the local change of area measure as we go from y-coordinates to x-coordinates. The change of variable formula can therefore be written

$$\int_\Omega f(x)\,dx = \int_{\tilde{\Omega}} f(g(y))|\det g'(y)|\,dy, \tag{64.23}$$

To prove this let $\tilde{\Omega}_i$ be a small subdomain of $\tilde{\Omega}$ and let $\Omega_i = g(\tilde{\Omega}_i)$ be the image of $\tilde{\Omega}_i$ under the mapping $x = g(y)$. If $g'(y)$ were constant over $\tilde{\Omega}_i$, and so $g(y)$ were linear on $\tilde{\Omega}_i$, then

$$d\Omega_i = |\det g'(y_i)|d\tilde{\Omega}_i,$$

where y_i is a point in $\tilde{\Omega}_i$, $d\Omega_i$ is the area of Ω_i, and $d\tilde{\Omega}_i$ is the area of $\tilde{\Omega}_i$. If $\{\tilde{\Omega}_i\}_{i=1}^n$ is a subdivision of $\tilde{\Omega}$ into subdomains $\tilde{\Omega}_i$ of maximal diameter d_n, we have

$$\int_\Omega f(x)\,dx \approx \sum_i f(x_i)d\Omega_i$$

$$\approx \sum_i f(g(y_i))|\det g'(y_i)|d\tilde{\Omega}_i \approx \int_{\tilde{\Omega}} f(g(y))|\det g'(y)|\,dy,$$

where $x_i = g(y_i)$ and the approximations are bounded by d_n times Lipschitz constants of the functions $f(x)$, $f(g(y))$ and $|\det g'(y)|$. The change of variables formula (64.23) follows by passing to the limit as n tends to infinity and d_n tends to 0.

We summarize:

Theorem 64.5 (Change of variables) *Assume $y \to x = g(y)$ maps a domain $\tilde{\Omega}$ in $\mathbb{R}^2$ onto a domain Ω in $\mathbb{R}^2$, where the Jacobian of g is Lipschitz continuous and let $f : \Omega \to \mathbb{R}$ be Lipschitz continuous. Then*

$$\int_\Omega f(x)\,dx = \int_{\tilde{\Omega}} f(g(y))|\det g'(y)|\,dy, \tag{64.24}$$

Example 64.6. Consider the mapping $x = g(y) = (2y_1 + y_2, y_1 - 2y_2)$ mapping the unit square $\tilde{\Omega} = [0,1] \times [0,1]$ onto the parallelogram Ω spanned by the vectors $(2, 1)$ and $(1, -2)$. We have $\det g'(y) = -5$, and thus

$$\int_\Omega f(x)\,dx = \int_{\tilde\Omega} f(2y_1 + y_2, y_1 - 2y_2)\,|-5|\,dy$$

$$= 5 \int_0^1 \int_0^1 f(2y_1 + y_2, y_1 - 2y_2)\,dy.$$

If $f(x) = x_2$ then

$$\int_\Omega f(x)\,dx = 5 \int_0^1 \int_0^1 (y_1 - 2y_2)\,dy = 5\left(\frac{1}{2} - 1\right) = -\frac{5}{2}.$$

Polar Coordinates

A particularly important change of variables is from rectangular coordinates to polar coordinates,

$$(x_1, x_2) = (r\cos(\theta), r\sin(\theta))$$

where $x = (x_1, x_2) \in \mathbb{R}^2$ and $r \geq 0$, $0 \leq \theta < 2\pi$, see Fig. 64.8.

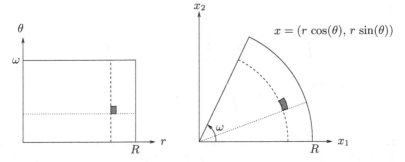

Fig. 64.8. Polar coordinates

The Jacobian of the mapping $(r, \theta) \to (x_1, x_2)$ is given by

$$\frac{d(x_1, x_2)}{d(r, \theta)} = \begin{pmatrix} \cos(\theta) & -r\sin(\theta) \\ \sin(\theta) & r\cos(\theta) \end{pmatrix},$$

and

$$\det \frac{d(x_1, x_2)}{d(r, \theta)} = r(\cos^2(\theta) + \sin^2(\theta)) = r.$$

Example 64.7. If $\Omega = \{x \in \mathbb{R}^2 : |x| \leq 1, x_1 \geq 0, x_2 \geq 0\}$ is the part of the unit circle in the positive quadrant, then the corresponding domain in polar coordinates takes the from $\tilde{\Omega} = \{(r, \theta) : 0 \leq r \leq 1, 0 \leq \theta \leq \frac{\pi}{2}\}$, and

$$\int_\Omega f(x_1, x_2)\, dx_1 dx_2 = \int_{\tilde{\Omega}} f(r\cos(\theta), r\sin(\theta))\, r dr\, d\theta.$$

In particular with $f(x) = 1$, we have

$$A(\Omega) = \int_\Omega dx_1 dx_2 = \int_{\tilde{\Omega}} r dr\, d\theta$$

$$= \int_0^{\frac{\pi}{2}} \int_0^1 r\, dr\, d\theta = \int_0^{\frac{\pi}{2}} \frac{1}{2}\, d\theta = \frac{\pi}{4}.$$

We have now computed the area of a quarter of a unit disc to be equal to $\frac{\pi}{4}$, so the area of a unit disc is π. A basic result of mathematics!

Example 64.8. Using polar coordinates, we have

$$\int_{\mathbb{R}^2} e^{-x_1^2 - x_2^2}\, dx = \int_0^{2\pi} \int_0^\infty e^{-r^2} r\, dr\, d\theta = 2\pi \left[-\frac{1}{2} e^{-r^2} \right]_0^\infty = \pi.$$

Since

$$\int_{\mathbb{R}^2} e^{-x_1^2 - x_2^2}\, dx = \int_{-\infty}^\infty e^{-x_1^2}\, dx_1 \int_{-\infty}^\infty e^{-x_2^2}\, dx_2,$$

we conclude that

$$\int_{-\infty}^\infty e^{-x^2}\, dx = \sqrt{\pi}. \tag{64.25}$$

Evidently, we did something magical: although we do not know a primitive function to e^{-x^2} we are able to obtain an analytic expression for $\int_{-\infty}^\infty e^{-x^2}\, dx$.

Chapter 64   Problems

64.1.   Compute with $\Omega = [0, 1] \times [0, 1]$ the unit square the integrals
(a) $\int_\Omega (x_1 + x_2)\, dx$ (b) $\int_\Omega x_1 x_2\, dx$ (c) $\int_\Omega \frac{dx}{x_1 + x_2}$ (d) $\int_\Omega \exp(-x_1 x_2)\, dx$

64.2. Compute with $\Omega = \{(x_1, x_2) : 0 \leq x_1 \leq x_2 \leq 1\}$ the integrals (a) $\int_\Omega \frac{x_1}{x_2}\, dx$
(b) $\int_\Omega \exp^{2x_2}\, dx$ (c) $\int_\Omega \exp^{x_2^2}\, dx$

64.3. Change the order of integration in the following integrals

1. $\int_{1/2}^{1} \int_{0}^{1-x_1} f(x_1, x_2)\, dx_2 dx_1$

2. $\int_{0}^{1} \int_{0}^{\sqrt{1-x_1^2}} f(x_1, x_2)\, dx_2 dx_1$

3. $\int_{0}^{1} \int_{x_2-1}^{0} f(x_1, x_2)\, dx_1 dx_2$

4. $\int_{0}^{1} \int_{1-x_1}^{1+x_1} f(x_1, x_2)\, dx_2 dx_1$

64.4. Evaluate the following integrals:

1. $\int_{\Omega}(x_1^2 + 2x_2^3)\, dx$, with Ω a triangle with vertices $(0,0)$, $(1,0)$, $(0,1)$.
2. $\int_{\Omega} x_1^2 x_2\, dx$, with $\Omega = \{x \in \mathbb{R}^2 : x_1^2 + x_2^2 \le 1, 0 \le x_2\}$.
3. $\int_{\Omega}(x_1+x_2)dx$, with Ω the tetrahedron with vertices $(0,0), (1,0), (2,1), (2,2)$.
4. $\int_{\Omega} |1 - x_1 - x_2|\, dx$, with Ω the unit square.

64.5. Find the volume under the graph of the following functions

1. $f(x) = e^{x_1} \cos(x_2)$, $0 \le x_1 \le 1$, $0 \le x_2 \le \frac{\pi}{2}$.
2. $f(x) = x_1^2 e^{-x_1 - x_2}$, $0 \le x_1 \le 1$, $0 \le x_2 \le 2$.
3. $f(x) = x_1^2 x_2$, $0 \le x_1 \le 1$, $x_1 + 1 \le x_2 \le x_1 + 2$.
4. $f(x) = \sqrt{x_1^2 - x_2^2}$, $x_1^2 - x_2^2 \ge 0$, $0 \le x_1 \le 1$.

64.6. A cylindrical hole of radius b is drilled symmetrically through a metal sphere of radius $a > b$. Find the volume of metal removed.

64.7. Evaluate

$$\int_{\Omega} \left(1 - \frac{x_1^2}{a_1^2} - \frac{x_2^2}{a_2^2}\right)^{3/2} dx$$

where Ω is the ellipse $\{x \in \mathbb{R}^2 : \frac{x_1^2}{a_1^2} + \frac{x_2^2}{a_2^2} \le 1\}$.

64.8. Evaluate

$$\int_{\Omega} \frac{x_1 + x_2}{x_1^2} e^{x_1+x_2}\, dx,$$

where $\Omega = \{x \in \mathbb{R}^2 : x_2 \le x_1 \le 2 - x_2, 0 \le x_2 \le 1\}$. Hint: Use the substitution $y_1 = x_1 + x_2$, $y_2 = \frac{x_2}{x_1}$.

64.9. Compute the area of one petal of the rose $0 \le r \le 3\sin(\theta)$ (polar coordinates).

64.10. Compute the area within the cardoid $r = 1 + \cos(\theta)$.

64.11. Compute the following double integrals:

1. $\int_\Omega x_1 \exp(x_1 x_2)\, dx$, for $\Omega = \{x : 0 \le x_1 \le 1, 0 \le x_2 \le 2\}$,
2. $\int_\Omega x_1 x_2 \exp(x_1 + x_2)\, dx$, for $\Omega = \{x : 1 \le x_1 \le 2 \le x_2 \le 3\}$,
3. $\int_\Omega x\, dx$, for $\Omega = \{x : 0 \le x_1 \le 1, 0 \le x_2 \le 1\}$.

64.12. Compute the following double integrals:

1. $\int_\Omega \exp(-x_1)\, dx$, for $\Omega = \{x : 0 \le x_1 \le 1, |x_2| \le x_1\}$,
2. $\int_\Omega x_1 x_2 \|x\|\, dx$, for $\Omega = \{x : 0 \le x_1 \le 1, 1 \le x_2 \le 2\}$,
3. $\int_\Omega \frac{x_1}{1+x_2}\, dx$, for $\Omega = \{x : 0 \le x_1 \le 1, 0 \le x_2 \le 1 - x_1\}$.

64.13. Compute the following double integrals by changing variables:

1. $\int_\Omega \|x\|^2\, dx$, for $\Omega = \{x : x_1^2 + x_2^2 - 2x_1 - 2x_2 \le 0\}$,
2. $\int_\Omega x_1 x_2\, dx$, for $\Omega = \{x : 3x_1^2 + x_2^2 - 2x_1 \le 0\}$,
3. $\int_\Omega \exp(-\|x\|^2)\, dx$, for $\Omega = \mathbb{R}^2$.

65
Surface Integrals

King Karl XII of Sweden (1682–1717) had an extraordinary talent for mathematics. He was by Swedenborg (the great Swedish Universal Genius, 1688–1772) considered equal if not better than Leibniz himself. King Karl XII could easily multiply large numbers without pen and paper, and proposed 64 as the right choice of basis of the natural numbers. Over night he constructed symbols and gave names to all the digits $0, 1, \ldots, 62, 63$. (from *The History of Sweden*, by Herman Lindquist).

65.1 Introduction

Previously, in Chapter *Curve integrals* we defined the notion of an integral computed over a curve or a curve integral. In this chapter, we use the same ideas to define an integral over a surface or a *surface integral*. We start with the surface integral representing *surface area*.

65.2 Surface Area

Let S be a surface in $\mathbb{R}^3$ parameterized by the mapping $s : \Omega \rightarrow \mathbb{R}^3$, where Ω is a domain in $\mathbb{R}^2$ with coordinates $y = (y_1, y_2) \in \mathbb{R}^2$, so that $s = s(y) = (s_1(y), s_2(y), s_3(y))$. We define the *area* $A(S)$ of the surface S

as the following integral over the parameter domain Ω,

$$A(S) = \int_\Omega \|s'_{,1} \times s'_{,2}\| \, dy, \qquad (65.1)$$

where

$$s'_{,1} = \begin{pmatrix} \frac{\partial s_1}{\partial y_1} \\ \frac{\partial s_2}{\partial y_1} \\ \frac{\partial s_3}{\partial y_1} \end{pmatrix}, \qquad s'_{,2} = \begin{pmatrix} \frac{\partial s_1}{\partial y_2} \\ \frac{\partial s_2}{\partial y_2} \\ \frac{\partial s_3}{\partial y_2} \end{pmatrix},$$

are the columns of the Jacobian

$$s' = \begin{pmatrix} \frac{\partial s_1}{\partial y_1} & \frac{\partial s_1}{\partial y_2} \\ \frac{\partial s_2}{\partial y_1} & \frac{\partial s_2}{\partial y_2} \\ \frac{\partial s_3}{\partial y_1} & \frac{\partial s_3}{\partial y_2} \end{pmatrix}.$$

Note all the coefficients are functions of $y \in \Omega$.

To motivate this definition, recall that the linearization of the mapping $s : \Omega \to \mathbb{R}^3$ at $\bar{y}$ is given by

$$y \to \hat{s}(y) = s(\bar{y}) + (y_1 - \bar{y}_1)s'_{,1}(\bar{y}) + (y_2 - \bar{y}_2)s'_{,2}(\bar{y}).$$

Consider a small square $R(\bar{y}, h) = [\bar{y}_1, \bar{y}_1 + h] \times [\bar{y}_2, \bar{y}_2 + h]$ in Ω of side length h and area h^2 with lower left-hand corner at the point $\bar{y} \in \Omega$. Here, we think of h as small. The linearization $\hat{s}(y)$ maps the square $R(\bar{y}, h)$ into a small parallelogram $P(s(\bar{y}), h)$ in the tangent plane of S through $s(\bar{y})$ spanned by the two vectors $s'_{,1}(\bar{y})$ and $s'_{,2}(\bar{y})$, with one of the corners of the parallelogram at $s(\bar{y})$. Recall now from Chapter *Analytic Geometry in $\mathbb{R}^2$* that the area of a parallelogram spanned by two vectors a and b in $\mathbb{R}^2$ is equal to $\|a \times b\|$. So, the area of $P(s(\bar{y}), h)$ is equal to

$$\|s'_{,1}(\bar{y}) \times s'_{,2}(\bar{y})\|h^2.$$

The change of scale of area is thus $\|s'_{,1}(\bar{y}) \times s'_{,1}(\bar{y})\|$. A small piece (square) of area h^2 at $\bar{y} \in \Omega$ in the parameter domain, thus corresponds to a small piece of the surface S at $s(\bar{y})$ of area approximately $\|s'_{,1}(\bar{y}) \times s'_{,2}(\bar{y})\|h^2$, where the approximation improves as h gets smaller.

Summing over all little pieces and letting h tend to zero, we are led to define the area $A(S)$ of the surface S by (65.1), which we write as

$$A(S) = \int_\Omega \|s'_{,1}(y) \times s'_{,2}(y)\| \, dy = \int_\Omega \|s'_{,1} \times s'_{,2}\| \, dy = \int_S ds.$$

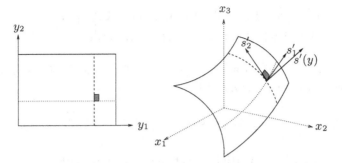

Fig. 65.1. The surface area scale

We thus write $ds = \|s'_{,1} \times s'_{,2}\|\, dy$, which expresses the change of scale. Of course, we assume that $\|s'_{,1} \times s'_{,2}\|$ is Lipschitz continuous to guarantee that the integral exists.

Example 65.1. Consider the surface S of a sphere of radius one centered at the origin. We describe this using spherical coordinates,

$$x = s(y_1, y_2) = (\sin(y_2)\cos(y_1), \sin(y_2)\sin(y_1), \cos(y_2))^{\top},$$

where $0 \leq y_1 < 2\pi$, $0 \leq y_2 < \pi$, see Fig. 66.3. We have

$$
\begin{aligned}
s'_{,1} &= (-\sin(y_2)\sin(y_1), \sin(y_2)\cos(y_1), 0)^{\top}, \\
s'_{,2} &= (\cos(y_2)\cos(y_1), \cos(y_2)\sin(y_1), -\sin(y_2))^{\top},
\end{aligned}
\tag{65.2}
$$

and thus by a direct computation $\|s'\| = \sin(y_2)$. We compute

$$A(S) = \int_0^{2\pi} \int_0^{\pi} \sin(y_2)\, dy_2\, dy_1 = \int_0^{2\pi} 2\, dy_1 = 4\pi,$$

and thus conclude that the surface area of a sphere of radius 1 is equal to 4π.

Example 65.2. We compute the area $A(S)$ of the surface S given by $s(y_1, y_2) = (2y_1 y_2, y_1^2, 2y_2^2)$ with $0 \leq y_1, y_2 \leq 1$. We have

$$s'(y) = (2y_2, 2y_1, 0) \times (2y_1, 0, 4y_2) = 4(2y_1 y_2, -2y_2^2, -y_1^2)$$

so that $\|s'(y)\| = 4(y_1^2 + 2y_2^2)$, and thus

$$A(S) = \int_0^1 \int_0^1 4(y_1^2 + 2y_2^2)\, dy_1 dy_2 = 4\left(\frac{1}{3} + \frac{2}{3}\right) = 4.$$

65.3   The Surface Area of a the Graph
of a Function of Two Variables

In the case S is given as the graph of a function $f : \Omega \to \mathbb{R}$, so that $s(y_1, y_2) = (y_1, y_2, f(y_1, y_2))$, then

$$A(S) = \int_S ds = \int_\Omega \|s'_{,1} \times s'_{,2}\| \, dy = \int_\Omega \sqrt{1 + f_{,1}^2 + f_{,2}^2} \, dy_1 dy_2, \quad (65.3)$$

where $f_{,i}$ denotes the partial derivative of f with respect to y_i. This follows from

$$s'_{,1} \times s'_{,2} = (1, 0, f_{,1}) \times (0, 1, f_{,2}) = (-f_{,1}, -f_{,2,}, 1).$$

Example 65.3. The surface S of a hemisphere of radius 1 and centered at the origin is given by $s(y_1, y_2) = (y_1, y_2, \sqrt{1 - y_1^2 - y_2^2})$ with $y \in \Omega = \{y \in \mathbb{R}^2 : y_1^2 + y_2^2 \le 1\}$. We have

$$A(S) = \int_\Omega \sqrt{1 + f_{,1}^2 + f_{,2}^2} \, dy_1 dy_2 = \int_\Omega \frac{1}{\sqrt{1 - y_1^2 - y_2^2}} \, dy$$

$$= \int_0^{2\pi} \int_0^1 \frac{1}{\sqrt{1 - r^2}} r \, dr \, d\theta = 2\pi \left[-\sqrt{1 - r^2} \right]_0^1 = 2\pi. \quad (65.4)$$

We retrieve the above result that the surface area of a sphere of radius 1 is equal to 4π.

65.4   Surfaces of Revolution

Surfaces of revolution occur in many practical applications. To generate a surface of revolution, we let $f : [a, b] \to \mathbb{R}$ be a given positive function and consider the surface S represented by

$$s(x_1, x_2) = (x_1, f(x_1) \cos(x_2), f(x_1) \sin(x_2)),$$

with $a \le x_1 \le b$ and $0 \le x_2 < 2\pi$, see Fig. 65.2. We use (x_1, x_2) as reference coordinates instead of (y_1, y_2). We have

$$s'_{,1} \times s'_{,2} = (1, f'(x_1) \cos(x_2), f'(x_1) \sin(x_2)) \times$$
$$(0, -f(x_1) \sin(\theta), f(x_1) \cos(\theta))$$

and thus by a direct computation

$$\|s'_{,1} \times s'_{,2}\| = f(x_1) \sqrt{1 + (f'(x_1))^2}. \quad (65.5)$$

The area $A(S)$ of S is given by:

$$A(S) = \int_0^{2\pi} \int_a^b f(x_1)\sqrt{1 + (f'(x_1))^2}\, dx_1 d\theta$$

$$= 2\pi \int_a^b f(x_1)\sqrt{1 + (f'(x_1))^2}\, dx_1. \quad (65.6)$$

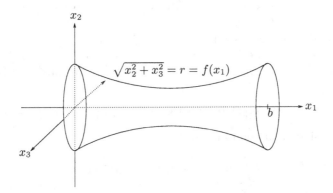

Fig. 65.2. A surface of revolution

Example 65.4. Consider the surface S of a parabolic reflector obtained by rotating the curve $f(x_1) = \sqrt{x_1}$ around the x_1-axis between $x_1 = 0$ and $x_1 = 1$. We have

$$A(S) = 2\pi \int_0^1 \sqrt{x_1}\sqrt{1 + \frac{1}{4x_1}}\, dx_1 = \pi \int_0^1 \sqrt{4x_1 + 1}\, dx_1 = \frac{\pi}{6}(5^{3/2} - 1).$$

65.5 Independence of Parameterization

We shall prove that if $t : \tilde{\Omega} \to \Omega$ is a one-to-one mapping of $\eta \in \tilde{\Omega} \subset \mathbb{R}^2$ onto $y = t(\eta) \in \Omega$, and $r(\eta) = s(t(\eta))$ maps $\tilde{\Omega}$ onto S, then

$$\int_S ds = \int_{\tilde{\Omega}} \|r'_{,1} \times r'_{,2}\|\, d\eta = \int_\Omega \|s'_{,1} \times s'_{,2}\|\, dy. \quad (65.7)$$

This shows that the surface area of the surface S is independent of the parametrization of S.

We need to show that with $y = t(\eta)$, we have

$$\|r'_{,1}(\eta) \times r'_{,2}(\eta)\| = \|s'_{,1}(y) \times s'_{,2}(y)\|\,|\det t'(\eta)|, \quad (65.8)$$

where $|\det t'|$ is the determinant of the Jacobian $t'(\eta)$ of $t(\eta)$. This follows after a lengthy computation that starts with differentiating $r(\eta) = s(t(\eta))$ using the Chain rule.

65.6 Surface Integrals

Let $S = s(\Omega)$ be a surface in $\mathbb{R}^3$ parameterized by the mapping $s : \Omega \to \mathbb{R}^3$, where Ω is a domain in $\mathbb{R}^2$, and let $u : S \to \mathbb{R}$ be a real-valued function defined on S. We assume that u, s and $\|s'_{,1} \times s'_{,2}\|$ are Lipschitz continuous. We define the integral of u over S to be

$$\int_S u \, ds = \int_\Omega u(s(y)) \|s'_{,1}(y) \times s'_{,2}(y)\| \, dy. \tag{65.9}$$

Example 65.5. Let $S = s(\Omega)$ be the "dome" given by $s(y_1, y_2) = (y_1, y_2, 1 - y_1^2 - y_2^2)$ and $\Omega = \{y \in \mathbb{R}^2 : y_1^2 + y_2^2 \le 1\}$, and $u(x) = (5x_1^2 + 5x_2^2 + x_3)^{1/2}$, so that $u(s(y)) = (5y_1^2 + 5y_2^2 + 1 - y_1^2 - y_2^2)^{1/2} = (1 + 4y_1^2 + 4y_2^2)^{1/2}$. We compute

$$\|s'_{,1}(y) \times s'_{,2}(y)\| = \|(1, 0, -2y_1) \times (0, 1, -2y_2)\| = (1 + 4y_1^2 + 4y_2^2)^{1/2},$$

and get using polar coordinates:

$$\int_S u \, ds = \int_\Omega u(s(y)) \|s'_{,1}(y) \times s'_{,2}(y)\| \, dy$$

$$= \int_\Omega (1 + 4y_1^2 + 4y_2^2)^{1/2}(1 + 4y_1^2 + 4y_2^2)^{1/2} \, dy$$

$$= 2\pi \int_0^1 (1 + 4r^2) r \, dr = \frac{3}{2}.$$

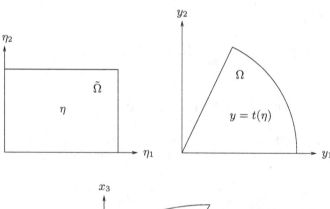

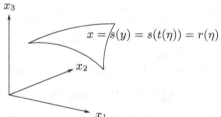

Fig. 65.3. Reparametrization $r(\eta) = s(t(\eta))$ of a surface given by $s(y)$

65.7   Moment of Inertia of a Thin Spherical Shell

The *moment of inertia* of a thin sphere $S = \{\|x\| = 1\}$ of (uniformly distributed) total mass m about the x_3-axis, is equal to

$$I = \frac{m}{4\pi} \int_S (x_1^2 + x_2^2)\, ds. \tag{65.10}$$

If the sphere rotates with angular speed ω around the x_3 axis, then the total *kinetic energy* is equal to

$$E = \frac{1}{2}\frac{m}{4\pi} \int_S \omega^2(x_1^2 + x_2^2)\, ds = \frac{1}{2}\omega^2 I. \tag{65.11}$$

Using spherical coordinates to compute gives

$$I = \frac{2m}{3}. \tag{65.12}$$

Chapter 65   Problems

65.1. (a) Verify that $\|s'_{,1}(y) \times s'_{,2}(y)\| = \sin(y_2)$ in (65.2). (b) Verify (65.5). (c) Prove (65.8).

65.2. Determine which famous building is defined by the *MATLAB*© code given below, and compute the surface area of its roof.

```
r=0:.1:1;
v=0:pi/20:2*pi;
[R,V]=meshgrid(r,v);
surf(10*cos(V),10*sin(V),R.*(5+cos(V).^ 2-sin(V).^ 2))
hold on
surf(10*R.*cos(V),10*R.*sin(V),5+(R.*cos(V)).^ 2-(R.*sin(V)).^ 2)
hold off
axis('equal')
```

65.3. Another famous building. What does it take to repaint it?

```
w=0:pi/20:3*pi/4;
v=0:pi/20:2*pi;
[W,V]=meshgrid(w,v);
h=surf(sin(W).*cos(V),sin(W).*sin(V),cos(W));
set(h,'FaceColor',[1 1 1])
axis('equal')
```

65.4. Motivate (65.11), and prove (65.12).

65.5. (a) Consider the surface $S = \{x : x = y_1 a + y_2 b + (1 - y_1 - y_2)c, y \in T\}$, where $a, b, c \in \mathbb{R}^3$ and $T = \{y \in \mathbb{R}^2 : y_1 + y_2 \leq 1, y_i \geq 0, i = 1, 2\}$. Give a geometric description of S and compute its area.
(b) Find a parametrization of the form $x = My + b$ of the (flat) triangular surface S with corners in $(1, 0, 0)$, $(0, 0, 3)$ and $(0, 3, -9)$, with parameter domain T as in (a), where b is a 3-vector and M a 3-by-2 matrix.
(b) Compute the area of S. Does the area depend on b? Interpret!
(c) Compute $\int_S (x_1 + 2x_2) \, dS$

65.6. Compute (a) $\int_S dS$ (b) $\int_S f(x) \, dS$ where $S = \{x : x = My, y \in Q\}$, Q is the unit square in $\mathbb{R}^2$ and M is the 3-by-2 matrix with columns $(1, 0, 1)^\top$ and $(0, 1, 2)^\top$, and $f(x) = x_3$. Also, plot the surface S and interpret (a) as the area of S. Compare the computation of (a) with the method for computing the area of a parallelogram using the cross product in Linear algebra.

65.7. Compute (a) $\int_S dS$ (b) $\int_S x_2 \, dS$ where $S = \{x : x = y_1(1 - y_2)(1, 0, 0) + (1 - y_1)(1 - y_2)(1, 2, 0) + (1 - y_1)y_2(0, 1, 1) + y_1 y_2(0, 0, 3), 0 \leq y_i \leq 1, i = 1, 2\}$. Plot the surface and describe its geometry.

65.8. Compute (a) $\int_S dS$ (b) $\int_S x_1 x_2 \, dS$ where $S = \{(y_1, y_2, y_1 y_2) : 0 \leq y_i \leq 1, i = 1, 2\}$.

65.9. Consider for given $r > 0$ and $h > 0$ the surface

$$S = \{x : x = (r \cos(v), r \sin(v), z), 0 \leq v \leq 2\pi, 0 \leq z \leq h\}.$$

(a) Give a geometrical description of S, and give corresponding parameterizations of the surfaces (b) $S = \{x \in \mathbb{R}^3 : x_2^2 + x_3^2 = 4, |x_1| \leq 5\}$ (c) $S = \{x \in \mathbb{R}^3 : x_2^2 + 4x_3^2 = 4, 0 \leq x_1 \leq x_2^2 + x_3^2\}$.

65.10. Compute $\int_S (x_1 + x_2 + x_3) \, dS$ where $S = \{(x_1, x_2, x_3) : x_1 = y_1 \cos(y_2), x_2 = y_1 \sin(y_2), x_3 = y_1(\cos(y_2) + \sin(y_2))\}$.

65.11. Compute $\int_S (x_1, x_2, x_3) \cdot n \, ds$ if S is the boundary of $\Omega = \{x : x_1 + x_2 + x_3 \leq 1, x_i \geq 0, i = 1, 2, 3\}$.

65.12. Compute $\int_S \frac{(x_1, x_2, x_3)}{\|x\|^2} \cdot n \, dS$ for the cylindrical shell $S = \{x \in \mathbb{R}^3 : x_1^2 + x_2^2 = 1, -a \leq x_3 \leq a\}$, and the corresponding limit as $a \to \infty$.

65.13. Compute the moment of inertia of the cylindrical shell $S = \{x \in \mathbb{R}^3 : x_1^2 + x_2^2 = 1, -1 \leq x_3 \leq 1\}$ with respect to the x_1-axis.

65.14. Compute $\int_S (x_1, 0, x_3) \cdot n \, dS$ where $S = \{(y_1 + y_2, y_1^2 - y_2^2, y_1 y_2) : 0 \leq y_1 \leq 1, 0 \leq y_2 \leq 1\}$, and n is the normal to S with $n_3 < 0$.

65.15. Compute the area of the torus (donut) in $\mathbb{R}^3$ given by

$$s(y_1, y_2) = \big((a + b\cos(y_2))\cos(y_1), (a + b\cos(y_2))\sin(y_1), b\sin(y_2)\big)$$

with $a > b$ constants and $0 \le y_1, y_2 < 2\pi$.

65.16. Plot and compute the area of the surface $S = \{(r\cos(v), r\sin(v), v) : 1 \le r \le 2, 0 \le v \le 4\pi\}$. In what type of buildings can one find constructions like this?

65.17. Describe/plot the surfaces (of rotation, if you wish) (a) $x_1^2 + x_2^2 = x_3^2, x_3 > 0$ (b) $5 + x_1^2 + x_2^2 = x_3^2 \le 9$, $x_3 > 0$ and compute its area.

66
Multiple Integrals

We met weekly, (sometimes at Dr Goddard's lodgings, sometimes at the Mitre in Wood Street near-by) at a certain hour, under a certain penalty, and a weekly contribution for the charge of experiments, with certain rules agreed among us. There, to avoid being diverted to other discourses and for some other reasons, we barred all discussion of Divinity, of State Affairs, and of news (other than what concerned our business of philosophy) confining ourselves to philosophical inquiries, and related topics; as medicine, anatomy, geometry, astronomy, navigation, statics, mechanics, and natural experiments. (Wallis about the formation of the Royal Society)

66.1   Introduction

We now consider *triple integrals* over domains in $\mathbb{R}^3$ and more generally *multiple integrals* over domains in $\mathbb{R}^n$ with $n > 3$.

66.2   Triple Integrals over the Unit Cube

A triple integral over the unit cube $Q = \{x \in \mathbb{R}^3 : 0 \leq x_i \leq 1, \ i = 1, 2, 3\}$ of a Lipschitz continuous function $f : Q \to \mathbb{R}$ takes the form

$$\int_Q f(x)\, dx = \int_0^1 \int_0^1 \int_0^1 f(x_1, x_2, x_3)\, dx_1\, dx_2\, dx_3.$$

This can be computed by iterated integration in any order, for example,

$$\int_Q f(x)\, dx = \int_0^1 \left(\int_0^1 \left(\int_0^1 f(x_1, x_2, x_3)\, dx_3 \right) dx_2 \right) dx_1.$$

The definition of the integral and the verification of the iterated integration formula is a direct generalization of the corresponding steps in the case of a double integral over the unit square.

Example 66.1. We compute the integral of $x_1^2 x_2 e^{x_1 x_2 x_3}$ over the unit cube Q,

$$\int_Q x_1^2 x_2 e^{x_1 x_2 x_3}\, dx = \int_0^1 \int_0^1 \left(\int_0^1 x_1^2 x_2 e^{x_1 x_2 x_3}\, dx_3 \right) dx_1 dx_2$$

$$= \int_0^1 \int_0^1 \left[x_1 e^{x_1 x_2 x_3} \right]_{x_3=0}^{x_3=1} dx_1 dx_2 = \int_0^1 \int_0^1 x_1 (e^{x_1 x_2} - 1)\, dx_1 dx_2.$$

which leaves a double integral that we know how to handle.

66.3   Triple Integrals over General Domains in $\mathbb{R}^3$

Let $\Omega = \{x \in \mathbb{R}^3 : \gamma_2(x_1, x_2) \le x_3 \le \gamma_1(x_1, x_2),\ (x_1, x_2) \in \omega\}$, where ω is a domain in $\mathbb{R}^2$ and $\gamma_1 : \omega \to \mathbb{R}$ and $\gamma_2 : \omega \to \mathbb{R}$ are given functions of (x_1, x_2), see Fig. 66.1. Let $f : \Omega \to \mathbb{R}$ Lipschitz continuous. We define the triple integral of $f(x)$ over Ω by

$$\int_\Omega f(x)\, dx = \int_\omega \left(\int_{\gamma_2(x_1,x_2)}^{\gamma_1(x_1,x_2)} f(x_1, x_2, x_3)\, dx_3 \right) dx_1 dx_2$$

via iterated integration first with respect to x_3 and then with respect to $(x_1, x_2) \in \omega$.

Expanding the double integral over ω into two one-dimensional integrals, assuming $\omega = \{(x_1, x_2, x_3) : \alpha_2 \le x_1 \le \alpha_1,\ \beta_2(x_1) \le x_2 \le \beta_1(x_1)\}$, we have

$$\int_\Omega f(x)\, dx = \int_{\alpha_2}^{\alpha_1} \left(\int_{\beta_2(x_1)}^{\beta_1(x_1)} \left(\int_{\gamma_2(x_1,x_2)}^{\gamma_1(x_1,x_2)} f(x)\, dx_3 \right) dx_2 \right) dx_1$$

$$= \int_{\alpha_2}^{\alpha_1} \left(\int_{\omega(x_1)} f(x_1, x_2, x_3)\, dx_2 dx_3 \right) dx_1,$$

where $\omega(x_1) = \{(x_2, x_3) : \beta_2(x_1) \le x_2 \le \beta_1(x_1),\ \gamma_2(x_1, x_2) \le x_3 \le \gamma_1(x_1, x_2)\}$ is the cross-section of the domain Ω with a plane with fixed x_1-coordinate. This way of splitting a triple integral into an one-dimensional

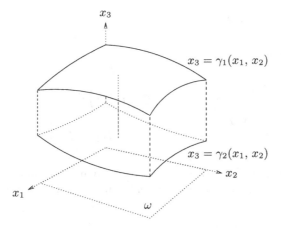

Fig. 66.1. Integration over a volume by first integrating in the x_3-direction

integral of double integrals over domain cross-sections corresponds to cutting a piece of bread or ham into slices.

We may define triple integrals similarly for more general domains by dividing the domain suitably into pieces.

66.4   The Volume of a Three-Dimensional Domain

We define the volume $V(\Omega)$ of a domain Ω in $\mathbb{R}^3$ as

$$V(\Omega) = \int_\Omega dx,$$

i.e. by integrating $f(x) = 1$ over $x \in \Omega$. If $\Omega = \{x \in [0,1] \times [0,1] \times \mathbb{R} : \gamma_2(x_1, x_2) \le x_3 \le \gamma_1(x_1, x_2)\}$, then

$$V(\Omega) = \int_0^1 \int_0^1 \left(\int_{\gamma_2(x_1, x_2)}^{\gamma_1(x_1, x_2)} dx_3 \right) dx_1 dx_2$$

$$= \int_0^1 \int_0^1 (\gamma_1(x_1) - \gamma_2(x_1)) \, dx_1 \, dx_2,$$

which conforms to the previous formula of the volume between the surfaces $\gamma_1(x_1, x_2)$ and $\gamma_2(x_1, x_2)$ as the integral of the difference $\gamma_1(x_1, x_2) - \gamma_2(x_1, x_2)$.

Example 66.2.   The volume of the pyramid Ω with corners at $(0,0,0)$, $(1,0,0)$, $(0,1,0)$, and $(0,0,1)$ described as $\{x \in \mathbb{R}^3 : 0 \le x_1 + x_2 + x_3 \le 1, x_1 x_2, x_3 \ge 0\}$, can be computed with $\omega = \{(x_1, x_2) : 0 \le x_1 \le 1, 0 \le$

$x_2 \leq 1 - x_1\}$ as follows

$$
\begin{aligned}
V(\Omega) &= \int_\omega \left(\int_0^{1-x_1-x_2} dx_3 \right) dx_1 dx_2 = \int_\Omega dx \\
&= \int_0^1 \left(\int_0^{1-x_1} \left(\int_0^{1-x_1-x_2} dx_3 \right) dx_2 \right) dx_1 \\
&= \int_0^1 \left(\int_0^{1-x_1} (1 - x_1 - x_2) \, dx_2 \right) dx_1 = \int_0^1 (1-x_1)^2/2 \, dx_1 = \frac{1}{6},
\end{aligned}
$$

which agrees with the earlier computation giving the volume of a pyramid as $\frac{1}{3}Bh$, where B is the area of the base and h the height, see Fig. 66.2.

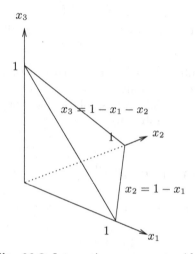

Fig. 66.2. Integration over a pyramid

66.5   Triple Integrals as Limits of Riemann Sums

We may also define integrals over domains in $\mathbb{R}^3$ as limits of Riemann sums

$$
\int_\Omega f(x) \, dx = \lim_{N \to \infty} \sum_{i=1}^N f(x_i) d\Omega_i, \tag{66.1}
$$

where $\{\Omega_i\}_{i=1}^N$ is a subdivision of the given domain Ω into pieces Ω_i with volume $V(\Omega_i) \leq d_N$ and quadrature points $x_i \in \Omega_i$, where d_N tends to 0 as N tends to infinity. The error estimate (64.20) for double integrals generalizes directly to three dimensions.

66.6   Change of Variables in a Triple Integral

We next prove an analog of the change of variable formula for two dimensional integrals. We thus want to make a change of variables in an integral

$$\int_{\Omega} f(x) \, dx = \int_{\Omega} f(x_1, x_2, x_3) \, dx_1 dx_2 dx_3, \tag{66.2}$$

where Ω is a given domain in $\mathbb{R}^3$ and the integration variable x runs over Ω. If $g : \tilde{\Omega} \to \Omega$ is a one-to-one mapping of $y \in \tilde{\Omega}$ onto $x = g(y)$ in Ω, we have the following change of variables formula

$$\int_{\Omega} f(x) \, dx = \int_{\tilde{\Omega}} f(g(y)) |\det g'(y)| \, dy, \tag{66.3}$$

where $|\det g'(y)|$ is the modulus of the determinant of the Jacobian $g'(y)$ of $g(y)$. Formally, we write $dx = |\det g'(y)| dy$ or $|\det g'(y)| = |\det \frac{dx}{dy}|$, and $|\det g'(y)|$ is the local change of volume measure as we go from y-coordinates to x-coordinates.

To prove this, let $\tilde{\Omega}_i$ be a small subdomain of $\tilde{\Omega}$ and let $\Omega_i = g(\tilde{\Omega}_i)$ be the image of $\tilde{\Omega}_i$ under the mapping $x = g(y)$. If $g'(y)$ were constant and $g(y)$ were linear over $\tilde{\Omega}_i$, then

$$d\Omega_i = |g'(y_i)| d\tilde{\Omega}_i, \tag{66.4}$$

where y_i is a point in $\tilde{\Omega}_i$, $d\Omega_i$ is the area of Ω_i, and $d\tilde{\Omega}_i$ is the area of $\tilde{\Omega}_i$. Thus,

$$\int_{\Omega} f(x) \, dx \approx \sum_i f(x_i) d\Omega_i$$

$$\approx \sum_i f(g(y_i)) |\det g'(y_i)| d\tilde{\Omega}_i \approx \int_{\tilde{\Omega}} f(g(y)) |\det g'(y)| \, dy,$$

where $x_i = g(y_i)$ and $\{\tilde{\Omega}_i\}_{i=1}^{N}$ is a subdivision of $\tilde{\Omega}$ of maximal diameter d_N. Assuming now that $f(x)$, $f(g(y))$ and $|\det g'(y)|$ are Lipschitz continuous, the formula (66.3) follows by passing to the limit as d_N tends to 0.

Spherical Coordinates

As a particular important change of variables, we consider spherical coordinates,

$$(x_1, x_2, x_3) = (r \sin(\varphi) \cos(\theta), r \sin(\varphi) \sin(\theta), r \cos(\varphi)),$$

where $x = (x_1, x_2, x_3) \in \mathbb{R}^3$ and $r \geq 0$, $0 \leq \theta < 2\pi$, $0 \leq \varphi < \pi$, see Fig. 66.3.

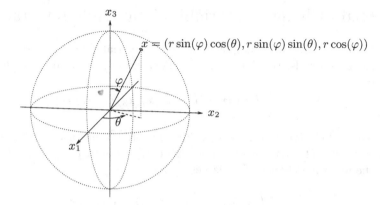

Fig. 66.3. Spherical coordinates

The Jacobian of the mapping $(r, \theta, \varphi) \to (x_1, x_2, x_3)$ is equal to

$$\frac{d(x_1, x_2, x_3)}{d(r, \theta, \varphi)} = \begin{pmatrix} \sin(\varphi)\cos(\theta) & -r\sin(\varphi)\sin(\theta) & r\cos(\varphi)\cos(\theta) \\ \sin(\varphi)\sin(\theta) & r\sin(\varphi)\cos(\theta) & r\cos(\varphi)\sin(\theta) \\ \cos(\varphi) & 0 & -r\sin(\varphi) \end{pmatrix}$$

and by a direct computation

$$\left| \det \frac{d(x_1, x_2, x_3)}{d(r, \theta, \varphi)} \right| = r^2 \sin(\varphi). \tag{66.5}$$

The change of variables formula from Cartesian x-coordinates to spherical coordinates takes the form

$$\int_\Omega f(x)\, dx$$
$$= \int_{\tilde{\Omega}} f\big(r\sin(\varphi)\cos(\theta), r\sin(\varphi)\sin(\theta), r\cos(\varphi)\big) r^2 \sin(\varphi)\, dr\, d\theta\, d\varphi,$$

where it is understood that $\tilde{\Omega}$ is a subdomain of $\{(r, \theta, \varphi) : 0 \le r, 0 \le \theta \le 2\pi, 0 \le \varphi \le \pi\}$, so that $(r, \theta, \varphi) \to x$ is a one-to-one mapping of $\tilde{\Omega}$ onto Ω.

Example 66.3. The unit ball $B = \{x \in \mathbb{R}^3 : |x| \le 1,\}$ is described in spherical coordinates as $\tilde{B} = \{(r, \theta, \varphi) : 0 \le r \le 1, 0 \le \theta < 2\pi, 0 \le \varphi < \pi\}$. The volume $V(B)$ of B is given by

$$B = \int_B dx = \int_{\tilde{B}} dr\, d\theta\, d\varphi = \int_0^\pi \int_0^{2\pi} \int_0^1 r^2 \sin(\varphi)\, dr\, d\theta\, d\varphi$$
$$= \int_0^\pi \sin(\varphi)\, d\varphi \int_0^{2\pi} d\theta \int_0^1 r^2\, dr = 2 \cdot 2\pi \frac{1}{3} = \frac{4\pi}{3}.$$

Note the way the triple integral splits into a product of three one-dimensional integrals because the limits of integration are fixed numbers in all the coordinate directions and the function to be integrated is a product of functions of the individual variables.

We have shown that the volume of the unit ball in $\mathbb{R}^3$ to be equal to $\frac{4\pi}{3}$. Another basic result of Calculus!

66.7   Solids of Revolution

To generate a *solid of revolution*, we let $f : [a, b] \rightarrow \mathbb{R}$ be a given (positive) function and consider the body B in $\mathbb{R}^3$ represented by $s(x_1, r, \theta) = (x_1, r\cos(\theta), r\sin(\theta))$ with $a \leq x_1 \leq b$, $0 \leq \theta < 2\pi$ and $0 \leq r \leq f(x_1)$, see Fig. 66.4. We have

$$\frac{d(x_1, x_2, x_3)}{d(x_1, r, \theta)} = \begin{pmatrix} 1 & 0 & 0 \\ 0 & \cos(\theta) & \sin(\theta) \\ 0 & -r\sin(\theta) & r\cos(\theta) \end{pmatrix}$$

and thus by a direct computation

$$\left| \det \frac{d(x_1, x_2, x_3)}{d(x_1, r, \theta)} \right| = r.$$

The coordinate system (x_1, r, θ) is an example of so called cylindrical coordinates suitable for data with rotational symmetry.

The volume $V(B)$ of B is given by:

$$V(B) = \int_0^{2\pi} \int_a^b \int_0^{f(x_1)} r\, dr dx_1 d\theta = \pi \int_a^b f^2(x_1)\, dx_1. \tag{66.6}$$

Example 66.4. Consider the body B obtained by rotating the parabola $f(x_1) = \sqrt{x_1}$ around the x_1-axis between $x_1 = 0$ and $x_1 = 1$. We have

$$V(B) = \pi \int_0^1 x_1\, dx_1 = \frac{\pi}{2}.$$

Example 66.5. The center of mass $\bar{x}$ of a body B of revolution obtained rotating a curve $f(x_1)$ around the x_1-axis from $x_1 = a$ to $x_1 = b$ is given by $\bar{x}_2 = \bar{x}_3 = 0$ (rotational symmetry) and

$$\bar{x}_1 = \frac{\pi}{V(B)} \int_a^b x_1 f^2(x_1)\, dx_1.$$

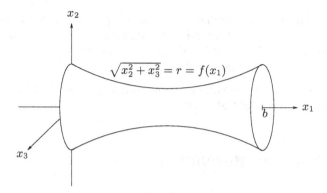

Fig. 66.4. A solid of revolution

66.8   Moment of Inertia of a Ball

The *moment of inertia* about the x_3-axis of the ball $B = \{\|x\| = 1\}$ of (uniformly distributed) total mass m, is equal to

$$I = \frac{m}{V(B)} \int_B (x_1^2 + x_2^2)\, dx. \qquad (66.7)$$

If the ball rotates with angular speed ω around the x_3 axis, then the total *kinetic energy* is equal to

$$E = \frac{1}{2}\frac{m}{V(B)} \int_B \omega^2(x_1^2 + x_2^2)\, dx = \frac{1}{2}\omega^2 I. \qquad (66.8)$$

Using spherical coordinates gives

$$I = \frac{2m}{5}. \qquad (66.9)$$

Chapter 66   Problems

66.1. Motivate (66.8) and prove (66.9).

66.2. Verify (66.5).

66.3. Compute the following triple integrals:

1. $\int_\Omega \|x\|^2\, dx$, for $\Omega = \{x \in \mathbb{R}^3 : 0 \le x_i \le 1, i = 1, 2, 3\}$,
2. $\int_\Omega \exp(x_1+x_2+x_3)\, dx$, for $\Omega = \{x \in \mathbb{R}^3 : 0 \le x_i \le 1, i = 1, 2, x_3 \le x_1+x_2\}$,
3. $\int_\Omega 1/\|x\|^2\, dx$, for $\Omega = \{x \in \mathbb{R}^3 : 1 \le \|x\| \le 2\}$.

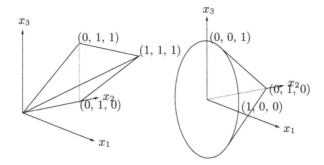

66.4. Compute with domains Ω as in Fig. 66.4 $\int_\Omega (1 - x_2)\, dx$.

66.5. Compute for $\Omega = \{(x_1, x_2, x_3) : x_1^2 + x_2^2 \le 1, |x_3| \le 1\}$

1. $\int_\Omega \frac{dx}{\|x\|^2}$
2. The moment of inertia of Ω with respect to the x_3-axis.
3. The moment of inertia of Ω with respect to the x_2-axis.

66.6. Compute the following multiple integrals:

1. $\int_\Omega \frac{\exp(-\|x\|)}{\|x\|}\, dx$, for $\Omega = \{x \in \mathbb{R}^3 : \|x\| > 1\}$,
2. $\int_\Omega x_1 + x_2 + x_3 + x_4\, dx$, for $\Omega = \{x \in \mathbb{R}^4 : 0 \le x_i \le 1, i = 1,2,3,4\}$,
3. $\int_\Omega x_1 + \ldots + x_n\, dx$, for $\Omega = \{x \in \mathbb{R}^n : 0 \le x_i \le 1, i = 1, \ldots, n\}$.

66.7. Compute the following multiple integrals:

1. $\int_\Omega x\, dx$,
2. $\int_\Omega \|x\|\, dx$,
3. $\int_\Omega \|x\|^2\, dx$,

where $\Omega = \{x \in \mathbb{R}^3 : \|x\| \le 1\}$.

66.8. Try to generalize the result in the previous exercise to $\mathbb{R}^n$, denoting the area of the unit sphere, $\{x \in \mathbb{R}^n : \|x\| = 1\}$, by S_n.

66.9. Compute the integral $\int_{\mathbb{R}^2} \exp(-\|x\|^2)\, dx$ and use the result to compute $\int_{\mathbb{R}^n} \exp(-\|x\|^2)\, dx$.

66.10. Find the moment of inertia of a unit cube with respect to its diagonal.

66.11. Let E_y be the domain in $\mathbb{R}^n$ where the absolute value of $f : \mathbb{R}^n \to \mathbb{R}$ is larger than y, i.e. $E_y = \{x \in \mathbb{R}^n : |f(x)| > y\}$, and let $g(y)$ be the volume (size, measure) of this domain, i.e. $g(y) = \int_{E_y} dx$. Show, by changing the order of integration, that

$$\int_{\mathbb{R}^n} |f(x)|\, dx = \int_0^\infty g(y)\, dy.$$

67

Gauss' Theorem and Green's Formula in $\mathbb{R}^2$

> Mathematics at its best: it looks impressive (incomprehensible), but is trivial for anyone who understands the notation. (R. Reagan)

67.1 Introduction

We now turn to two of the corner stones of calculus in several dimensions, namely *Gauss' theorem* and *Green's formula*, beginning with two dimensions. We shall see that these famous (and useful) results are direct consequences of the fundamental formula,

$$\int_\Omega \frac{\partial u}{\partial x_2} \, dx = \int_\Gamma u \, n_2 \, ds, \tag{67.1}$$

where Ω is a domain in $\mathbb{R}^2$ with boundary Γ and $n(x) = (n_1(x), n_2(x))$ is the *outward unit normal* to Γ at $x \in \Gamma$, that is $n(x)$ is orthogonal to the tangent to Γ at x and points out of Ω and $\|n(x)\| = 1$, see Fig. 67.2. We shall see that this formula is an analog of the Fundamental Theorem

$$\int_a^b \frac{du}{dx} \, dx = u(b) - u(a), \tag{67.2}$$

stating that the integral over an interval $[a, b]$ of the derivative $\frac{du}{dx}$ of a function u is equal to the difference between the end-point values $u(b)$ and $u(a)$.

67.2 The Special Case of a Square

To see the connection between (67.1) and (67.2), we first assume that Ω is the unit square $[0,1] \times [0,1]$. In this case $n_2 = 1$ on the top Γ_1 of the square and $n_2 = -1$ on the bottom Γ_3, and $n_2 = 0$ on the vertical sides Γ_2 and Γ_4, see Fig. 67.2. Therefore,

$$\int_\Gamma u\, n_2 \, ds = \int_0^1 u(x_1, 1)\, dx_1 - \int_0^1 u(x_1, 0)\, dx_1$$

if we parameterize Γ_1 by $s(x_1) = (x_1, 1)$ and Γ_3 by $s(x_1) = (x_1, 0)$. On the other hand, integrating first with respect to x_2 and then with respect to x_1 and using (67.2), we have

$$\int_\Omega \frac{\partial u}{\partial x_2}\, dx = \int_0^1 \left(\int_0^1 \frac{\partial u}{\partial x_2}(x_1, x_2)\, dx_2 \right) dx_1 = \int_0^1 (u(x_1, 1) - u(x_1, 0))\, dx_1$$

$$= \int_0^1 u(x_1, 1)\, dx_1 - \int_0^1 u(x_1, 0)\, dx_1 = \int_\Gamma u\, n_2 \, ds,$$

which proves (67.1) when Ω is a square. We see that (67.1) results from using (67.2) with $\frac{du}{dx}\, dx$ replaced by $\frac{\partial u}{\partial x_2} dx_2$, followed by an integration with respect to x_1. The net result is that the integral of $\frac{\partial u}{\partial x_2} dx_2$ over Ω is replaced by a curve integral of $u n_2$ over the boundary Γ of Ω.

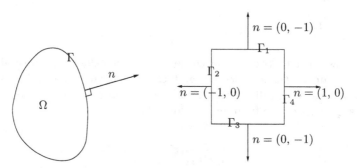

Fig. 67.1. To the *left*: A domain Ω with boundary Γ and normal n. To the *right*: A special case

67.3 The General Case

We now consider a domain Ω bounded by two curves Γ_1 parameterized by $s_1(x_1) = (x_1, \gamma_1(x_1))$ and Γ_2 parameterized by $s_2(x_1) = (x_1, \gamma_2(x_1))$ with $a \le x_1 \le b$, and $n = (n_1, n_2)$ is the outward normal to Γ, see Fig. 67.2.

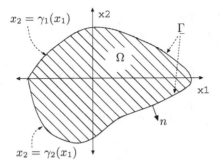

Fig. 67.2. A domain Ω with two curves defining the boundary Γ

The proof of (67.1) depends on the key observation that

$$\left\|\frac{ds_1}{dx_1}\right\| = \sqrt{1+(\gamma_1')^2}, \quad n_2 = \frac{1}{\sqrt{1+(\gamma_1')^2}},$$

$$\left\|\frac{ds_2}{dx_1}\right\| = \sqrt{1+(\gamma_2')^2}, \quad n_2 = -\frac{1}{\sqrt{1+(\gamma_2')^2}}.$$

Formally, $n_2 ds_1 = dx_1$ and $n_2 ds_2 = -dx_1$, see Fig. 67.3.

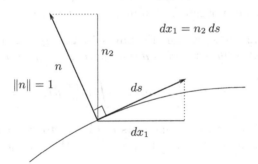

Fig. 67.3. The key observation that $\frac{dx_1}{ds} = \frac{n_2}{1}$ by similarity

Note that n_2 is positive on the upper boundary curve s_1 and negative on the lower boundary curve s_2. We thus have

$$\int_{\Gamma_1} u n_2 \, ds_1 = \int_a^b u(x_1, \gamma_1(x_1)) n_2 \left\|\frac{ds_1}{dx_1}\right\| \, dx_1 = \int_a^b u(x_1, \gamma_1(x_1)) \, dx_1,$$

$$\int_{\Gamma_2} u n_2 \, ds_2 = \int_a^b u(x_1, \gamma_2(x_1)) n_2 \left\|\frac{ds_2}{dx_1}\right\| \, dx_1 = -\int_a^b u(x_1, \gamma_1(x_1)) \, dx_1.$$

Secondly, integrating first with respect to x_2 and then with respect to x_1 and using the Fundamental Theorem, we see that

$$\int_\Omega \frac{\partial u}{\partial x_2}\, dx = \int_\Omega \frac{\partial u}{\partial x_2}\, dx_2\, dx_1 = \int_a^b \left(\int_{\gamma_2(x_1)}^{\gamma_1(x_1)} \frac{\partial u}{\partial x_2}\, dx_2 \right) dx_1$$

$$= \int_a^b u(x_1, \gamma_1(x_1))\, dx_1 - \int_a^b u(x_1, \gamma_2(x_1))\, dx_1.$$

Since

$$\int_\Gamma u\, n_2\, ds = \int_{\Gamma_1} u n_2\, ds_1 + \int_{\Gamma_2} u n_2\, ds_2,$$

the desired formula (67.1) now follows. The proof generalizes to arbitrary domains bounded by smooth curves with Lipschitz continuous tangents. We summarize in the following basic theorem:

Theorem 67.1 *If Ω is a domain in $\mathbb{R}^2$ with boundary Γ with outward unit normal (n_1, n_2) and $u : \Omega \to \mathbb{R}$ is differentiable, then*

$$\int_\Omega \frac{\partial u}{\partial x_i}\, dx = \int_\Gamma u\, n_i\, ds, \quad i = 1, 2. \tag{67.3}$$

Applying (67.3) to the product vw of two functions v and w, we obtain the following analog of integration by parts in two dimensions:

Theorem 67.2 (Integration by parts in 2d) *If Ω is a domain in $\mathbb{R}^2$ with boundary Γ with outward unit normal (n_1, n_2) and $v, w : \Omega \to \mathbb{R}$, then*

$$\int_\Omega \frac{\partial v}{\partial x_i} w\, dx = \int_\Gamma vw\, n_i\, ds - \int_\Omega v \frac{\partial w}{\partial x_i}\, dx, \quad i = 1, 2. \tag{67.4}$$

Applying (67.3) to the components u_i of a vector valued function $u = (u_1, u_2)$ and summing over $i = 1, 2$, we obtain the *Divergence theorem*, or *Gauss' theorem*

$$\int_\Omega \nabla \cdot u\, dx = \int_\Gamma u \cdot n\, ds, \tag{67.5}$$

where $u \cdot n = u_1 n_1 + u_2 n_2$ and

$$\nabla \cdot u = \left(\frac{\partial}{\partial x_1}, \frac{\partial}{\partial x_2} \right) \cdot (u_1, u_2) = \frac{\partial u_1}{\partial x_1} + \frac{\partial u_2}{\partial x_2}.$$

Applying (67.4) with w replaced by $\frac{\partial w}{\partial x_i}$ and summing over $i = 1, 2$, we obtain *Green's formula*:

$$\int_\Omega \nabla v \cdot \nabla w\, dx = \int_\Gamma v \partial_n w\, ds - \int_\Omega v \Delta w\, dx, \tag{67.6}$$

where

$$\partial_n w = \nabla w \cdot n = \frac{\partial w}{\partial x_1} n_1 + \frac{\partial w}{\partial x_2} n_2, \tag{67.7}$$

is the *outward normal derivative* of w on Γ. We often use Green's formula in the form

$$\int_\Omega v \Delta w \, dx - \int_\Omega \Delta v \, w \, dx = \int_\Gamma v \, \partial_n w \, ds - \int_\Gamma \partial_n v \, w \, ds, \tag{67.8}$$

which results after applying (67.6) twice and using

$$\Delta w = \text{div grad} w = \nabla \cdot \nabla w$$
$$= \left(\frac{\partial}{\partial x_1}, \frac{\partial}{\partial x_2} \right) \cdot \left(\frac{\partial w}{\partial x_1}, \frac{\partial w}{\partial x_2} \right) = \frac{\partial}{\partial x_1} \left(\frac{\partial w}{\partial x_1} \right) + \frac{\partial}{\partial x_2} \left(\frac{\partial w}{\partial x_2} \right),$$

which can be written succinctly as $\Delta w = \frac{\partial^2 w}{\partial x_1^2} + \frac{\partial^2 w}{\partial x_2^2}$.

We also note the following analog of the Divergence theorem:

$$\int_\Omega \nabla \times u \, dx = \int_\Gamma n \times u \, ds, \tag{67.9}$$

where $u : \Omega \to \mathbb{R}^2$ and $\nabla \times u = \frac{\partial u_2}{\partial x_1} - \frac{\partial u_1}{\partial x_2}$, and $n \times u = u_2 n_1 - u_1 n_2$. This is just a restatement of

$$\int_\Omega \left(\frac{\partial u_2}{\partial x_1} - \frac{\partial u_1}{\partial x_2} \right) dx = \int_\Gamma \left(u_2 n_1 - u_1 n_2 \right) ds \tag{67.10}$$

and therefore follows from (67.3). We further note that $\tau = (-n_2, n_1)$ is a unit tangent to Γ, since $n = (n_1, n_2)$ is a unit normal and $(-n_2, n_1) \cdot (n_1, n_2) = 0$, and $\tau = (-n_2, n_1)$ is directed in the counter-clockwise direction of Γ, see Fig. 67.4.

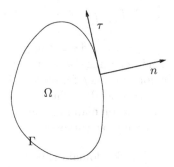

Fig. 67.4. The unit tangent $\tau = (-n_2, n_1)$ to Γ expressed in terms of the normal $n = (n_1, n_2)$

We often write

$$\int_\Gamma u_2 n_1 - u_1 n_2 \, ds = \int_\Gamma u \cdot \tau \, ds = \int_\Gamma u \cdot ds,$$

interpreting ds in the last integral as the *vector* τds with the old use of ds as the element of curve length. This is consistent with the notation

$$\int_\Gamma u \cdot ds = \int_a^b u(s(t)) \cdot s'(t) \, dt,$$

where $s : [a, b] \rightarrow \mathbb{R}^2$ represents Γ, which was introduced in Chapter *Curve Integrals*. Caution: we here use "ds" with two different interpretations: as the element of curve length (a scalar), and as an element of the tangent vector (a vector).

We summarize the basic results derived in this chapter as follows:

Theorem 67.3 *If Ω is a domain in $\mathbb{R}^2$ with boundary Γ with outward unit normal (n_1, n_2), and $u : \Omega \rightarrow \mathbb{R}^2$ and $v, w : \Omega \rightarrow \mathbb{R}$, then*

$$\int_\Omega \frac{\partial v}{\partial x_i} \, dx = \int_\Gamma v \, n_i \, ds, \quad i = 1, 2, \tag{67.11}$$

$$\int_\Omega \frac{\partial v}{\partial x_i} w \, dx = \int_\Gamma vw \, n_i \, ds - \int_\Omega v \frac{\partial w}{\partial x_i} \, dx, \quad i = 1, 2, \tag{67.12}$$

$$\int_\Omega \nabla \cdot u \, dx = \int_\Gamma u \cdot n \, ds, \tag{67.13}$$

$$\int_\Omega \nabla \times u \, dx = \int_\Gamma n \times u \, ds = \int_\Gamma u \cdot ds, \tag{67.14}$$

$$\int_\Omega \nabla v \cdot \nabla w \, dx = \int_\Gamma v \partial_n w \, ds - \int_\Omega v \Delta w \, dx, \tag{67.15}$$

$$\int_\Omega v \Delta w \, dx - \int_\Omega \Delta v \, w \, dx = \int_\Gamma v \, \partial_n w \, ds - \int_\Gamma \partial_n v \, w \, ds. \tag{67.16}$$

Example 67.1. For $u(x_1, x_2) = x_1$ and $i = 1$ in (67.11), we obtain $\int_\Omega dx = \int_\Gamma x_1 n_1 \, ds = \int_\Gamma x_1 \, dx_2$. An interesting observation from this is that you may compute the area $\int_\Omega dx$, for example of a piece of land, simply by walking its boundary and computing $\int_\Gamma x_1 \, dx_2$. The *planetometer* is a mechanical devise for computing the area of plane domains built on this principle, which has been used extensively by Surveyors.

Example 67.2. If $\nabla \times u = 0$ in the domain Ω between two curves Γ_1 and Γ_2 that both start at the point a and end at the point b, then $\int_{\Gamma_1} u \cdot ds = \int_{\Gamma_2} u \cdot ds$, where ds is the vector tangential to the curves in the direction from a to b of length equal to the element of curve length. This follows from the fact that $\int_{\Gamma_1 \cup \Gamma_2^-} u \cdot ds = \int_{\Gamma} u \cdot ds = 0$, by (67.14), where Γ_2^- denotes the curve Γ_2 with the direction of ds reversed. We conclude that curve integrals of a field $u = (u_1, u_2)$ with $\nabla \times u = 0$, that is of an *irrotational* field, is independent of the particular "path" of the curve from a to b. The integral of $u \cdot ds$ only depends on the two end-points a and b of the integration. Fields $u = (u_1, u_2)$ of this type are called *conservative*. As we shall see below, such fields are given by a *potential*, that is, they are the gradient field of some scalar potential $\varphi = \varphi(x)$ so $u = \nabla \varphi$.

Furthermore, $\int_{\gamma} u \cdot ds = \varphi(b) - \varphi(a)$ for a curve γ from a to b. For example, the field $u = (x_2, x_1)$ has $u_1(x_1, x_2) = x_2$ and $u_2(x_1, x_2) = x_1$ and thus $\nabla u = \frac{\partial u_2}{\partial x_1} - \frac{\partial u_1}{\partial x_2} = 1 - 1 = 0$. We find easily that $u = \nabla \varphi$ for $\varphi(x) = x_1 x_2$, and the integral of $u \cdot ds$ from a point $a = (a_1, a_2)$ to $b = (b_1, b_2)$ is given by $b_1 b_2 - a_1 a_2$.

Chapter 67 Problems

67.1. Derive (67.4), (67.5), (67.6) and (67.8) from (67.3).

67.2. (a) Explain why (67.1) is valid also for a domain like $\{(x_1, x_2) : x_1 \le |x_2|, x_1^2 + x_2^2 \le 1\}$. (b) Verify by direct computation of $\int_{\Omega} \frac{\partial u}{\partial x_2} dx$ and $\int_{\Gamma} u n_2 ds$ that (67.1) is valid for $u = r^{1/4} \sin(v/4)$ and $\Omega = \{(r\cos(v), r\sin(v)) : 0 < r < 1, 0 < v < 2\pi\}$, where $r = \sqrt{x_1^2 + x_2^2}$ and $v = \text{arccot}(x_1/x_2)$ for $x_2 > 0$, $v = \text{arccot}(x_1/x_2) + \pi$ for $x_2 < 0$ are the usual polar coordinates. Recall that by the chain rule you may express $\frac{\partial u}{\partial x_2}$ in terms of $\frac{\partial u}{\partial r}$ and $\frac{\partial u}{\partial v}$ if you like.

67.3. Assume $u = (u_1, u_2)$ is divergence free in Ω with boundary Γ. What can be said about (a) $\int_{\Gamma} u \cdot n \, ds$, (b) $u(x) \cdot n(x)$ for points x on Γ.

67.4. Assume $\int_{\Gamma} u \cdot n \, ds = 0$, where Γ is the boundary of a domain Ω with exterior unit normal n. What can be said about $\nabla \cdot u$ in Ω? (Before you give a too definite answer you may want to consider for example the case $u = (x_1^2, x_2^2)$ with Ω the unit disc.) Assume $\int_{\gamma} u \cdot n \, ds = 0$ for all closed curves γ in Ω, and the derivatives of u_i are Lipschitz. What can then be said about u in Ω?

67.5. Consider a "deformation" of $\mathbb{R}^2$ where the points $x = (x_1, x_2)$ are displaced to new positions $x + u(x)$, $u = (u_1, u_2)$, $u_i = u_i(x)$. We call $u(x)$ the displacement field and the Jacobian $u'(x)$ of $u(x)$ the deformation tensor (matrix). Consider for simplicity the case $u_i(x) = a_i x_i$, $i = 1, 2$, and assume the displacement is "area preserving", corresponding to "incompressibility" of the deformed material. Show that for small deformations, one has div $u \approx 0$. Hint: Consider $x \to x + u(x)$ as a change of variables and use an established fact about the Jacobian of area preserving maps.

67.6. Consider the vector field $u(x) = x/\|x\|^2$. Let Ω be the disc $\{x \in \mathbb{R}^2 : \|x - a\| \le 1\}$, and Γ its boundary with exterior unit normal n. Compute $\int_{\Gamma} u \cdot n \, ds$

for $a = (2,0)$ and $a = 0$. Do the results conform with the Divergence theorem? Make an "arrow plot" of u in the (x_1, x_2)-plane. Can you see a connection the eruption of a volcano? Does the Divergence theorem apply in the case $a = (0,0)$?

67.7. Show that if $\nabla \cdot u = 0$ and Γ and $\bar{\Gamma}$ are curves with normals n and $\bar{n}$ as in Fig. 67.7, then $\int_\Gamma u \cdot n \, ds = \int_{\bar{\Gamma}} u \cdot \bar{n} \, ds$.

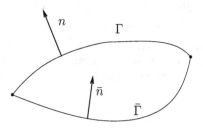

67.8. For $u = \left(\frac{2x_1 x_2}{1+x_2^2}, -\log(1+x_2^2) \right)$ and Γ the curve (a) $\{(x_1, x_2) : x_1^2 + x_2^2 = 1, x_i \geq 0, i = 1, 2\}$ (b) $\{(x_1, x_2) : x_1 = 2 - (x_2 - 1)^2, x_1 \geq 1\}$, compute $\int_\Gamma u \cdot n \, ds$. Hint: Close the curves and use the Divergence theorem.

67.9. Show that the field $u = e^{x_1 x_2}(1 + x_1 x_2, x_1^2)$ is irrotational, and find a potential φ such that $u = \nabla \varphi$.

67.10. Evaluate the integrals in (67.16) for w a solution to the *differential equation* $-\Delta w = f$ in $\Omega = \mathbb{R}^2$ and $v = -\frac{1}{2\pi}\log(x - \bar{x})$, assuming w and $\partial_n w$ vanish for $\|x\|$ large. Show that this gives a formula for $w(\bar{x})$ in terms of f and v. Hint: Take $\Omega = \{x \in \mathbb{R}^2 : \|x - \bar{x}\| > \epsilon\}$ and let ϵ tend to zero.

67.11. Let w be the solution to $-\Delta w = f$ in the upper half plane $x_2 > 0$, $-\frac{\partial w}{\partial x_2} = g$ for $x_2 = 0$, and assume w and ∇w vanish for $\|x\|$ large. Show that for $\bar{x} = (\bar{x}_1, 0)$ on $\Gamma = \{(x_1, x_2) : x_2 = 0\}$, one has $\frac{1}{2}w(\bar{x}) = \int_{\{x : x_2 > 0\}} vf \, dx + \int_{\{x : x_2 = 0\}} vg \, ds$, where $v = -\frac{1}{2\pi}\log(x - \bar{x})$. Hint: Take $\Omega = \{x \in \mathbb{R}^2 : x_2 > 0, \|x - \bar{x}\| > \epsilon\}$ in (67.16), and let ϵ tend to zero.

67.12. Show that for *harmonic functions* v and w, that is with $\Delta v = 0$ and $\Delta w = 0$, one has $\int_\Gamma \partial_n vw \, ds = \int_\Gamma v\partial_n w \, ds$ for a closed curve Γ.

67.13. Find the area of the domain enclosed by the curve

$$\Gamma = \{(r\cos(v), r\sin(v)) : r = 2 + \sin(v), 0 \leq v < 2\pi\}.$$

Hint: Integrals of the form $\int \sin^4(v)\, dv$ and $\int \cos^4(v)\, dv$ may be computed using integration by parts, as follows:

$$I = \int \cos^4(v)\, dv = \int (1 - \sin^2(v)) \cos(v) \cdot \cos(v)\, dv = \{\text{int. by parts}\}$$

$$= \left(\sin(v) - \frac{1}{3}\sin^3(v) \right) \cdot \cos(v) - \int \left(\sin(v) - \frac{1}{3}\sin^3(v) \right) (-\sin(v))\, dv$$

$$= \left(\sin(v) - \frac{1}{3}\sin^3(v) \right) \cdot \cos(v) + \int \sin^2(v)\, dv - \frac{1}{3}\int (1 - \cos^2(v))^2\, dv$$

$$= \left(\sin(v) - \frac{1}{3}\sin^3(v) \right) \cos(v) + \int \sin^2(v)\, dv - \frac{1}{3}\int (1 - 2\cos^2(v))\, dv - \frac{1}{3}I,$$

from which I can be computed.

68

Gauss' Theorem and Green's Formula in $\mathbb{R}^3$

Of those who with me have written something about these matters, either I alone am mad, or I alone am not mad. No third option can be maintained, unless (as perchance it may seem to some) we are all mad. (Hobbes to Wallis)

If he is mad, he is not likely to be convinced by reason; on the other hand, if we be mad, we are in no position to attempt it. (Wallis to Hobbes)

We now extend the results of the previous chapter to three dimensions. The basic result is the following analog of (67.1): If Ω is a domain in $\mathbb{R}^3$ with boundary Γ, then

$$\int_\Omega \frac{\partial u}{\partial x_3}\, dx = \int_\Gamma u\, n_3\, ds, \tag{68.1}$$

where (n_1, n_2, n_3) is the outward normal to Γ. To prove this, we assume that Γ is composed of the two surfaces Γ_1 given by $s_1(x_1, x_2) = (x_1, x_2, \gamma_1(x_1, x_2))$ and Γ_2 given by $s_2(x_1, x_2) = (x_1, x_2, \gamma_2(x_1, x_2))$, where $(x_1, x_2) \in \omega$ and the parameter domain ω is a domain in $\mathbb{R}^2$, and we assume that $\Omega = \{x \in \mathbb{R}^3 : (x_1, x_2) \in \omega, \gamma_2(x_1, x_2) < x_3 < \gamma_1(x_1, x_2)\}$, see Fig. 68.1. We have $s'_{i,1} \times s'_{i,2} = (1, 0, \gamma_{i,1}) \times (0, 1, \gamma_{i,2})$ for $i = 1, 2$, where $\gamma_{i,j} = \frac{\partial \gamma_i}{\partial x_j}$, and thus on Γ_1

$$\|s'_{1,1} \times s'_{1,2}\| = \sqrt{1 + (\gamma'_{1,1})^2 + (\gamma'_{1,2})^2}, \quad n_3 = \frac{1}{\sqrt{1 + (\gamma'_{1,1})^2 + (\gamma'_{1,2})^2}}$$

and on Γ_2

$$\|s'_{2,1} \times s'_{2,2}\| = \sqrt{1 + (\gamma'_{2,1})^2 + (\gamma'_{2,2})^2}, \quad n_3 = -\frac{1}{\sqrt{1 + (\gamma'_{2,1})^2 + (\gamma'_{2,2})^2}}.$$

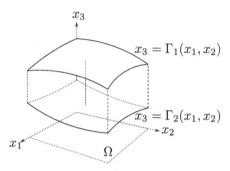

Fig. 68.1. A domain Ω bounded by two graphs Γ_1 and Γ_2

Integrating first with respect to x_3 and using the Fundamental Theorem we get

$$\int_\Omega \frac{\partial u}{\partial x_3} \, dx = \int_\Omega \frac{\partial u}{\partial x_3} \, dx_3 dx_1 dx_2$$

$$= \int_\omega u(x_1, x_2, \gamma_1(x_1, x_2)) \, dx_1 dx_2 - \int_\omega u(x_1, x_2, \gamma_2(x_1, x_2)) \, dx_1 dx_2$$

$$= \int_\omega u \, n_3 \|s'_{1,1} \times s'_{1,2}\| dx_1 dx_2 + \int_\omega u \, n_3 \|s'_{2,1} \times s'_{2,2}\| dx_1 dx_2$$

$$= \int_{\Gamma_1} u \, n_3 \, ds + \int_{\Gamma_2} u \, n_3 \, ds = \int_\Gamma u \, n_3 \, ds,$$

which proves (68.1). Note that $n_3 = 0$ on the "vertical" parts of Γ! This result generalizes to

$$\int_\Omega \frac{\partial u}{\partial x_i} \, dx = \int_\Gamma u \, n_i \, ds, \quad i = 1, 2, 3, \tag{68.2}$$

for a general domain Ω in $\mathbb{R}^3$ with boundary Γ with outward unit normal (n_1, n_2, n_3).

Applying (68.2) to the product vw of two functions v and w, we obtain the analog of integration by parts in three dimensions,

$$\int_\Omega \frac{\partial v}{\partial x_i} w \, dx = \int_\Gamma vw \, n_i \, ds - \int_\Omega v \frac{\partial w}{\partial x_i} \, dx, \quad i = 1, 2, 3. \tag{68.3}$$

Applying (68.3) to the components u_i of a vector valued function $u = (u_1, u_2, u_3)$ with $w = 1$ and summing over i, we obtain the *Divergence theorem*, or *Gauss' theorem* in three dimensions

$$\int_\Omega \nabla \cdot u \, dx = \int_\Gamma u \cdot n \, ds, \qquad (68.4)$$

where $\nabla \cdot u = (\frac{\partial}{\partial x_1}, \frac{\partial}{\partial x_2}, \frac{\partial}{\partial x_3}) \cdot (u_1, u_2, u_3) = \frac{\partial u_1}{\partial x_1} + \frac{\partial u_2}{\partial x_2} + \frac{\partial u_3}{\partial x_3} = \sum_{i=1}^{3} \frac{\partial u_i}{\partial x_i}$, and $u \cdot n = u_1 n_1 + u_2 n_2 + u_3 n_3$ is the component of u in the direction of the normal n. If u represents a *flux* of some quantity, like heat flux or water flux, then $u(x) \cdot n(x)$ at a point $x \in \Gamma$ represents the flux through Γ (out of Ω), or *normal flux*, and thus

$$\int_\Gamma u \cdot n \, ds$$

represents the *total flux* through Γ.

We also directly obtain the following analog of Gauss' theorem for a function $u : \Omega \to \mathbb{R}^3$:

$$\int_\Omega \nabla \times u \, dx = \int_\Gamma n \times u \, ds, \qquad (68.5)$$

which is now a vector equation!!

Another consequence of (68.3) is *Green's formula*:

$$\int_\Omega \nabla v \cdot \nabla w \, dx = \int_\Gamma v \partial_n w \, ds - \int_\Omega v \Delta w \, dx, \qquad (68.6)$$

where $\partial_n v = \nabla v \cdot n = \frac{\partial v}{\partial x_1} n_1 + \frac{\partial v}{\partial x_2} n_2 + \frac{\partial v}{\partial x_3} n_3$ is the outward normal derivative of v on Γ, and now $\Delta w = \frac{\partial^2 w}{\partial x_1^2} + \frac{\partial^2 w}{\partial x_2^2} + \frac{\partial^2 w}{\partial x_3^2}$. We often use Green's formula in the form

$$\int_\Omega v \Delta w \, dx - \int_\Omega \Delta v \, w \, dx = \int_\Gamma v \, \partial_n w \, ds - \int_\Gamma \partial_n v \, w \, ds, \qquad (68.7)$$

which results after applying (68.6) twice.

We summarize the basic results derived in this chapter as follows:

Theorem 68.1 *If Ω is a domain in $\mathbb{R}^3$ with boundary Γ with outward unit normal $n = (n_1, n_2, n_3)$, and $u : \Omega \to \mathbb{R}^3$ and $v, w : \Omega \to \mathbb{R}$, then*

$$\int_\Omega \frac{\partial v}{\partial x_i} \, dx = \int_\Gamma v \, n_i \, ds, \quad i = 1, 2. \qquad (68.8)$$

$$\int_\Omega \frac{\partial v}{\partial x_i} w \, dx = \int_\Gamma v w \, n_i \, ds - \int_\Omega v \frac{\partial w}{\partial x_i} \, dx, \quad i = 1, 2. \qquad (68.9)$$

$$\int_\Omega \nabla \cdot u \, dx = \int_\Gamma u \cdot n \, ds \qquad (68.10)$$

$$\int_\Omega \nabla \times u \, dx = \int_\Gamma n \times u \, ds, \tag{68.11}$$

$$\int_\Omega \nabla v \cdot \nabla w \, dx = \int_\Gamma v \partial_n w \, ds - \int_\Omega v \Delta w \, dx, \tag{68.12}$$

$$\int_\Omega v \Delta w \, dx - \int_\Omega \Delta v \, w \, dx = \int_\Gamma v \, \partial_n w \, ds - \int_\Gamma \partial_n v \, w \, ds. \tag{68.13}$$

Example 68.1. We compute the total flow of the vector field $u(x) = (x_1 + x_2^5, x_2 + x_3 x_1, x_3 + x_1 x_2)$ out of the boundary S of the unit ball $B = \{x \in \mathbb{R}^3 : \|x\| = 1\}$, that is the integral,

$$\int_S u \cdot n \, ds = \int_S ((x_1 + x_2^5)x_1 + (x_2 + x_3 x_1)x_2 + (x_3 + x_1 x_2)x_3) \, ds, \tag{68.14}$$

where we used that the outward unit normal n to S at $x \in S$ is given by $n(x) = x$. Since div $u(x) = 3$ for $x \in \mathbb{R}^3$, we have by Gauss's theorem

$$\int_S u \cdot n \, ds = \int_B 3 \, dx = 3V(B) = 4\pi,$$

which gives a quick way of computing the quite difficult integral (68.14).

68.1 George Green (1793–1841)

George Green, a millers son and self-taught mathematician (he left school at age 9 after two years of study), published 1827 on his own "An Essay on the Application of Mathematical Analysis to the Theories of Electricity and Magnetism" introducing in particular so-called *Green's functions* forming the basis of the modern theory of partial differential equations. His importance in mathematics was only recognized after his death in the work by in particular Maxwell on electromagnetics.

Chapter 68 Problems

68.1. Write out and verify (68.5) from (68.2).

68.2. (a) Prove Green's formula (68.6) using (68.3). (b) Prove (68.13).

68.3. Compute the integral $\int_\Gamma \frac{x}{\|x\|^3} \cdot n \, ds$, where $\Gamma = \{x \in \mathbb{R}^3 : x_1^2 + x_2^2 + (x_3 - ja)^2 = a^2\}$ for $a > 0$ and $j = 0, 1, 2$, respectively, where n is the exterior unit normal to Γ. Interpret the results.

68.4. Compute the integral $\int_\Gamma \frac{1}{x_1^2 + x_2^2} \frac{(-x_2, x_1)}{x_1^2 + x_2^2} \times ds$, where $\Gamma = \{x \in \mathbb{R}^3 : x_1^2 + x_2^2 + x_3 = 1\}$.

68.5. Let Γ be the unit sphere in $\mathbb{R}^3$ with exterior unit normal n and compute the following integrals:

1. $\int_\Gamma x \cdot n \, ds$,
2. $\int_\Gamma x \times n \, ds$,
3. $\int_\Gamma \frac{1}{\|x\|^2} \frac{x}{\|x\|} \cdot n \, ds$,
4. $\int_\Gamma \frac{1}{\|x\|^2} \frac{x}{\|x\|} \times n \, ds$.

68.6. Verify that for a radial field $F(x) = \|x\|^\alpha \frac{x}{\|x\|}$ one has div $F = (\alpha + 2)\|x\|$.

68.7. What is the smallest possible value of the integral $\int_\Gamma F \cdot n \, ds$, where $F(x) = (x_1 x_2^2 - 4x_1 x_2, 4x_2 x_3^2 + 8x_2 x_3 + 5x_2, x_1^2 x_3 - 2x_1 x_3)$ and Γ is a closed surface in $\mathbb{R}^3$, and n its exterior unit normal? Hint: Enclose all the "sinks" of F, that is, consider the domain where div $F \leq 0$.

68.8. Compute the surface integral $\int_\Gamma F \cdot n \, ds$, where

$$F(x) = (x_2^2, x_1 x_2 (\cos(x_1))^2 + x_1 x_2^3 + \exp(\cos(x_1 x_3^2)), x_1 x_3 (\sin(x_1))^2 - 3x_1 x_2^2 x_3),$$

and Γ is the part of the sphere $\|x\| = 2$ with positive x_3-coordinate, and n its normal with also positive x_3-component. Hint: The function F is chosen *seemingly* difficult only to confuse you.

68.9. Let $\{x_1, x_2, \ldots, x_N\}$ be a set of points in $\mathbb{R}^n$, and let

$$F(x) = \sum_{j=1}^N \frac{1}{4\pi \|x - x_j\|^2} \frac{x - x_j}{\|x - x_j\|}.$$

Compute the surface integral $\int_\Gamma F \cdot n \, ds$ for any closed surface Γ containing $k \leq N$ of the points $\{x_1, x_2, \ldots, x_N\}$, with n the exterior unit normal as usual.

68.10. Show, as you did in Chapter Newton's Nightmare, that the gravitation from a sphere is the same as if all the mass of the sphere was concentrated to its center, but now using Gauss' theorem to make things easier. Use as starting point that the divergence of the gravitational field is (proportional to) the density, i.e.

$$\nabla \cdot F = \rho/c$$

for some constant c, and assume spherical symmetry, i.e. the direction of the gravitational field is in the radial direction from the center of the sphere.

68.11. Show that if $-\Delta u = f$ in Ω, then for any function v that is zero on Γ, the boundary of Ω, one has

$$\int_\Omega \nabla u \cdot \nabla v = \int_\Omega fv.$$

Also prove that if $\partial_n u = g$ on Γ, then for all functions v,

$$\int_\Omega \nabla u \cdot \nabla v = \int_\Omega fv + \int_\Gamma gv \, ds.$$

69
Stokes' Theorem

I too feel that I have been thinking too much of late, but in a different way, my head running on divergent series, the discontinuity of arbitrary constants,... I often thought that you would do me good by keeping me from being too engrossed by those things. (Stokes asking Mary Susanna Robinson to marry him 1857)

69.1   Introduction

Stokes' theorem states that if $u : \mathbb{R}^3 \to \mathbb{R}^3$ is differentiable, then

$$\int_S (\nabla \times u) \cdot n \, ds = \int_\Gamma u \cdot ds \,, \qquad (69.1)$$

where S is a surface in $\mathbb{R}^3$ bounded by a closed curve Γ, n is a unit normal to S, and Γ is oriented in a clockwise direction following the positive direction of the normal n, see Fig. 69.1. The integral

$$\int_\Gamma u \cdot ds$$

is called the *circulation* of u around Γ. The integral

$$\int_S (\nabla \times u) \cdot n \, ds$$

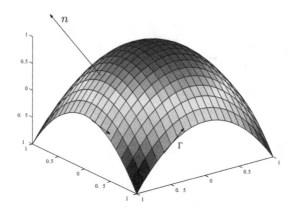

Fig. 69.1. A Stokes surface S with boundary curve Γ

Fig. 69.2. Stokes at age 22: "After taking my degree I continued to reside in College and took private pupils. I thought I would try my hand at original research..."

is the *total flow of the rotation* $\nabla \times u$ *across the surface* S. Stokes' theorem states that the total flow of $\nabla \times u$ across S is equal to the circulation of u around the boundary Γ of S.

Stokes (1819–1903), an Irish mathematician/physicist and professor in Cambridge 1849, gave basic contributions to the theory of viscous fluid flow modeled by the Navier-Stokes equations, see Fig. 69.2.

69.2   The Special Case of a Surface in a Plane

We start by considering the special case of a plane surface $\bar{S}$ in the plane $\{x \in \mathbb{R}^3 : x_3 = 0\}$ with normal $\bar{n} = (0, 0, 1)$ and with boundary Γ, see Fig. 69.3. In this case, Stokes' theorem takes the form

$$\int_{\bar{S}} (\nabla \times u) \cdot \bar{n} \, ds = \int_{\bar{S}} \left(\frac{\partial u_2}{\partial x_1} - \frac{\partial u_1}{\partial x_2} \right) dx_1 \, dx_2$$

$$= \int_{\Gamma} u \cdot ds = \int_{\Gamma} (u_2 n_1 - u_1 n_2) \, ds. \qquad (69.2)$$

By identifying the plane $\{x_3 = 0\}$ with $\mathbb{R}^2$, this is (67.10) and is a direct consequence of (67.3). This result is often referred to as *Green's formula in two dimensions*. We have thus proved Stokes' theorem in the case of a plane surface S in the plane $\{x_3 = 0\}$.

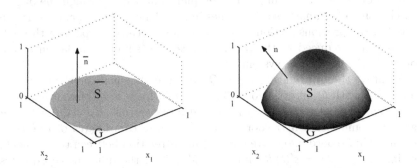

Fig. 69.3. Two special cases: A plane surface $\bar{S}$ with normal $\bar{n}$ and boundary curve Γ, and a curved surface S with normal n and a plane boundary curve Γ

Note that the unit tangent direction is given by $\tau = (-\tilde{n}_2, \tilde{n}_1)$, where $\tilde{n} = (\tilde{n}_1, \tilde{n}_2)$ is the outward normal direction to Γ in the plane $\{x : x_3 = 0\}$ with a counter clockwise orientation when viewed from the top of the normal $\bar{n} = (0, 0, 1)$ of $\bar{S}$. The orientation is consistent with the specification that τ should be oriented clockwise when following the direction of the normal to $\bar{S}$.

Example 69.1. Let $S = \{x \in \mathbb{R}^3 : \|x\| \le 1, x_3 = 0\}$ be the unit disc in the plane $\{x_3 = 0\}$ bounded by the curve Γ parameterized by $s(t) = (\cos(t), \sin(t), 0)$, $0 \le t \le 2\pi$. Choose $n = (0, 0, 1)$ and let $u(x) = (-x_2, x_1, 0)$ so that $\nabla \times u(x) = (0, 0, 2)$. We compute

$$\int_S (\nabla \times u)\, ds = 2\pi, \quad \int_\Gamma u \cdot ds = \int_0^{2\pi} (\cos^2(t) + \sin^2(t))dt = 2\pi,$$

in accordance with Stokes' theorem.

Example 69.2. Ampere's law states the $\nabla \times H = J$, where H is the magnetic field and J the electric current. Stokes' theorem states that the circulation of H around a closed curve Γ bounding a surface S is equal to the total current through the surface S. Stokes' theorem is thus one of the cornerstones of electromagnetic field theory.

69.3   Generalization to an Arbitrary Plane Surface

We shall now verify that both the left and right hand side of Stokes' equality

$$\int_S (\nabla \times u) \cdot n\, ds = \int_\Gamma u \cdot ds,$$

are invariant under orthogonal coordinate transformations. We thus obtain a proof of Stokes' theorem for a given plane surface S through the origin, by choosing coordinates so that S lies in the plane $\{x_3 = 0\}$, and using the proof of the previous section. The case of a surface S not passing through the origin is reduced to the previous case by a simple translation of the origin of the coordinate system.

To prove the invariance, let $x = Q\bar{x}$ be an orthogonal coordinate transformation with Q an orthogonal 3×3 matrix from a set of coordinates $\bar{x}$ to x. The dependent vector variable u also transforms as $u = Q\bar{u}$, where u are the components in x-coordinates and $\bar{u}$ the coordinates of the same quantity in $\bar{x}$-coordinates. We have a similar relation between the elements of integration $ds = s'(t)dt$ and $d\bar{s} = \bar{s}'(t)dt$ in the different coordinates since $s'(t) = Q\bar{s}'(t)$, that is $ds = Qd\bar{s}$. Therefore,

$$\int_\Gamma u \cdot ds = \int_\Gamma Q\bar{u} \cdot Qd\bar{s} = \int_\Gamma Q^\top Q\bar{u} \cdot d\bar{s} = \int_\Gamma \bar{u} \cdot d\bar{s},$$

and the invariance of the right hand side of (69.1) follows.

To prove the invariance of the left hand side of (69.1), we use the Chain rule to obtain the following relation between the gradient ∇ with respect to x and the gradient $\bar{\nabla}$ with respect to $\bar{x}$,

$$\nabla = Q\bar{\nabla}.$$

A direct computation shows that

$$(\nabla \times u) \cdot n = (Q\bar{\nabla} \times Q\bar{u}) \cdot Q\bar{n} = (\bar{\nabla} \times \bar{u}) \cdot \bar{n}, \qquad (69.3)$$

which proves the invariance since $d\bar{x} = dx$. Note that (69.3) is analogous to the the the relation

$$(Qa \times Qb) \cdot Qc = (a \times b) \cdot c$$

for $a, b, c \in \mathbb{R}^3$. This expresses the invariance of the volume spanned by three vectors a, b and c under orthogonal coordinate transformations.

69.4 Generalization to a Surface Bounded by a Plane Curve

Suppose that S is a surface bounded by a curve Γ contained in the plane $\{x_3 = 0\}$, see Fig. 69.3. We do not assume that S is contained in $\{x_3 = 0\}$. Let $\bar{S}$ be the surface in the plane $\{x_3 = 0\}$ with the boundary Γ and let Ω be the volume bounded by the surface S and the plane surface $\bar{S}$. Since $\nabla \cdot (\nabla \times u) = 0$, the Divergence theorem implies

$$0 = \int_\Omega \nabla \cdot (\nabla \times u) \, dx = \int_S \nabla \times u \cdot n \, ds + \int_{\bar{S}} \nabla \times u \cdot n \, ds, \qquad (69.4)$$

where n is the outward unit normal to the boundary $\partial\Omega$ of Ω. If n is a normal to S and $\bar{n} = -n$ is a normal to $\bar{S}$, then (69.4) implies

$$\int_S \nabla \times u \cdot n \, ds = \int_{\bar{S}} \nabla \times u \cdot \bar{n} \, ds.$$

Applying Stokes' theorem to $\bar{S}$, we obtain

$$\int_S \nabla \times u \cdot n \, ds = \int_{\bar{S}} \nabla \times u \cdot \bar{n} \, ds = \int_\Gamma u \cdot ds,$$

which proves Stokes' theorem for the surface S bounded by the plane curve Γ.

A proof of Stokes' theorem for the case of a general curve is outlined in Problem 69.1. We now summarize:

Theorem 69.1 (Stokes' theorem). *If S is a surface in $\mathbb{R}^3$ with unit normal n, and Γ is the boundary of S oriented clockwise following the direction of n, then*

$$\int_S (\nabla \times u) \cdot n \, ds = \int_\Gamma u \cdot ds.$$

We state the following important direct consequence of Stokes' theorem:

Theorem 69.2 *If $u : \Omega \to \mathbb{R}^3$ with Ω a domain in $\mathbb{R}^3$ is a differentiable vector field such that*

$$\int_\Gamma u \cdot ds = 0$$

for all closed curves Γ in Ω, then $\nabla \times u = 0$ in Ω.

Chapter 69 Problems

69.1. Prove Stokes theorem for a curve Γ given by $s(t) = (x_1(t), x_2(t), f(x_1(t), x_2(t)))$, $t \in [a, b]$, where $f : \mathbb{R}^2 \to \mathbb{R}$, bounding a surface $\Omega = \{x \in \mathbb{R}^3 : x_3 - f(x_1, x_2) = 0\}$ in $\mathbb{R}^3$. Hint: The projection of Γ on the x_1x_2-plane is the curve $\tilde{\Gamma}$ represented by $\tilde{s}(t) = (x_1(t), x_2(t), 0)$ which bounds the domain $\tilde{\Omega}$ in the x_1x_2-plane. Show that, writing $u_i = u_i(x_1, x_2, f(x_1, x_2))$,

$$\int_\Gamma u \cdot ds = \int_a^b \left(u_1 x_1' + u_2 x_2' + u_3 \left(\frac{\partial f}{\partial x_1} x_1' + \frac{\partial f}{\partial x_2} x_2' \right) \right) dt$$

$$= \int_a^b \left(\left(u_1 + u_3 \frac{\partial f}{\partial x_1} \right) x_1' + \left(u_2 + u_3 \frac{\partial f}{\partial x_2} \right) x_2' \right) dt$$

$$= \int_{\tilde{\Gamma}} \left(u_1 + u_3 \frac{\partial f}{\partial x_1}, u_2 + u_3 \frac{\partial f}{\partial x_2} \right) \cdot ds = I.$$

Then use the Stokes theorem for a plane curve established above, to show that

$$I = \int_{\tilde{\Omega}} \left(\frac{\partial}{\partial x_1} \left(u_2 + u_3 \frac{\partial f}{\partial x_2} \right) - \frac{\partial}{\partial x_2} \left(u_1 + u_3 \frac{\partial f}{\partial x_1} \right) \right) dx,$$

and prove by performing the differentiations and direct computation that

$$I = \int_\Omega (\nabla \times u) \cdot n \, ds,$$

where $n \, ds = (-\frac{\partial f}{\partial x_1}, -\frac{\partial f}{\partial x_2}, 1) \, dx$.

69.2. Give a proof of the equality $\int_\Omega \nabla \times u \, dx = \int_\Gamma n \times u \, ds$, where Ω is a subset of $\mathbb{R}^3$ with boundary Γ with outward unit normal n, by applying the divergence theorem to $u \times a$ with a an arbitrary constant vector.

69.3. Study the relation between Green's formula (67.9), in the form (67.10), and the divergence theorem applied to the two-dimensional domain S with boundary Γ:

$$\int_S \left(\frac{\partial v_1}{\partial x_1} + \frac{\partial v_2}{\partial x_2} \right) dx_1 \, dx_2 = \int_\Gamma (v_1 n_1 + v_1 n_2) \, ds,$$

with the identification $(u_1, u_2) = (-v_2, v_1)$ corresponding to counter clockwise rotation of the vector (v_1, v_2) by $\pi/2$. Explain how the clockwise direction in Stokes' theorem becomes a counter clockwise direction in (67.9).

69.4. Compute the integral

$$\int_\Gamma (x_1^2 x_2, -x_1^3)/\|x\|^4 \cdot ds,$$

where Γ is the curve in the $x_1 x_2$-plane from $(1,0)$ to $(0,1)$ defined by $(x_1(t), x_2(t)) = (\cos(t)^{15}, \sin(t)^{17})$, $0 \le t \le \pi/2$.

69.5. Compute the integral

$$\int_\Gamma \frac{1}{x_1^2 + x_2^2} \frac{(-x_2, x_1, x_3)}{\|x\|} \cdot ds,$$

where Γ is a curve traversing the unit circle in the $x_1 x_2$-plane five times counter-clockwise, then two times clockwise, and then again four times counterclockwise, as viewed from the positive x_3-axis.

69.6. Use Stokes' theorem to prove that

$$\int_\Gamma v\, ds = \int_S n \times \nabla v\, ds,$$

where S is a surface in $\mathbb{R}^3$ bounded by the closed curve Γ. Hint: Use Stokes' theorem with $u = va$ and a is a arbitrary vector in $\mathbb{R}^3$.

69.7. Verify by direct computation Stokes' theorem for (a) S the hemisphere $\{x \in \mathbb{R}^3 : \|x\| = 1, x_3 \ge 0\}$ and $u = (x_2, 2x_3, 3x_1)$, (b) $S = \{x \in \mathbb{R}^3 : x_3 = 1 - x_1^2 - x_2^2, x_3 \ge 0\}$.

69.8. (a) Let Ω be a domain in $\mathbb{R}^2$ with boundary Γ. Show that the area $A(\Omega)$ is given by the formula

$$A(\Omega) = \frac{1}{2} \int_\Gamma u \cdot ds,$$

where $u(x) = (-x_2, x_1)$ and Γ is oriented counter-clockwise. Use this result to show that the area bounded by the ellipse $x = (a\cos(t), b\sin(t))$, $0 \le t \le 2\pi$, with half-axes a and b, is equal to πab. (b) Try to design a mechanical instrument for measuring the area of a domain in $\mathbb{R}^2$ (planimeter).

70
Potential Fields

He is a rather tall, lanky-looking man, with moustache and beard about to turn grey with a somewhat harsh voice and rather deaf. He was unwashed, with his cup of coffee and cigar. One of his failings is forgetting time, he pulls his watch out, finds it past three, and runs out without even finishing the sentence. (Thomas Hirst about Dirichlet 1850)

70.1 Introduction

We know from Chapter Curve integrals that potential force fields play an important role in mechanics. Let $u : \Omega \to \mathbb{R}^3$ be a given vector function, where Ω is a domain in $\mathbb{R}^3$. How can we check if $u(x)$ is a *potential field*, that is, if there is a scalar function or scalar *potential*, φ such that

$$u(x) = \nabla\varphi(x) \quad \text{for } x \in \Omega? \tag{70.1}$$

We recall that if $u = \nabla\varphi$ is a potential field and Γ is a curve parameterized by $s : [0, 1] \to \mathbb{R}^3$ from $a = s(0)$ to $b = s(1)$, then the work of u along Γ is given by

$$\int_\Gamma u \cdot ds = \int_0^1 \nabla\varphi(s(t)) \cdot s'(t) \, dt = \int_0^1 \frac{d\varphi(s(t))}{dt} \, dt = \varphi(b) - \varphi(a).$$

In particular, the work is the same along all curves from a to b, and if the curve is closed with $\varphi(1) = \varphi(0)$ then the work performed when moving around the curve is zero. A field with the property that the work along a closed curve is zero is referred to as a *conservative field*. A potential field is thus a conservative field.

A basic example of a gradient field is the gravitational field of a mass m at the origin,

$$u(x) = -m\frac{x}{\|x\|^3} = \nabla\left(\frac{m}{\|x\|}\right),$$

normalizing units so the gravitational constant is one. The electrical field of a charge m at the origin has the same form. In that case, the potential $\varphi(x) = m/\|x\|$ represents *potential energy* (gravitational or electrical), and curve integrals $\int_\Gamma u \cdot ds = \varphi(b) - \varphi(a)$ represents the work performed by a unit mass or charge when moved from a to b along Γ.

70.2   An Irrotational Field Is a Potential Field

We saw earlier that a potential field $u = \nabla\varphi$ is *irrotational*, that is $\nabla \times u = \nabla \times (\nabla\varphi) = 0$. This follows from a direct computation using $\frac{\partial^2\varphi}{\partial x_i\partial x_j} = \frac{\partial^2\varphi}{\partial x_j\partial x_i}$. In other words, $\nabla \times u = 0$ in Ω is a *necessary condition* for u to be a potential field in Ω.

We shall now prove that, the condition $\nabla \times u = 0$ in Ω is a *sufficient condition* for u to be a potential field in Ω, under the assumption that Ω is convex. We recall that Ω is *convex* if for any two points x and $\bar{x}$ in Ω, the entire line segment $\bar{x} + t(x - \bar{x}), 0 \le t \le 1$ between $\bar{x}$ and x, is also in Ω, see Fig. 70.1. Convexity implies in particular that Ω has "no holes". We thus conclude that u is a potential field in a convex domain Ω if and only u is irrotational in Ω. In other words, $u = \nabla\varphi$ in Ω for some potential φ if and only if $\nabla \times u = 0$ in Ω.

Fig. 70.1. One convex and two non-convex domains

We carry out the proof by *constructing* a potential φ such that $\nabla\varphi = u$ for a given irrotational field $u(x)$ in the convex domain Ω. For the construction, we choose a fixed point $\bar{x}$ in Ω. For each point x, we let Γ_x be a curve in Ω connecting $\bar{x}$ to x and we define

$$\varphi(x) = \int_{\Gamma_x} u \cdot ds. \tag{70.2}$$

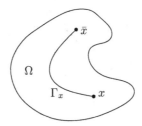

Fig. 70.2. A curve Γ_x in Ω joining $\bar{x}$ and x

We first prove that $\varphi(x)$ is independent of the choice of curve Γ_x from $\bar{x}$ to x. Assume that Γ_x and $\tilde{\Gamma}_x$ are two curves from $\bar{x}$ to x. Together they form a closed curve Γ bounding a surface S so Stokes theorem implies

$$\int_\Gamma u \cdot ds = \pm \int_S (\nabla \times u) \cdot n \, ds = 0,$$

since $\nabla \times u = 0$ on S. Now

$$\int_\Gamma u \cdot ds = \int_{\Gamma_x} u \cdot ds - \int_{\tilde{\Gamma}_x} u \cdot ds$$

if we orient Γ in the same direction as Γ_x and thus in the opposite direction as $\tilde{\Gamma}_x$. We conclude that

$$\int_{\tilde{\Gamma}_x} u \cdot ds = \int_{\Gamma_x} u \cdot ds,$$

and the independence of the choice of curve connecting $\bar{x}$ with x follows.

Next, we prove that the function $\varphi(x)$ defined by (70.2) satisfies $\nabla\varphi(x) = u(x)$ for $x \in \Omega$. We do this by choosing a curve Γ_x to connect to x along the x_1-axis, the x_2-axis, or the x_3-axis. Letting Γ_x connect along the x_1-axis according to Fig. 70.3, for $\hat{x}$ close to x we have

$$\varphi(x) - \varphi(\hat{x}) = \int_{\hat{x}_1}^{x_1} u_1(t, x_2, x_3) \, dt,$$

and the Fundamental Theorem implies

$$\frac{\partial\varphi}{\partial x_1}(x) = u_1(x).$$

Similarly, we obtain $\frac{\partial\varphi}{\partial x_i}(x) = u_i(x)$ for $i = 2, 3$. We summarize:

Theorem 70.1 *If $u : \Omega \to \mathbb{R}^d$, with Ω being a convex domain in $\mathbb{R}^d$ for $d = 2, 3$, satisfies $\nabla \times u(x) = 0$ for all $x \in \Omega$, then there is a function $\varphi : \Omega \to \mathbb{R}$ such that $u(x) = \nabla\varphi(x)$ for $x \in \Omega$.*

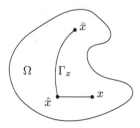

Fig. 70.3. A curve Γ_x in Ω connecting to x along the x_1-axis

70.3 A Counter-Example for a Non-Convex Ω

Consider the function $u : \Omega \to \mathbb{R}^2$, defined by $u(x) = (-x_2, x_1)/\|x\|^2$ with $\Omega = \{x \in \mathbb{R}^2 : x \neq 0\}$. This function satisfies

$$\nabla \times u(x) = \frac{\partial u_2}{\partial x_1} - \frac{\partial u_1}{\partial x_2} = \frac{-2x_1 x_2}{\|x\|^4} - \frac{-2x_1 x_2}{\|x\|^4} = 0 \quad \text{for } x \in \Omega.$$

Nevertheless, $u(x)$ cannot be written in the form $u(x) = \nabla\varphi(x)$ for $x \in \Omega$. This follows by noting that if, for example, Γ is the closed circle given by $s(t) = r(\cos(t), \sin(t))$, $0 \leq t < 2\pi$, then

$$\int_\Gamma u \cdot ds = \int_0^{2\pi} \frac{1}{r^2} r^2 \, dt = 2\pi,$$

while if $u(x) = \nabla\varphi(x)$, $\int_\Gamma u \cdot ds = 0$ since Γ is closed. The reason is that in this case Ω *is not convex*. The point $x = 0$ does not belong to Ω and thus Ω has a "hole". We cannot extend Ω to include $x = 0$ since the function $u(x)$ is singular at $x = 0$ and in particular not Lipschitz continuous at $x = 0$.

Chapter 70 Problems

70.1.   If possible, find a potential φ for (a) $u(x) = (x_1, x_2, x_3)$ (b) $u(x) = (x_3, x_1, x_2)$ (c) $u(x) = (x_2^2 - x_3, 2x_1 x_2, 3x_3^2 - x_1)$.

70.2.   We recall from above that $\nabla \times u = 0$ if and only if $u = \nabla\varphi$ for some φ.

We now ask the question if $\nabla \cdot u = 0$ if and only if $u = \nabla \times \psi$ for some (vector) potential ψ. Recall that we already know that the "if-part" of this is true, namely that $\nabla \cdot u = 0$ if $u = \nabla \times \psi$ for some ψ for some ψ.

It turns out that also the "only if-part" is true, that is, if $\nabla \cdot u = 0$ we may construct a (vector) potential ψ such that $u = \nabla \times |psi$. Verify this, using the construction $\psi(x) = \int_0^1 u(tx) \times tx \, dt$, and assuming $\nabla \cdot u$ in all of $\mathbb{R}^3$ for simplicity.

70.3.   Extend the above counterexample to the function $u : \mathbb{R}^3 \to \mathbb{R}^3$ given by $u(x) = (-x_2, x_1, 0)/\|x\|^2$ representing the magnetic field around a current along the x_3-axis.

71

Center of Mass and Archimedes' Principle*

The simplest schoolboy is now familiar with facts for which Archimedes would have sacrificed his life. (Ernest Renan)

71.1 Introduction

We now turn to a study of the stability of floating bodies, including the question of how to design a big ship or a sailing boat so that it does not tip over. An example of an unfortunate design is given by the warship Vasa, which tipped over on its maiden voyage on August 10, 1628 in the harbor of Stockholm and sank along with 50 of the crew of 150 people. In the resulting trial, it was decided that the ship was "well built, but badly proportioned" and no-one was held guilty for the disaster. The ship can now be studied at the Vasa museum in Stockholm.

Evidently, the stability properties of Vasa came as a surprise. Vasa had a new design with two gundecks with heavy artillery instead of one and the planned ballast of stone was not sufficient as a counterbalance. The old rules of ship design apparently did not apply to the new design and Calculus and scientific computing at that time was too primitive for trustworthy predictions.

Let's see what we can do today with a little bit of Calculus. We start with the concept of *center of mass*, pass on to *Archimedes principle* and the question of stability of floating bodies.

Fig. 71.1. Vasa goes unstable on August 10, 1628

71.2　Center of Mass

Consider a body B occupying the volume V in $\mathbb{R}^3$. Suppose the *density* of
the body at x is given by $\rho(x)$. The total mass $m(B)$ of the body is

$$m(B) = \int_V \rho(x)\, dx.$$

The *center of mass* $\bar{x} = (\bar{x}_1, \bar{x}_2, \bar{x}_3) \in \mathbb{R}^3$ of the body B is defined by

$$\bar{x}_i \int_V \rho(x)\, dx = \int_V x_i \rho(x)\, dx\,, \quad i = 1, 2, 3,$$

that is

$$\bar{x}_i = \frac{\int_V x_i \rho(x)\, dx}{\int_V \rho(x)\, dx}\,, \quad i = 1, 2, 3.$$

In vector form, this is

$$\bar{x} = \frac{\int_V x \rho(x)\, dx}{\int_V \rho(x)\, dx}.$$

We now explain the relevance of the concept of center of mass using the
concept of *torque*. Assume the body B is acted upon by a vertical gravity

force field $-e_3$ of unit strength with the coordinate direction e_3 oriented vertically upward. The torque about a point $\bar{x}$ of a force F acting at x is equal to

$$(x - \bar{x}) \times F = -F \times (x - \bar{x}),$$

see Fig. 71.2. In other words, the torque is a vector that is perpendicular to the plane generated by the direction of the force F and the lever arm $x - \bar{x}$, with modulus equal to the modulus of F times the distance of the point $\bar{x}$ from the line of action of F.

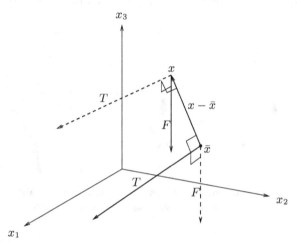

Fig. 71.2. The torque $T = (x - \bar{x}) \times F$ about the point $\bar{x}$ of F acting at x

The torque of the gravity field (assuming the acceleration of gravity $g = 1$) acting on an element of mass $\rho(x)\, dx$ at position x about a given point $\bar{x}$ is equal to

$$\rho(x)\, dx\, e_3 \times (x - \bar{x}).$$

The total torque T of the gravity field $-e_3$ on the body B about $\bar{x}$ is thus equal to

$$T = e_3 \times \int_V \rho(x)(x - \bar{x})\, dx = 0,$$

by the definition of the center of mass $\bar{x}$. The torque about $\bar{x}$ thus vanishes which means that body will balance if supported at $\bar{x}$, see Fig. 71.3.

More precisely,

$$T = e_3 \times \left(\int_V \rho(x)x\, dx - \bar{x} \int_V \rho(x)\, dx \right) = 0 \qquad (71.1)$$

if and only if

$$\bar{x}_i = \frac{\int_V x_i \rho(x)\, dx}{\int_V \rho(x)\, dx},$$

for $i = 1, 2$. This means that the body will balance if supported at a point $x = (x_1, x_2, x_3)$ with $x_1 = \bar{x}_1$ and $x_2 = \bar{x}_2$, while x_3 may be chosen arbitrarily, see Fig. 71.3. Thus, if the body is supported at its center of mass $\bar{x}$ then it will balance independently of its orientation. If the body is supported at a point x different from the center of mass $\bar{x}$, then it will balance only if $\bar{x} - x$ is parallel to the direction of the gravity field.

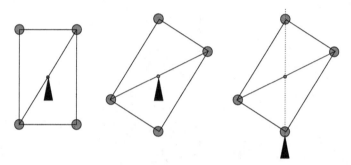

Fig. 71.3. A body supported at its center of mass, in two stable positions, and a body supported at a boundary point, balanced but unstable

Example 71.1. We compute the center of mass $\bar{x}$ of a thin triangular plate of uniform thickness occupying the region $\Omega = \{x \in \mathbb{R}^2 : 0 \le x_1, x_2, x_1 + x_2 \le 1\}$ in the plane. We get

$$\bar{x}_i = \frac{\int_\Omega x_i\, dx}{\int_\Omega dx} = \frac{1/6}{1/2} = \frac{1}{3}.$$

Example 71.2. We compute the center of mass of the half-ball $\Omega = \{x \in \mathbb{R}^3 : \|x\| \le 1, x_3 \ge 0\}$. By symmetry $\bar{x}_1 = \bar{x}_2 = 0$. For $\bar{x}_3$ we get using spherical coordinates

$$\int_\Omega x_3\, dx = \int_0^{2\pi} \int_0^{\pi/2} \int_0^1 r\cos(\varphi)\, r^2 \sin(\varphi)\, drd\varphi d\theta$$

$$= \int_0^{2\pi} \int_0^{\pi/2} \frac{1}{2}\sin(2\varphi) \left[\frac{1}{4} r^4\right]_0^1 d\varphi d\theta$$

$$= \frac{1}{4} \int_0^{2\pi} \left[\frac{1}{4}\cos(2\varphi)\right]_0^{\pi/2} d\theta = \frac{1}{4}\int_0^{2\pi} d\theta = \pi/4,$$

that is, $\bar{x}_3 = \int_\Omega x_3\, dx / \int_\Omega dx = \frac{\pi/4}{2\pi/3} = \frac{3}{4}$.

71.3 Archimedes' Principle

Archimedes principle states that (i) the *buoyancy force* acting on a body B totally immersed in a liquid is equal to the weight of the displaced liquid and (ii) acts along a vertical line through the center of mass of the displaced fluid, which we refer to as the *center of bouyancy c_b*. We shall now prove this fact using vector Calculus. The force from the fluid acting on an element $ds = ds(x)$ of the surface S of the body B at position x is equal to $-p(x)n(x)\,ds$, where $p(x)$ is the pressure of the liquid and $n(x)$ is the outward (from B) unit normal to S at x. The total pressure force on B is thus

$$F = - \int_S p(x)n(x)\,ds(x) = - \int_S pn\,ds.$$

Since

$$\int_V \frac{\partial p}{\partial x_i}\,dx = \int_S pn_i\,ds, \quad i = 1, 2, 3,$$

where V is the volume occupied by B, we have

$$F = - \int_V \nabla p(x)\,dx$$

The *pressure $p(x)$* in a fluid at rest, called the *hydro-static pressure*, is given by

$$p(x) = \rho z(x) + p_0,$$

where $z(x)$ is the depth, ρ is the constant density of the fluid and p_0 is the pressure on the surface of the fluid, see Fig. 71.4.

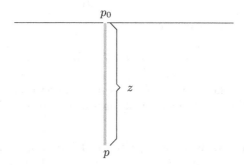

Fig. 71.4. Hydrostatic pressure $p(x) = \rho z(x) + p_0$

The pressure force at a point x is equal in all directions and its modulus $p(x)$ equal is to the weight $\rho z(x)$ of the column of fluid above the point x

plus the surface pressure p_0 from the atmosphere. We conclude that

$$\nabla p(x) = -\rho e_3,$$

where we assume that the coordinate direction e_3 is vertical and pointed upwards. Therefore,

$$F = \int_V \rho \, dx \, e_3 \equiv W e_3,$$

where $W = \int_V \rho \, dx$ is the total weight of the displaced fluid. This proves the first part of Archimedes principle.

Next, the total torque T from the fluid pressure forces on S about a point $\bar{x}$ is given by

$$T = \int_S (x - \bar{x}) \times (-p(x)n(x)) \, ds(x) = \int_S n(x) \times p(x)(x - \bar{x}) \, ds.$$

Recalling that

$$\int_S n \times F \, ds = \int_V \nabla \times F \, dx,$$

we find that

$$T = \int_V \nabla \times (p(x)(x - \bar{x})) \, dx.$$

But,

$$\nabla \times (p(x)(x - \bar{x})) = \nabla p \times (x - \bar{x}) + p \nabla \times (x - \bar{x}).$$

Since $\nabla \times (x - \bar{x}) = 0$, it follows that

$$T = \int_V \nabla p \times (x - \bar{x}) \, dx = -\int_V \rho(x - \bar{x}) \, dx \times e_3$$

and the torque T vanishes if $\bar{x}$ satisfies

$$\bar{x}_i \int_V \rho \, dx = \int_V x_i \rho \, dx \quad \text{for } i = 1, 2.$$

We conclude that the buoyancy force is vertical upward and is acting along a vertical line through the center of mass of the displaced fluid. We have now proved:

Theorem 71.1 (Archimedes' principle) *The buoyancy force acting on a body immersed in a liquid is equal to the weight of the displaced liquid and acts along a vertical line through the center of mass of the displaced fluid.*

We can directly extend Archimedes' principle to a partially immersed body assuming the pressure on the surface of the fluid to be is zero.

71.4   Stability of Floating Bodies

The stability of a floating body B is of central importance in all forms of boating, from canoes to big ships. The question of stability can be reduced to a question of the relative position of (i) the center of mass c_m of the body B and (ii) the center of buoyancy c_b of B according to the following discussion. Consider the body in rest position with the gravity force acting vertically downward from the center of gravity, and the buoyancy force acting vertically upward from the center of buoyancy. We assume the body is in equilibrium with the gravity force and the buoyancy force balancing and acting along the vertical line through the centers of gravity and buoyancy, see Fig. 71.5.

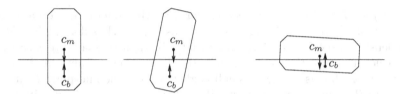

Fig. 71.5. Floating bodies with centers of gravity and buoyancy

Assume that the body is tilted a small angle so that the centers of gravity and buoyancy are displaced horizontally, see Fig. 71.5. Let T be the resulting torque from the pair of gravity and buoyancy forces. The sign of T will govern the stability! If T acts in the same direction as the tilting, then the tendency of tilting will be enforced and the body will depart from its equilibrium position and eventually tilt over, see Fig. 71.5. This happens if the center of gravity is displaced horizontally in the direction of tilting more quickly than the center of buoyancy. Conversely, if the center of gravity is displaced more slowly, then the resulting torque T will be negative and act as a restoring force seeking to bring back the body to the rest position, see Fig. 71.5. We now consider two examples with simple geometry.

Example 71.3. Consider a space capsule with hemispherical base and conical top floating in the Pacific and waiting to be recovered. Will the capsule float upright or not? Assuming the capsule is floating upright with a part of the hemispherical base immersed into the water, see Fig. 71.6.

The resultant of the buoyancy forces is directed upward and acts through the center C of the hemisphere, see Fig. 71.6. If the capsule is tilted a little, the resultant of the buoyancy forces is still directed upwards through C and the torque from the gravity force will be de-stabilizing if the center of mass c_m of the capsule is positioned above C, and stabilizing if c_m is below C, for c_m on the symmetry axis of the capsule, see Fig. 71.6.

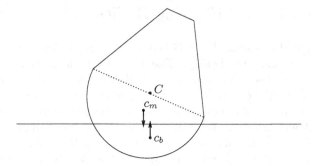

Fig. 71.6. Space capsule floating upright

Example 71.4. Consider a rectangular box with square horizontal cross section of width $2w$ and height $2h$ and density $\bar{\rho}$ which is floating in a fluid of density ρ, see Fig. 71.7. Suppose that $\bar{\rho}$ is small compared to ρ so that it penetrates into the fluid only slightly. To test the stability of the box, suppose the box is rotated a small angle θ around the mid-point C at the bottom. The de-stabilizing torque about C resulting from the gravity force through the center of gravity is equal to $g\bar{\rho}(2w)^2 2hh\sin(\theta)$, see Fig. 71.7. The stabilizing torque from the change of buoyancy forces caused by the rotation is equal to

$$2\frac{2}{3}www\sin\theta\frac{1}{2}\rho gw,$$

because the area of the triangle CAB is equal to $ww\sin\theta\frac{1}{2}$, and the center of gravity of CAB is at horizontal distance $2\frac{2}{3}w$ from C. The position is stable if

$$2\frac{2}{3}w^4\sin\theta\frac{1}{2}\rho g > g\bar{\rho}8w^2h^2\sin(\theta),$$

that is, if

$$w^2\rho > 12h^2\bar{\rho}.$$

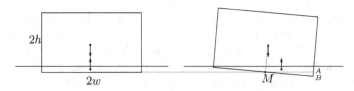

Fig. 71.7. Floating box

Chapter 71 Problems

71.1. The density of ice is 0.917 times the density of water (at $-4°C$). How large a part of an iceberg is visible above the water surface?

71.2. How does a log float? Why does it not want to float in an upright position?

71.3. Understand why catamaran ships have good stability properties.

71.4. Find the stable floating position of a "log" with a quadratic cross-section and density $\bar{\rho} = \frac{1}{2}\rho$. Find the stable positions as a function of the ratio $\bar{\rho}/\rho$ (we know from above that for $\bar{\rho}/\rho$ sufficiently small it will float as the box in Fig. 71.7). May there be more than one stable position (disregarding symmetric ones). Discuss! May the conclusion depend on the shape of the cross-section?

71.5. How does a (perfect) ice cube float? How does a barrel (cylinder) float, given hight/diameter ratio and density?

71.6. Study the design of sailing boats from the stability point of view. Study in particular modern designs with good form stability (wide and flat bottom), and classical designs with a narrow deep hull. Connect to the discussion above.

71.7. Extend Archimedes principle to a body immersed into a system of two layers of different fluids on top of each other.

72

Newton's Nightmare*

God does not care about mathematical difficulties. He integrates
empirically. (Einstein)

Newton's theory of gravitation states that the gravitational force
field $F(x)$ generated by a point mass m at the origin is the potential field

$$F(x) = -m\frac{x}{\|x\|^3} = \nabla\left(\frac{m}{\|x\|}\right), \qquad (72.1)$$

corresponding to the potential $\varphi(x) = m/\|x\|$, in units where the gravitational constant is one. This means the gravitational force from the mass m at the origin on a unit point mass at position x is equal to $F(x)$. Taking norms gives

$$\|F(x)\| = \frac{m}{\|x\|^2},$$

which is known as *Newton's Inverse Square Law* . More generally, the gravitational force field of a mass m a position y is given by

$$F(x) = -m\frac{x-y}{\|x-y\|^3}, \qquad (72.2)$$

with $F(x)$ being the force on a unit point mass a position x and the corresponding potential $\varphi(x) = m/\|x-y\|$.

Over a long period, Newton tried to show one consequence of his new theory of gravitation: the gravitational force between two solid balls is the

Fig. 72.1. Isaac Newton 1689: "I have not been able to discover the cause of those properties of gravity from phenomena, and I frame no hypotheses; for whatever is not deduced from the phenomena is to be called a hypothesis, and hypotheses, whether metaphysical or physical, whether of occult qualities or mechanical, have no place in experimental philosophy"

same as if the total mass of each ball was concentrated at the center of mass of each ball. This result has important practical implications. For example, it would allow the modeling of the solar system as 9 small point masses representing the planets orbiting around one fixed big point mass representing the Sun, that is as a 9-body system. Without the simplifying basic result, we would have to take into account the gravitational attraction between the parts of each of the bodies and we would end up with a very complicated model. The practicality of Newton's gravitational theory could easily be questioned by anyone having some interest in that direction, like the Church. Lacking this basic result, Newton delayed the publication of his monumental *Principia Mathematica* many years. Newton states that he purposely made Principia difficult to read "to avoid being bated by little smatterers of mathematics". Newton did not like critics.

It fact, even a 9-body system of point masses may be far beyond comprehension or mathematical analysis. Luckily, the solar system is a very special 9-body system in which the motion of each planet can be viewed to good approximation as a 1-body system, i.e. as each planet orbiting undisturbed around one heavy Sun. Such 1-body systems have a full analytical solution available, as we saw in the Chapter *Lagrange and the Principle of Least Action*.

The basic result that Newton finally succeeded in proving can be phrased as follows: Consider a thin spherical shell S of radius r and uniform thickness centered at the origin and assume the total mass of the shell is m. Let $F(x)$ be the gravitational force field generated by the spherical shell so that $F(x)$ is the gravitational force of the shell on a unit point mass at position x outside the sphere. Newton proved that

$$F(x) = -m \frac{x}{\|x\|^3} \quad \text{for } \|x\| > r,$$

which says that the gravitational field generated by the sphere on a point outside the sphere is the same as the field generated by a point mass m at the center of the sphere.

The gravitational field $F(x)$ of the shell/sphere is the sum of the gravitational fields of all the little pieces $ds(y)$ of the surface of mass $dm(y)$ at position y making up the sphere S, that is

$$F(x) = \int_S f(y)ds(y),$$

where

$$f(y)ds(y) = -dm(y)\frac{x - y}{\|x - y\|^3}$$

is the gravitational field of the piece of surface $ds(y)$ of mass $dm(y)$ at position y. We note that

$$dm(y) = \frac{mds(y)}{4\pi r^2},$$

since the area of the sphere is $4\pi r^2$ and the total mass is m, and thus

$$f(y) = -\frac{m}{4\pi r^2}\frac{x - y}{\|x - y\|^3}. \tag{72.3}$$

Newton thus wanted to verify that

$$\int_S f(y)\,ds(y) = -m\frac{x}{\|x\|^3} \quad \text{for } \|x\| > r, \tag{72.4}$$

where $f(y)$ is given by (72.3). Once this basic result for a sphere is established, the corresponding result for a solid ball follows by simply viewing the ball as the union of a collection of thin spheres of varying radii. The desired final result for two solid balls follows similarly.

We now prove (72.4) giving the gravitational field of a thin spherical shell S of radius r and total mass m centered at the origin. We assume that $x = (R, 0, 0)$ with $R > r$. By symmetry, this covers the general situation. We note that the components $F_2(x)$ and $F_3(x)$ of the gravitational force vanish because the gravitational force is directed from $(R, 0, 0)$ towards the origin, and we have simply to verify that

$$F_1(x) = -\frac{m}{4\pi r^2}\int_S \frac{R - y_1}{\|x - y\|^3}ds(y) = -\frac{m}{R^2}.$$

To compute the surface integral, we use spherical coordinates

$$y = (r\cos(\varphi), r\sin(\varphi)\cos(\theta), r\sin(\varphi)\sin(\theta))$$

with $0 \le \varphi \le \pi$ and $0 \le \theta \le 2\pi$, see Fig. 72.2, and recall from the Chapter Surface integrals that

$$ds(y) = r^2 \sin(\varphi) d\varphi \, d\theta.$$

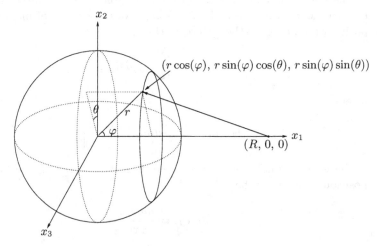

Fig. 72.2. Newtons nightmare

We have according to Fig. 72.2,

$$F_1(x) = -\frac{m}{4\pi r^2} \int_S \frac{R - y_1}{\|x - y\|^3} ds(y)$$

$$= -\frac{m}{4\pi} \int_0^\pi \int_0^{2\pi} \frac{(R - r\cos(\varphi))\sin(\varphi)}{((R - r\cos(\varphi))^2 + (r\sin(\varphi))^2)^{3/2}} d\theta d\varphi$$

$$= -\frac{m}{2} \int_0^\pi \frac{(R - r\cos(\varphi))\sin(\varphi)}{((R - r\cos(\varphi))^2 + (r\sin(\varphi))^2)^{3/2}} d\varphi$$

where we performed the integration with respect to θ using the fact that the integrand is independent of θ. We thus need to verify that

$$I = \int_0^\pi \frac{(R - r\cos(\varphi))\sin(\varphi) \, d\varphi}{((R - r\cos(\varphi))^2 + (r\sin(\varphi))^2)^{3/2}} = \frac{2}{R^2}. \qquad (72.5)$$

To this end, we change variables to set $t = \cos(\varphi)$ and we use $dt = -\sin(\varphi)d\varphi$ to get

$$I = \int_{-1}^1 \frac{(R - rt) \, dt}{(R^2 + r^2 - 2Rrt)^{3/2}} = \frac{1}{R^2} \int_{-1}^1 \frac{(1 - at) \, dt}{(1 + a^2 - 2at)^{3/2}},$$

where $a = \frac{r}{R} < 1$. By a routine computation, we find that $a < 1$ implies

$$\int_{-1}^{1} \frac{(1-at)\,dt}{(1+a^2-2at)^{3/2}}$$

$$= \int_{-1}^{1} \frac{(\frac{1+a^2}{2} - at)\,dt}{(1+a^2-2at)^{3/2}} - \int_{-1}^{1} \frac{(\frac{1+a^2}{2} - 1)\,dt}{(1+a^2-2at)^{3/2}}$$

$$= \frac{1}{2a}\left[-(1+a^2-2at)^{1/2}\right]_{-1}^{1} - \frac{a^2-1}{2a}\left[(1+a^2-2at)^{-1/2}\right]_{-1}^{1}$$

$$= \frac{1}{2a}(1+a-(1-a)) - \frac{a^2-1}{2a}\left(\frac{1}{1-a} - \frac{1}{1+a}\right) = 1+1 = 2,$$

and the desired result follows:

$$F_1(x) = -\frac{m}{R^2} \quad \text{if } x = (0,0,R), \quad R > r.$$

Below, we give a much shorter proof of this result using some tools of Calculus to be developed in the next chapters.

Chapter 72 Problems

72.1. Prove that the gravitational field from a thin sphere is equal to zero *inside* the sphere.

72.2. Compute the gravitational field $F(x)$ for $x \in \mathbb{R}^3$ of a solid ball of total mass m and radius r centered at the origin

72.3. Compute the gravitational field of a "black hole" with mass density $\frac{exp(-r)}{r}$, $r = \|x\|$.

72.4. Determine the gravitational field generated by a thin straight uniform rod.

72.5. Determine the gravitational field generated by a thin circular flat (a) ring (b) disc.

72.6. (a) Consider a particle cloud of uniform density in the form of a ball. Assume the particles attract each other according to Newton's Law of gravitation. Compute the evolution of the cloud for $t > 0$ assuming the particles are at rest at $t = 0$. (b) Do the same with a cloud in the form of the volume between two concentric spheres. (c) Extend to clouds of variable density.

73

Laplacian Models

... on aura donc $\Delta u = 0$; cette équation remarquable nous sera de la plus grande utilité ... (Laplace)

If one has to stick to this damned quantum jumping, then I regret ever having been involved in this thing. I don't like it (quantum mechanics), and I'm sorry I ever had anything to do with it. (Schrödinger)

73.1 Introduction

In this chapter, we present some basic models involving the Laplacian, including models for heat conduction, elasticity, electromagnetics, fluid mechanics, and gravitation. In deriving these models, we make use of the basics of Calculus in several dimensions including Gauss' and Stokes' theorems, and we get a quick and easy introduction to some of mysteries of the mechanics and physics of "continuous media". We also make connections to linear algebra when discretizing the Laplacian using the 5-point scheme and variants of "Svensson's formula".

73.2 Heat Conduction

We first model *heat conduction* in a heat-conducting material occupying the volume Ω in $\mathbb{R}^3$ with boundary Γ, over a time interval $I = [0, T]$. We let $u(x,t)$ denote the *temperature* and $q(x,t)$ the *heat flux* at the point x at time t. The heat flux is a vector $q = (q_1, q_2, q_3)$, where q_i is the heat flux, or rate of heat flowing in the direction x_i. We let $f(x,t)$ denote the rate of heat (per unit of volume) supplied at (x,t) by a *heat source*.

We derive the model using a basic *conservation law* expressing *conservation of heat* in the following form: for any fixed domain V in Ω with boundary S, the rate of the total heat introduced in V by the external source is equal to the rate of the total heat accumulated in V plus the total heat flux through S. This is based on the conviction that the heat introduced in V by the external source can choose between two options only: (i) flow out of V or (ii) be accumulated in V. With S denoting the boundary of V and n denoting the outward unit normal to S, see Fig. 73.1, the conservation law can be expressed as

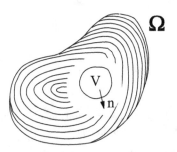

Fig. 73.1. An arbitrary subset V of a heat conducting body Ω

$$\int_V f \, dx = \frac{\partial}{\partial t} \int_V \lambda u \, dx + \int_S q \cdot n \, ds, \qquad (73.1)$$

where $\lambda(x,t)$ is the *heat capacity coefficient* giving the amount heat per unit of volume needed to raise the temperature one unit, and all functions are evaluated at a specific time $t \in I$. By the Divergence theorem,

$$\int_S q \cdot n \, ds = \int_V \nabla \cdot q \, dx,$$

and combined with (73.1), this implies that

$$\int_V \left(\frac{\partial}{\partial t}(\lambda u) + \nabla \cdot q \right) dx = \int_V f \, dx,$$

where the time derivative could be moved under the integral sign because V does not depend on time t. Since V is arbitrary, assuming the integrands are Lipschitz continuous, it follows that

$$\frac{\partial}{\partial t}(\lambda u)(x,t) + \nabla \cdot q(x,t) = f(x,t) \quad \text{for all } x \in \Omega, \, 0 < t \le T, \qquad (73.2)$$

which is a differential equation describing *conservation of heat* involving two unknowns: the temperature $u(x, t)$ and the heat flux $q(x, t)$. We thus have one equation and two unknowns and we need yet another equation.

The second equation is a *constitutive equation* that couples the heat flux q to the temperature gradient ∇u. *Fourier's law* states that heat flows from warm to cold regions with the heat flux proportional to the temperature gradient:

$$q(x, t) = -a(x, t) \nabla u(x, t) \quad \text{for } x \in \Omega, \, 0 < t \leq T \qquad (73.3)$$

where the factor of proportionality $a(x, t)$ is the coefficient of heat conductivity. Note the minus sign indicating that the heat flows from warm to cold regions, and that the heat conductivity $a(x, t)$ is positive. Combining (73.2) and (73.3), we obtain the basic differential equation describing heat conduction:

$$\frac{\partial}{\partial t}(\lambda u) - \nabla \cdot (a \nabla u) = f \quad \text{in } \Omega \times (0, T], \qquad (73.4)$$

where $a(x, t)$ and $\lambda(x, t)$ are given positive coefficients depending on (x, t) and $f(x, t)$ is a given heat source, and the unknown $u(x, t)$ represents the temperature.

To define the solution uniquely, the differential equation is complemented by initial and boundary conditions. The complete model with *Dirichlet boundary conditions* reads

$$\begin{cases} \frac{\partial}{\partial t}(\lambda u) - \nabla \cdot (a \nabla u) = f & \text{in } \Omega \times (0, T], \\ u = u_b & \text{on } \Gamma \times (0, T], \\ u(x, 0) = u_0(x) & \text{for } x \in \Omega, \end{cases} \qquad (73.5)$$

where u_0 is the initial temperature and u_b is the boundary temperature. The Dirichlet boundary condition corresponds to immersing the body Ω in a large reservoir with a specified temperature u_b and assuming that the boundary acts as a perfect thermal conductor so that the temperature of the body on the boundary is equal to the specified outside reservoir temperature u_b. Note that the given boundary temperature $u_b = u_b(x, t)$ may vary with (x, t).

Other commonly encountered boundary conditions are *Neumann* and *Robin* boundary conditions. A Neumann boundary condition corresponds to prescribing the heat flux $q \cdot n$ across (out of) the boundary:

$$q \cdot n = -a \nabla u \cdot n = -a \frac{\partial u}{\partial n} = -a \partial_n u = g \quad \text{on } \Gamma,$$

with g given. A *homogeneous Neumann boundary condition* with $g = 0$ corresponds to a perfectly insulating boundary with the heat flux across the boundary being zero. A homogenous Robin boundary condition is intermediate with the boundary neither being perfectly conducting nor perfectly insulated, with the heat flux through the boundary being proportional to the difference of the temperature u inside and a given temperature u_b outside Ω:

$$-a\partial_n u = \kappa(u - u_b)$$

with κ a positive coefficient representing the heat conductivity of the boundary.

Partitioning the boundary Γ into disjoint pieces Γ_1, Γ_2 and Γ_3 with different types of boundary conditions, the *general initial boundary value problem* IBVP for the heat equation has the form,

$$\begin{cases} \frac{\partial}{\partial t}(\lambda u) - \nabla \cdot (a\nabla u) = f & \text{in } \Omega \times (0, T], \\ u = u_b & \text{on } \Gamma_1 \times (0, T], \\ -a\partial_n u = g & \text{on } \Gamma_2 \times (0, T], \\ a\partial_n u + \kappa(u - u_b) = 0 & \text{on } \Gamma_3 \times (0, T], \\ u(x, 0) = u_0(x) & \text{for } x \in \Omega, \end{cases} \qquad (73.6)$$

where u_b represents a given "exterior" boundary temperature, and g represents a given outward normal heat flux on the boundary.

We note that in a stationary situation with $\frac{\partial}{\partial t}(\lambda u) = 0$ and with the heat source $f = 0$, the equation (73.2) expressing conservation of heat, takes the form

$$\nabla \cdot q = 0. \qquad (73.7)$$

If heat is neither produced nor accumulated, then conservation of heat is expressed by the equation $\nabla \cdot q = 0$, that is, the heat flux q is *divergence-free*. Below we shall meet several other examples of divergence-free fields.

73.3   The Heat Equation

We refer to the special case of (73.6) with $\lambda = a = 1$ as the *heat equation*. In the case with homogeneous Dirichlet boundary conditions, we get the model

$$\begin{cases} \frac{\partial u}{\partial t} - \Delta u = f & \text{in } \Omega \times (0, T], \\ u = 0 & \text{on } \Gamma \times (0, T], \\ u(x, 0) = u_0(x) & \text{for } x \in \Omega, \end{cases}$$

where u_0 is the initial temperature, and $\Delta u = \nabla \cdot (\nabla u)$ is the Laplacian. The heat equation serves as a basic prototype of a *parabolic problem*.

73.4 Stationary Heat Conduction: Poisson's Equation

The stationary analog of (73.6) reads

$$\begin{cases} -\nabla \cdot (a\nabla u) = f & \text{in } \Omega, \\ u = u_b & \text{on } \Gamma_1, \\ -a\partial_n u = g & \text{on } \Gamma_2, \\ a\partial_n u + \kappa(u - u_b) = 0 & \text{on } \Gamma_3. \end{cases} \tag{73.8}$$

Choosing $a = 1$ leads to the *Poisson equation*:

$$\begin{cases} -\Delta u = f & \text{in } \Omega, \\ u = u_1 & \text{on } \Gamma_1, \\ -a\partial_n u = g_2 & \text{on } \Gamma_2, \\ a\partial_n u + \kappa(u - u_b) = g_3 & \text{on } \Gamma_3. \end{cases} \tag{73.9}$$

In the case of homogeneous Dirichlet boundary conditions on the whole of the boundary, the Poisson equation reads

$$\begin{cases} -\Delta u = f & \text{in } \Omega, \\ u = 0 & \text{on } \Gamma. \end{cases} \tag{73.10}$$

Poisson's equation serves as a basic model of an *elliptic* problem and has numerous applications in physics and mechanics. We present the basic applications below. Poisson's equation $-\Delta u = f$ with $f = 0$ is referred to as *Laplace's equation*: $\Delta u = 0$.

Fig. 73.2. Poisson (1781–1840): "Life is good for only two things: to study mathematics and to teach it"

We now give a couple of analytic solutions to the heat equation in simple situations:

Example 73.1. The stationary temperature u in a heat conduction unit cube Q with heat production and conduction coefficient equal to one, zero boundary temperature for $x_1 = 0, 1$, and zero heat flux for $x_2, x_3 = 0, 1$, is given by

$$u(x) = \frac{1}{2} x_1 (1 - x_1).$$

We see that the temperature is maximal for $x_1 = 0.5$ and drops off quadratically towards the Dirichlet boundary, see Fig. 73.3 for a plot in the corresponding case in two dimension in the unit square.

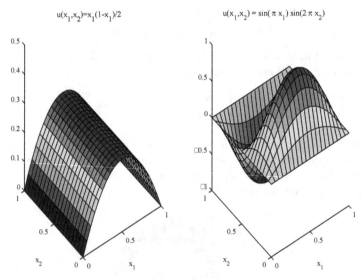

Fig. 73.3. The functions $\frac{1}{2}(x_1(1-x_1))$ and $\sin(\pi x_1) \sin(2\pi x_2)$

Example 73.2. Consider the homogenous heat equation in the unit square Q with $f = 0$ and homogenous Dirichlet boundary conditions: the function

$$u(x, t) = e^{-(n^2 + m^2)t} \sin(nx_1) \sin(mx_2)$$

with $m, n = 1, 2, 3, \ldots$ is a solution of the homogenous heat equation $\frac{\partial u}{\partial t} - \Delta u = 0$ with initial value $u_0(x_1, x_2) = \sin(nx_1) \sin(mx_2)$, see Fig. 73.3. We see that the temperature $u(x, t)$ decays exponentially in time very quickly if n and/or m is only moderately large. This corresponds to the fact that a temperature oscillating in space, is quickly levelled out.

Example 73.3. The stationary temperature $u(x)$ between the two planes $\{x_3 = 0\}$ and $\{x_3 = 1\}$ bounding a heat conducting layer with heat conductivity coefficient equal to one, zero heat source, and the temperature $u = 1$ on $\{x_3 = 1\}$ and $u = 0$ on $\{x_3 = 0\}$, is given by $u(x) = x_3$ displaying a linear variation of the temperature between the plates. No surprise of course.

73.5   Convection-Diffusion-Reaction

The heat equation models the physical phenomenon of *diffusion*, and we now extend this model to include the phenomena of *convection* and *reaction*. We obtain a scalar *convection-diffusion-reaction* equation, which is another basic model in science. We consider a typical case where u represents the concentration of a certain chemical species subject to convection in a given velocity field $\beta(x,t)$, diffusion with diffusion coefficient $\epsilon(x,t)$ and reaction with reaction rate $\alpha(x,t)$. For example, u may represent the concentration of a contaminant in a volume of water moving with the velocity $\beta(x,t)$.

The model results from a principle of conservation of mass together with a constitutive equation generalizing Fourier's law expressing the *flow rate q* of the chemical species in terms of ∇u and βu. Conservation of mass is expressed by

$$\dot{u} + \nabla \cdot q + \alpha u = f,$$

where f represents a source term, and the constitutive law takes the form

$$q = \beta u - \epsilon \nabla u.$$

which says that the total flow rate q is the sum of a convective rate βu and a diffusive rate $-\epsilon \nabla u$. The model thus takes the form:

$$\dot{u} + \nabla \cdot (\beta u) + \alpha u - \nabla \cdot (\epsilon \nabla u) = f \quad \text{in } \Omega \times (0, T], \tag{73.11}$$

together with initial and boundary conditions, where Ω is domain in space and $[0, T]$ a given time interval. We shall meet this model and generalizations thereof in several different contexts below.

73.6   Elastic Membrane

Consider a horizontal elastic net covering the unit square $Q = \{x \in \mathbb{R}^2 : 0 \leq x_i \leq 1, i = 1, 2\}$ formed by elastic strings tied together at nodes $a_{ij} \in \mathbb{R}^2$ in a uniform quadrilateral mesh with mesh size $h = 1/N$, so that $a_{ij} = (ih, jh)$, $i, j = 0, 1, \ldots, N$, where N is the number of cells in each coordinate direction. Assume that the net is stretched so that the tension in each string is equal to h, corresponding to the tension being equal to one per unit of length. Note the normalization introduced says that the tension in each string decreases as the number of strings increases. We refer to the situation in which all the nodes lie in the plane of the square and there is no external load on the net as the unloaded reference configuration of the net.

Suppose the net is subject to a set of downward vertical loads of size $f_{ij}h^2$ at the nodes a_{ij}. The net will deform under the loads and the nodes will be displaced from the initial unloaded reference configuration. Let the vertical displacement of node a_{ij} be denoted by $u_{i,j}$. If the displacements are small, then (recalling the Chapter String theory) the vertical upward force from the net on node a_{ij} is equal to

$$(u_{i,j} - u_{i-1,j}) + (u_{i,j} - u_{i+1,j}) + (u_{i,j} - u_{i,j-1}) + (u_{i,j} - u_{i,j+1}),$$

with contributions from the four pieces of string meeting at a_{ij}. This is because the vertical slope of the line between for example node (i, j) and $(i - 1, j)$ is equal to $(u_{i,j} - u_{i-1,j})/h$ and the tension is h. We thus obtain the following vertical equilibrium equation for each node a_{ij}:

$$-\frac{u_{i-1,j} - 2u_{i,j} + u_{i+1,j}}{h^2} - \frac{u_{i,j-1} - 2u_{i,j} + u_{i,j+1}}{h^2} = f_{ij}.$$

Passing to the limit as h tends to zero, and recalling that Taylor's theorem implies

$$\lim_{h \to 0} \frac{v(x - h) - 2v(x) + v(x + h)}{h^2} = v''(x) = \frac{d^2 v}{dx^2}(x)$$

if $v : \mathbb{R} \to \mathbb{R}$ is twice differentiable, we are led to the equation

$$-\Delta u(x) = f(x).$$

This equation expresses the equilibrium of a horizontal membrane made by an elastic fabric and carrying a vertical load of intensity (force per unit area) $f(x)$, where $u(x)$ is the vertical displacement of the membrane at x and we assume that the membrane in its unloaded plane reference configuration is prestressed to uniform tension in all directions.

We can generalize to a horizontal membrane covering a general domain Ω in $\mathbb{R}^2$. Assuming the membrane is fixed at the boundary Γ of Ω, so that the vertical displacement $u(x)$ is zero at Γ, we thus obtain Poisson's equation

$$-\Delta u = f \quad \text{in } \Omega, \, u = 0 \quad \text{on } \Gamma \tag{73.12}$$

as a model for the vertical deflection of a horizontal elastic membrane spanned over the boundary Γ of a domain Ω in $\mathbb{R}^2$, subject to a vertical load of intensity $f(x)$. This is a basic model of elasticity theory.

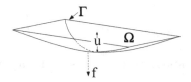

Fig. 73.4. An elastic membrane under a load $f(x)$ and supported at Γ

Example 73.4. If $\Omega = \{x \in \mathbb{R}^2 : \|x\| < 1\}$ is the unit disc, and the load f is radially symmetric, then the deflection u will also be radially symmetric. Recalling the form of the Laplacian in polar coordinates from the Chapter The divergence, rotation and Laplacian, we can write (73.12) in the form

$$-\Delta u = -\frac{1}{r}\frac{\partial}{\partial r}\left(r\frac{\partial u}{\partial r}\right) = f(r) \quad \text{for } 0 < r < 1,\, u(1) = 0,\, \frac{\partial u}{\partial r}(0) = 0.$$

Note the boundary condition $\frac{\partial u}{\partial r}(0) = 0$, which has no counterpart in x-coordinates, says that $u(x)$ is differentiable at $x = 0$. If $\frac{\partial u}{\partial r}(0) \neq 0$, then $u(x)$ has a a a conical "to" at $x = 0$ and thus is not differentiable at $x = 0$. If $f(r) = 1$, then the solution is given by

$$u(r) = \frac{1}{4}(1 - r^2) \quad \text{for } 0 \le r \le 1.$$

73.7 Solving the Poisson Equation

Suppose we would like to numerically solve the Poisson equation

$$-\Delta u = f \quad \text{in } Q, \quad u = u_b \quad \text{on } \Gamma$$

where Q is the unit square with boundary Γ and $f(x)$ a given function on Q. Recalling the derivation of the model $-\Delta u = f$ from the previous section, we are led to computing approximations $U_{i,j}$ of $u(ih, jh)$ for $i, j = 0, 1, \ldots, N$, where $h = 1/N$, from the system of equations

$$-\frac{U_{i-1,j} - 2U_{i,j} + U_{i+1,j}}{h^2} - \frac{U_{i,j-1} - 2U_{i,j} + U_{i,j+1}}{h^2} = f(ih, jh),$$
$$i, j = 1, \ldots, N - 1,$$

that is

$$4U_{i,j} - U_{i-1,j} - U_{i+1,j} - U_{i,j-1} - U_{i,j+1} = h^2 f(ih, jh),$$
$$i, j = 1, \ldots, N - 1, \quad (73.13)$$

where $U_{i,j} = u_b(ih, jh)$ if i or j is equal to 0 or N. We see that this is an $m \times m$ system of equations with $m = (N - 1) \times (N - 1)$ in the unknowns $U_{i,j}$ where $i, j = 1, \ldots, N-1$. This is the famous *5-point scheme* for the Poisson's equation, where the unknown $U_{i,j}$ is coupled to its four neighbors $U_{i-1,j}$, $U_{i+1,j}$, $U_{i,j-1}$, $U_{i,j+1}$.

If $f = 0$, then the 5-point scheme takes the form ("Svensson's formula")

$$U_{i,j} = \frac{1}{4}(U_{i-1,j} + U_{i+1,j} + U_{i,j-1} + U_{i,j+1}),$$

stating that each value $U_{i,j}$ is the mean value of the neighboring values (reflecting a basic feature of the Swedish national character).

Note that (73.13) is a linear system of equations for the values of U that requires some work to solve. For example, we may try to solve (73.13) by fixed point iteration as follows with $k = 0, 1, \ldots$

$$U_{i,j}^{k+1} = U_{i,j}^k$$
$$- \alpha\big(4U_{i,j}^k - U_{i-1,j}^k - U_{i+1,j}^k - U_{i,j-1}^k - U_{i,j+1}^k - h^2 f(ih, jh)\big), \quad (73.14)$$

for $i, j = 1, \ldots, N-1$, with $U_{i,j}^{k+1} = u_b(ih, jh)$ if i or j is equal to 0 or N. Here, $U_{i,j}^k$ is an approximation of $U_{i,j}$ after k steps starting with an initial approximation U_{ij}^0 and α is a positive constant. It turns out that if α is sufficiently small, then the iteration converges, see Problem 73.9, although the convergence gets slower as the step size h decreases.

Example 73.5. Assuming x_2 independence, we are led to the model

$$-u''(x) = f(x) \quad \text{for } 0 < x < 1, \quad u(0) = u_0, \ u(1) = u_1,$$

where $u'(x) = \frac{du}{dx}$. The corresponding discrete model takes the form

$$-(U_{i-1} - 2U_i + U_{i+1}) = h^2 f(ih),$$
$$i = 1, \ldots, N-1, \ U_0 = u_0, \ U_N = u_1, \quad (73.15)$$

with U_i representing an approximation of $u(ih)$. Assuming for simplicity $u_0 = u_1 = 0$, the discrete model can be written in the form

$$AU = b,$$

with $U = (U_1, \ldots, U_{N-1})$, $b = (b_1, \ldots, b_{N-1})$ with $b_i = h^2 f(ih)$, $A = (a_{ij})$ an $(N-1) \times (N-1)$ matrix with $a_{ii} = 2$, $a_{i,i-1} = a_{i-1,i} = -1$ and $a_{ij} = 0$ if $|i - j| > 1$. The fixed point iteration described above can be written

$$U^{k+1} = U^k - \alpha(AU^k - b),$$

and the criterion of convergence is $\|I - \alpha A\| < 1$, which we prove in Problem 73.9 to be valid if $\alpha > 0$ is sufficiently small. Here, $\|I - \alpha A\|$ is the Euclidean norm of the matrix $I - \alpha A$ and the Spectral theorem implies

$$\|I - \alpha A\| = \max_i |1 - \alpha \lambda_i|,$$

where the λ_i are the eigenvalues of the symmetric matrix A.

73.8   The Wave Equation: Vibrating Elastic Membrane

We now model the dynamic motion of the elastic membrane considered above in the static. We complement the given exterior force $f(x,t)$, which now may be dependent on time, by a dynamic force, which according to Newton's law, takes the form $m\ddot{u}$, with m representing mass per unit area and $\ddot{u}$ representing the acceleration of vertical displacement u. This to leads to the *wave equation*, modeling a vibrating membrane subject to an exterior load,

$$\begin{cases} \ddot{u} - \Delta u = f & \text{in } \Omega \times (0,T], \\ u = 0 & \text{on } \Gamma \times (0,T], \\ u(x,0) = u^0(x),\ \dot{u}(x,0) = \dot{u}^0(x) & \text{for } x \in \Omega, \end{cases} \tag{73.16}$$

where Ω denotes a domain in $\mathbb{R}^d$ with boundary Γ, u^0 is a given initial displacement, $\dot{u}^0$ is a given initial displacement velocity, and we assume homogeneous Dirichlet boundary conditions for simplicity. Other boundary conditions, notably periodic boundary conditions, are also relevant for this model.

73.9   Fluid Mechanics

Fluid flow opens a rich field for mathematical modelling. We think of a fluid as a collection of many small "fluid particles" and we seek to describe the fluid flow resulting from the motion of all these fluid particles. We work under the assumption that the particles are so small and there are so many, that we can treat the fluid as a continuum. Usually, we use an *Eulerian* mode of description in which we describe the flow in terms of the *velocity* $u(x,t) \in \mathbb{R}^3$ of the fluid particles at position $x \in \mathbb{R}^3$ at time t, or simply the velocity of the fluid at (x,t). This corresponds to attaching an observer to each fixed point x for the purpose of observing the velocity $u(x,t)$ of the fluid particles that happen to be at position x at time t. The observer thus sits at position x and watches the fluid particles swirl by.

Alternatively, in a *Lagrangian* mode of description, an observer is attached to each fluid particle with the purpose of observing the change of velocity of the fluid particle with time. In this case, the observer follows the particle. The different modes of description are both useful and may also be used together, see the chapters on convection-diffusion in [10].

The Equation of Mass Conservation

We consider the flow of a fluid within a certain volume $\Omega \in \mathbb{R}^3$ using an Eulerian description with $u(x, t)$ representing the velocity of the fluid at x at time t. The velocity u is a vector $u = (u_1, u_2, u_3)$.

Let $\rho(x, t)$ denote the *density* of a fluid at (x, t) measuring the mass of the fluid particles per unit of volume. Let V be a fixed volume with boundary S. The total mass of the fluid in V at time t is given by

$$\int_V \rho(x, t)\, dx.$$

The mass of fluid at time t passing out through the boundary S per unit of time is given by

$$\int_S \rho(x, t) u(x, t) \cdot n(x)\, ds(x) = \int_V \nabla \cdot (\rho u)(x, t)\, dx,$$

where we used the Divergence theorem. The rate of change of mass in V plus the rate of mass flow through the boundary must be zero if we assume that no fluid is added or removed, which leads to the following expression of *mass conservation*,

$$\frac{\partial}{\partial t} \int_V \rho(x, t)\, dx + \int_V \nabla \cdot (\rho u)(x, t)\, dx = 0.$$

If ρ varies smoothly, then $\frac{\partial}{\partial t}$ may be moved under the integral sign and since V was arbitrarily, we are led to the differential equation expressing *mass conservation*,

$$\frac{\partial \rho}{\partial t} + \nabla \cdot (\rho u) = 0, \tag{73.17}$$

Of course this is a basic equation of mathematical modelling. Performing the differentiation with respect to x, we can express mass conservation in the form

$$\frac{\partial \rho}{\partial t} + u \cdot \nabla \rho + \rho \nabla \cdot u = 0. \tag{73.18}$$

Particle Paths and Streamlines

Let the velocity of a fluid be given by the function $u(x, t)$. Consider the IVP

$$\frac{d}{dt} x(t) = u(x(t), t) \quad \text{for } t > 0, \ x(0) = x_0.$$

The solution $x(t)$ represents the curve, or path or trajectory, followed by a fluid particle that starts at position x_0 at time $t = 0$ and moves with velocity $u(x(t), t)$ for $t > 0$. If the velocity $u(x, t) = u(x)$ is independent of time t, then particle paths are also referred to as *streamlines*.

Incompressible Flow

If the fluid velocity $u(x, t)$ satisfies

$$\nabla \cdot u(x, t) = \left(\frac{\partial u_1}{\partial x_1} + \frac{\partial u_2}{\partial x_2} + \frac{\partial u_3}{\partial x_3} \right)(x, t) = 0 \quad \text{for } x \in \Omega\, t > 0,$$

then the flow is said to be *incompressible* in Ω for $t > 0$.

If the flow is incompressible, the equation (73.18) of mass conservation takes the form

$$\frac{\partial \rho}{\partial t} + u \cdot \nabla \rho = 0. \tag{73.19}$$

Since $\frac{dx}{dt} = u$ for a $x(t)$ particle path, the Chain rule implies

$$\frac{\partial}{\partial t} \rho(x(t), t) = \frac{\partial \rho}{\partial t} + u \cdot \nabla \rho = 0.$$

This says that the density is constant along particle paths, or in other words the volume occupied by a certain set of fluid particles is constant. So, the fluid cannot be compressed. It is common to assume that the density of an incompressible fluid is constant.

Water is very nearly incompressible; it is very difficult to change the total volume of a bucket of water. Air is compressible; the air tank of a diver contains a huge volume of air at normal pressure compressed and stored in a small volume at high pressure. But to get the air into the tank consumes energy.

Incompressible Potential Flow

In so-called *stationary flow*, the velocity $u(x, t)$ is independent of time and thus the fluid velocity $u(x)$ is a function of $x \in \Omega$. Note that in a stationary flow the fluid particles at x are moving if $u(x) \neq 0$, but the velocity of the fluid particles at x does not change with time.

The velocity field $u(x)$ of *rotation-free* fluid flow satisfies $\nabla \times u = 0$, which implies $u = \nabla \varphi$ for a scalar *velocity potential* φ under appropriate convexity assumptions. If the fluid is *incompressible*, then $\nabla \cdot u = 0$, and we obtain the Laplace equation $\Delta \varphi = 0$ for the potential of a rotation-free incompressible flow. At a solid boundary, through which the fluid cannot penetrate, the normal velocity of the fluid is zero, which translates into a homogeneous Neumann boundary condition $\partial_n \varphi = 0$ for the potential φ.

We now give some basic examples of incompressible potential flow. For simplicity, we consider situations in which the velocity $u(x)$ is independent of the x_3-coordinate.

Example 73.6. The potential

$$\varphi(x_1, x_2) = x_1^2 - x_2^2$$

satisfies $\Delta\varphi = 0$ and the corresponding flow velocity $u = \nabla\varphi$ is given by

$$u(x) = (2x_1, -2x_2).$$

This represents stationary flow in a corner, see Fig. 73.5. A streamline $x(t)$ satisfies $\frac{dx}{dt} = (2x_1, -2x_2)$, which is a separable equation with solutions satisfying

$$x_1(t)x_2(t) = c,$$

where c is a constant, see Fig. 73.5. We check by computing $\frac{d}{dt}x_1 x_2 = \dot{x}_1 x_2 + x_1 \dot{x}_2 = 2x_1 x_2 - 2x_1 2x_2 = 0$.

Example 73.7. The potential

$$\varphi(x) = \log(\|x\|)$$

satisfies $\Delta\varphi(x) = 0$ for $x \neq 0$, and the corresponding flow velocity $u = \nabla\varphi$ is given by $u(x) = \frac{x}{\|x\|^2}$, see Fig. 73.5.

Example 73.8. We consider incompressible potential flow around an infinite circular cylinder along the x_3-axis with cross-section $\Omega = \{x = (x_1, x_2) \in \mathbb{R}^2 : \|x\| < 1\}$ from left to right, see Fig. 73.5. The potential φ is given in polar coordinates $x = r(\cos(\theta), \sin(\theta))$ by

$$\varphi(x) = \varphi(r, \theta) = \left(r + \frac{1}{r}\right)\cos(\theta),$$

corresponding to a flow from right to left sweeping around Ω and approaching $u(x) = (1, 0)$ for $\|x_1\|$ large and x_2 bounded. We note that $\Delta\varphi = 0$ for $r \neq 0$, and that $\frac{\partial\varphi}{\partial r} = 1 - 1/r^2 = 0$ for $r = 1$ and thus the flow is tangential to the boundary of Ω.

Note that fluid flow is rarely rotation-free in the whole region occupied by the fluid. In particular, if the fluid is viscous then rotation is generated at solid boundaries.

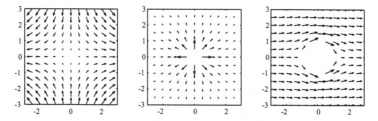

Fig. 73.5. Examples of incompressible potential flow

Incompressible Flow With Rotation

We now consider basic examples of incompressible flow in two dimensions with non-zero rotation. We assume $u(x) = (u_1(x), u_2(x))$ satisfies $\nabla \cdot u = 0$, where $x = (x_1, x_2)$. Defining $v = (-u_2, u_1)$ this equation reads $\nabla \times v = 0$ and under appropriate convexity assumptions, there is a potential φ with $v = \nabla \varphi$. Thus,

$$u = (v_2, -v_1) = (\frac{\partial \varphi}{\partial x_2}, -\frac{\partial \varphi}{\partial x_1}) = \nabla \times \varphi.$$

With the rotation $\nabla \times u = f$ given, we are led to the Poisson equation for φ,

$$f = \nabla \times u = \nabla \times (\nabla \times \varphi) = -\Delta \varphi.$$

Example 73.9. Given $f = 4$ we find the corresponding solution $\varphi(x) = -\|x\|^2$ with $u(x_1, x_2) = (-2x_2, 2x_1)$, see Fig. 73.6. Choosing $\varphi(x) = \log(\|x\|)$ corresponds to $f(x) = 0$ for $x \neq 0$, and the corresponding velocity $u(x_1, x_2) = \|x\|^{-2}(x_2, -x_1)$, see Fig. 73.6.

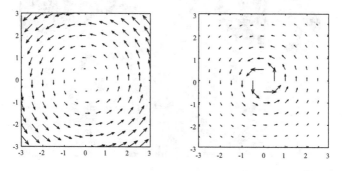

Fig. 73.6. Incompressible flow with rotation

The Euler and Navier-Stokes Equations

The *Euler equations* for an incompressible *inviscid* fluid with constant density equal to one, take the form

$$\frac{\partial u}{\partial t} + (u \cdot \nabla)u + \nabla p = f, \quad \nabla \cdot u = 0, \tag{73.20}$$

where $u(x,t)$ is the velocity and $p(x,t)$ the *pressure* of the fluid at the point x at time t, and f is an applied volume force like a gravitational force. In an inviscid fluid, the *viscosity* is zero and the only interior force acting between the fluid particles is the pressure force that is equal in all directions and acts normal to any surface. The equation $\nabla \cdot u = 0$ expresses the incompressibility of the flow. The first equation expresses Newton's law stating that the acceleration $\frac{d}{dt}u(x(t),t)$, where $x(t)$ is the trajectory followed by a fluid particle satisfying $\frac{dx}{dt} = u(x(t),t)$, is equal to the force $-\nabla p + f$, consisting of the pressure force $-\nabla p$ and the applied force f. We see this by computing with the the Chain rule and the equation $\frac{dx}{dt} = u(x(t),t)$ to get

$$\frac{d}{dt}u_i(x(t),t) = \frac{\partial u_i}{\partial t} + \frac{dx}{dt} \cdot \nabla u_i = \frac{\partial u_i}{\partial t} + (u \cdot \nabla)u_i,$$

which leads to the vector form (73.20). The *Navier-Stokes equations* are modifications of the Euler equations with an additional viscous force term $-\nu\Delta u$, where ν is the viscosity coefficient. In a fluid with non-zero viscosity, there are also tangential (shear) forces acting on a surface.

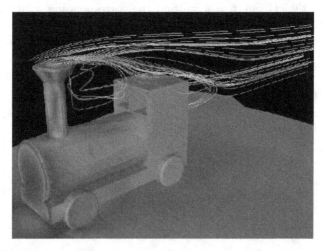

Fig. 73.7. Flow of air and pressure around a toy train

73.10   Maxwell's Equations

The interaction between electric and magnetic fields are described by *Maxwell's equations*:

$$\begin{cases} \dfrac{\partial B}{\partial t} + \nabla \times E = 0, \\[2mm] -\dfrac{\partial D}{\partial t} + \nabla \times H = J, \\[2mm] \nabla \cdot B = 0, \quad \nabla \cdot D = \rho, \\[2mm] B = \mu H, \quad D = \epsilon E, \quad J = \sigma E, \end{cases} \tag{73.21}$$

where E is the *electric field*, H is the *magnetic field*, D is the *electric displacement*, B is the *magnetic flux*, J is the *electric current*, ρ is the *charge*, μ is the *magnetic permeability*, ϵ is the *dielectric constant* of *electric permittivity*, and σ is the *electric conductivity*. The first equation is referred to as *Faraday's law*, the second is *Ampère's law*, $\nabla \cdot D = \rho$ is *Coulomb's law*, *Gauss law* $\nabla \cdot B = 0$ expresses the absence of "magnetic charge", and $J = \sigma E$ is *Ohm's law*. Maxwell, see Fig. 73.8, included the term $\partial D/\partial t$ for purely mathematical reasons and then used Calculus to predict the existence of electromagnetic waves before these had been observed experimentally.

Fig. 73.8. Maxwell (1831–1879), inventor of the mathematical theory of electromagnetism: "We can scarcely avoid the conclusion that light consists in the transverse undulations of the same medium which is the cause of electric and magnetic phenomena"

Typical boundary conditions include various combinations of $E \cdot n$ (perfect insulator), $E \times n$ (perfect conductor), $H \cdot n$ and $H \times n$.

Maxwell's equations describe the whole world of electromagnetic phenomena with an astounding economy of notation and accuracy of modelling. Our modern information society builds on electromagnetic waves.

We shall now pick out a couple of Laplace equation models from Maxwell's equations by considering some basic particular cases.

Electrostatics

A basic problem in *electrostatics* is to describe the stationary electric field $E(x)$ in a volume Ω in $\mathbb{R}^3$ containing *charges* of density $\rho(x)$ and enclosed by a perfectly conducting surface Γ. Faraday's law states that

$$\nabla \times E = 0 \quad \text{in } \Omega,$$

since we assume that $\frac{\partial B}{\partial t} = 0$. Recalling Chapter Potential fields, it follows that the electric field E is the gradient of a scalar *electric potential* φ, i.e. $E = \nabla\varphi$. Coulomb's law says

$$\nabla \cdot E = \rho \quad \text{in } \Omega,$$

so we are led to the Poisson equation for the potential φ,

$$\Delta\varphi = \nabla \cdot \nabla\varphi = \rho \quad \text{in } \Omega.$$

The boundary condition $E \times n = 0$ on the boundary Γ of Ω with n denoting the outward unit normal, says that the tangential component of E vanishes on the boundary. This models a perfectly conducting boundary in which differences in the electric field are leveled out. This means that $E = \nabla\varphi$ is normal to the boundary, so the boundary is a level surface of φ and the potential φ is constant on the boundary. Since φ is undetermined up to a constant, we may assume that $\varphi = 0$ on the boundary and we arrive at Poisson's equation $-\Delta\varphi = f$ with $f = -\rho$ in Ω with homogenous Dirichlet boundary conditions $\varphi = 0$ on Γ.

The potential $\varphi(x)$ of a point charge at the origin is given by

$$\varphi(x) = \frac{c}{\|x\|}$$

with the corresponding electric field

$$E(x) = -\frac{cx}{\|x\|^3}$$

and c a suitable constant. We shall return to this solution below.

Example 73.10. Let $\Omega = \{x \in \mathbb{R}^2 : \|x\| < 1, x_1 < 0 \text{ or } x_2 > 0\}$ be a circular disc with a piece cut out and a *reentrant* corner of angle $\omega = \frac{3\pi}{2}$, see Fig. 73.9. By a direct computation we can verify that the function

$$\varphi(x) = r^\alpha \sin(\alpha\theta)$$

expressed in polar coordinates $x = r(\cos(\theta), \sin(\theta))$, where $\alpha = \frac{\pi}{\omega} = \frac{2}{3}$, satisfies the Laplace equation $\Delta\varphi = 0$ in Ω and the boundary condition $\varphi = 0$ in the straight parts of the boundary meeting at the origin. Letting φ represent an electric potential, the corresponding electric field $E(x) = \nabla\varphi(x)$ satisfies

$$\frac{\partial E}{\partial r} = \alpha r^{\alpha-1} \sin(\alpha\theta)$$

and thus since $\alpha < 1$, is singular (infinite) at the corner where $r = 0$. This means that the electric field is very strong close to the corner, and the sharper the corner (α smaller) the stronger is the field. This may support the observation that an electric lightening is more likely to hit the pointed tower of church than a smooth hill, or the design of an electronic scanner where electrons pop out of the pin of a sharp needle.

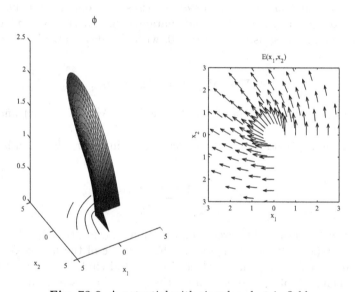

Fig. 73.9. A potential with singular electric field

Example 73.11. The potential φ of the electric field between two concentric spheres $S_1 = \{x \in \mathbb{R}^3 : \|x\| < r_1\}$ and $S_2 = \{x \in \mathbb{R}^3 : \|x\| < r_2\}$ with $r_2 > r_1$, is given by

$$\varphi(x) = \frac{1}{\|x\|}$$

if we assume that $\varphi = 1/r_i$ on S_i, $i = 1, 2$.

Example 73.12. The function

$$\varphi(x_1, x_2) = \arctan\left(\frac{x_2}{x_1}\right)$$

defined for $x_1 > 0$ satisfies $\Delta\varphi(x) = 0$ for $x_1 > 0$, and is constant $= \arctan(c)$ on rays $x_2 = cx_1$ through the origin of slope c. The corresponding electric field $E(x) = \nabla\varphi(x)$ given by

$$E(x) = \frac{(-x_2, x_1)}{\|x\|^2}.$$

We see that $E(x)$ is singular at $x = 0$.

Magnetostatics

The basic problem in *magnetostatics* arises by combining Gauss' law $\nabla \cdot H = 0$, assuming μ constant and guaranteeing that $H = \nabla \times \varphi$ for some vector potential ψ satisfying $\nabla \cdot \psi = 0$, with Faraday's law $\nabla \times H = J$ to give

$$\nabla \times (\nabla \times \psi) = -\Delta\psi = J,$$

where we use the facts that $\nabla \times (\nabla \times \psi) = -\Delta\psi + \nabla(\nabla \cdot \psi)$ and that $\nabla \cdot \varphi = 0$.

The magnetic field around a unit current J in the x_3-direction is given by

$$H(x) = \frac{1}{2\pi}\frac{(-x_2, x_1, 0)}{\|x\|^2},$$

which can be verified by direct computation showing that $\nabla \cdot H(x) = 0$ and $\nabla \times H(x) = 0$ for $(x_1, x_2) \neq 0$. The presence of the factor $\frac{1}{2\pi}$ makes $\int_\Gamma H \cdot ds = 1$ for any counter-clockwise oriented circle in the $x_1 x_2$-plane, from which by Stokes theorem follows that $\nabla \times H = J$, see the next section on Gravitation and delta functions.

Time-Dependent Magnetics

In low frequency applications, the term $\frac{\partial D}{\partial t}$ so cleverly introduced by Maxwell, plays a minor role and can be discarded. Let's see where this leads. Since $\nabla \cdot B = 0$, we can write B as $B = \nabla \times \psi$, where ψ is a magnetic vector potential. Inserting this into Faraday's law gives

$$\nabla \times \left(\frac{\partial\psi}{\partial t} + E\right) = 0,$$

from which it follows that

$$\frac{\partial \psi}{\partial t} + E = \nabla \varphi,$$

for some scalar potential φ. Multiplying by σ and using the laws of Ohm and Ampère, we obtain a vector equation for the magnetic potential ψ:

$$\sigma \frac{\partial \psi}{\partial t} + \nabla \times \left(\mu^{-1} \nabla \times \psi \right) = \sigma \nabla \varphi.$$

This system reduces to a scalar equation in two variables if we assume that $B = (B_1, B_2, 0)$ is independent of x_3. It then follows that ψ has the form $\psi = (0, 0, u)$ for some scalar function u that depends only on x_1 and x_2, so that $B_1 = \partial u / \partial x_2$ and $B_2 = -\partial u / \partial x_1$. We end up with a scalar equation for the scalar magnetic potential u in the form

$$\sigma \frac{\partial u}{\partial t} - \nabla \cdot \left(\mu^{-1} \nabla u \right) = f, \tag{73.22}$$

for some function $f(x_1, x_2)$. Choosing $\sigma = \mu = 1$ leads to the heat equation,

$$\begin{cases} \frac{\partial}{\partial t} u(x, t) - \Delta u(x, t) = f(x, t) & \text{for } x \in \Omega, \, 0 < t \leq T, \\ u(x, t) = 0 & \text{for } x \in \Gamma, \, 0 < t \leq T, \\ u(x, 0) = u_0(x) & \text{for } x \in \Omega, \end{cases} \tag{73.23}$$

where $\Omega \subset \mathbb{R}^2$ with boundary Γ, and we posed homogeneous Dirichlet boundary conditions. In the stationary case, we again obtain Poisson's equation with Dirichlet boundary conditions.

73.11   Gravitation

In his famous treatise *Mécanique Céleste* in five volumes published during 1799–1825, Laplace extended Newton's theory of gravitation and in particular developed a theory for describing gravitational fields based on using gravitational potentials that satisfy Laplace's equation, or more generally Poisson's equation.

We consider a gravitational field in $\mathbb{R}^3$ with gravitational force $F(x)$ at position x, generated by a distribution of mass of density $\rho(x)$. We recall that the work of a unit mass, moving along a curve Γ is given by

$$\int_\Gamma F \cdot ds,$$

If the curve Γ is closed, then the total work performed by the gravitational forces should be zero. Stokes' theorem implies that a gravitational field F

should satisfy $\nabla \times F = 0$. Using the basic result of the Chapter Potential fields, we conclude that F is the gradient of a scalar potential φ, i.e.

$$F(x) = \nabla \varphi(x). \tag{73.24}$$

Laplace proposed the following relation between the gravitational field F and the mass distribution ρ,

$$-\nabla \cdot F(x) = \rho(x), \tag{73.25}$$

assuming the gravitational constant is normalized to one. This is analogous to Coulomb's law $\nabla \cdot E = \rho$ in electrostatics and also to the energy balance equation $\nabla \cdot q = f$ for stationary heat conduction, where q is the heat flux and f a heat source. In particular, (73.25) states that $\nabla \cdot F(x) = 0$ at points x where there is no mass so that $\rho(x) = 0$. Combining (73.24) and (73.25), we obtain Poisson's equation $-\Delta \varphi = \rho$ for the gravitational potential φ.

Since the origin and property of gravitation of "acting at a distance" is still lacking a convincing physical explication, the equation $-\nabla \cdot F(x) = \rho(x)$ including $\nabla \cdot F = 0$ in empty space, should be viewed as a basic postulate on the nature of a gravitational field. Of course it seems very difficult to motivate that $\nabla \cdot F$ should be something different from zero in empty space, but a real a "proof" that $\nabla \cdot F$ must be zero in empty space seems to be missing.

Newton considered gravitational fields generated by *point masses*. Mathematically, a unit point mass at a point $z \in \mathbb{R}^3$ is represented by the so-called *delta function* δ_z at z, defined by the property that for any smooth function v,

$$\int_{\mathbb{R}^3} \delta_z v \, dx = v(z), \tag{73.26}$$

where the integration is to be interpreted in a generalized sense. We could think of a δ_z as a limit of positive functions $\varphi_h(x)$ such that $\varphi_h(x) = 0$ if $\|x - z\| > h$ and

$$\int_{\mathbb{R}^3} \varphi_h(x) \, dx = 1,$$

as h tends to zero. For example we may choose

$$\varphi_h(x) = \frac{3}{4\pi h^3} \quad \text{if } \|x - z\| < h$$

and $\varphi_h(x) = 0$ elsewhere. If $v(x)$ is Lipschitz continuous at z, then

$$lim_{h \to 0} \int_{\mathbb{R}^3} \varphi_h(x) v(x) \, dx = v(z),$$

which gives (73.26) its meaning. The function $\varphi_h(x)$ represents a very tall and narrow "hump" around z with volume one.

We expect that the gravitational potential $\Phi(x)$ corresponding to a unit point mass at the origin, to satisfy

$$-\Delta\Phi = \delta_0 \quad \text{in } \mathbb{R}^3, \tag{73.27}$$

assuming the gravitational constant to be equal to one. To give a precise meaning to this equation involving the somewhat mysterious delta function δ_0 at 0, we first formally multiply by a smooth *test function* v vanishing outside a bounded set to get

$$-\int_{\mathbb{R}^3} \Delta\Phi(x)v(x)\,dx = v(0). \tag{73.28}$$

Next, we integrate the left-hand side by parts formally using Green's formula to move the Laplacian from E to v, noting that the boundary terms disappear since v vanishes outside a bounded set. We may thus reformulate (73.27) as seeking a potential $E(x)$ satisfying

$$-\int_{\mathbb{R}^3} \Phi(x)\Delta v(x)\,dx = v(0), \tag{73.29}$$

for all smooth functions $v(x)$ vanishing outside a bounded set. We may view this as the concrete interpretation of (73.27), which is perfectly well defined since now the Laplacian acts on the smooth function $v(x)$ and the potential Φ is assumed to be integrable. We also require the potential $\Phi(x)$ to decay to zero as $\|x\|$ tends to infinity, which corresponds to a "zero Dirichlet boundary condition at infinity".

In the Chapter The divergence, rotation and Laplacian, we showed that the function $1/\|x\|$ satisfies Laplace's equation $\Delta u(x) = 0$ for $0 \neq x \in \mathbb{R}^3$, while it is singular at $x = 0$. We shall prove that the following scaled version of this function satisfies (73.29):

$$\Phi(x) = \frac{1}{4\pi} \frac{1}{\|x\|}. \tag{73.30}$$

We refer to this function as the *fundamental solution* of $-\Delta$ in $\mathbb{R}^3$. We conclude in particular that the gravitational field in $\mathbb{R}^3$ created by a unit point mass at the origin is given by

$$F(x) = \nabla\Phi(x) = -\frac{1}{4\pi} \frac{x}{\|x\|^3},$$

which is precisely Newton's inverse square law of gravitation. Laplace thus gives a motivation why the exponent should be two, which Newton did not (and therefore was criticized by Leibniz). Of course, it still remains to motivate (73.25). In the context of heat conduction, the fundamental solution $E(x)$ represents the stationary temperature in a homogeneous body with heat conductivity equal to one filling the whole of $\mathbb{R}^3$, subject to a concentrated heat source of strength one at the origin and with the temperature tending to zero as $\|x\|$ tends to infinity.

We now prove that the function $\Phi(x)$ defined by (73.30) satisfies (73.29). We first note that since Δv is smooth and vanishes outside a bounded set and $\Phi(x)$ is integrable over bounded domains,

$$\int_{\mathbb{R}^3} \Phi \Delta v \, dx = \lim_{a \to 0^+} \int_{D_a} \Phi \Delta v \, dx, \qquad (73.31)$$

where $D_a = \{x \in \mathbb{R}^3 : a < \|x\| < a^{-1}\}$, with a small positive, is a bounded region obtained from $\mathbb{R}^3$ by removing a little sphere of radius a with boundary surface S_a and also points further away from the origin than a^{-1}, see Fig. 73.10. We now use Green's formula on D_a with $w = \Phi$. Since v is zero for $\|x\|$ large, the integrals over the outside boundary vanish when a is sufficiently small. Using the fact that $\Delta\Phi = 0$ in D_a, $\Phi = 1/(4\pi a)$ on S_a and $\partial\Phi/\partial n = 1/(4\pi a^2)$ on S_a with the normal pointing in the direction of the origin, we obtain

$$-\int_{D_a} \Phi \Delta v \, dx = \int_{S_a} \frac{1}{4\pi a^2} v \, ds - \int_{S_a} \frac{1}{4\pi a} \frac{\partial v}{\partial n} \, ds = I_1(a) + I_2(a),$$

with the obvious definitions of $I_1(a)$ and $I_2(a)$. Now, $\lim_{a \to 0} I_1(a) = v(0)$ because $v(x)$ is continuous at $x = 0$ and the surface area of S_a is equal to $4\pi a^2$, while $\lim_{a \to 0} I_2(a) = 0$. The desired equality (73.29) now follows recalling (73.31).

The corresponding fundamental solution of $-\Delta$ in $\mathbb{R}^2$ is given by

$$\Phi(x) = \frac{1}{2\pi} \log\left(\frac{1}{\|x\|}\right). \qquad (73.32)$$

In this case the fundamental solution is not zero at infinity.

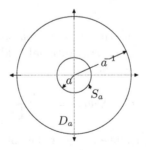

Fig. 73.10. A cross-section of the domain D_a

Replacing 0 by an arbitrary point $z \in \mathbb{R}^3$, (73.29) becomes

$$-\int_{\mathbb{R}^3} \Phi(z-x)\Delta v(x)\, dx = v(z), \qquad (73.33)$$

which leads to a solution formula for Poisson's equation in $\mathbb{R}^3$. For example, if u satisfies the Poisson equation $-\Delta u = f$ in $\mathbb{R}^3$ and $|u(x)| = O(\|x\|^{-1})$ as $\|x\| \to \infty$, then u may be represented in terms of the fundamental solution Φ and the right-hand side f as follows:

$$u(z) = \int_{\mathbb{R}^3} \Phi(z-x) f(x)\, dx = \frac{1}{4\pi} \int_{\mathbb{R}^3} \frac{f(x)}{\|z-x\|}\, dx. \qquad (73.34)$$

We see that $u(z)$ is a mean value of f centered around z weighted so that the influence of the values of $f(x)$ is inversely proportional to the distance from z.

Similarly, the potential u resulting from a distribution of mass of density $\rho(x)$ on a (bounded) surface Γ in $\mathbb{R}^3$ is given by

$$u(z) = \frac{1}{4\pi} \int_{\Gamma} \frac{\rho(x)}{\|z-x\|}\, ds(x). \qquad (73.35)$$

Formally, we obtain this formula by simply adding the potentials from all the different pieces of mass on Γ. One can show that the potential u defined by (73.35) is continuous in $\mathbb{R}^3$ if ρ is bounded on Γ, and of course u satisfies Laplace's equation away from Γ. Suppose now that we would like to determine the distribution of mass ρ on Γ so that the corresponding potential u defined by (73.35) is equal to a given potential u_0 on Γ, that is we seek in particular a function u solving the boundary value problem $\Delta u = 0$ in Ω and $u = u_0$ on Γ, where Ω is the volume enclosed by Γ. This leads to the following *integral equation*: given u_0 on Γ find the function ρ on Γ such that

$$\frac{1}{4\pi} \int_{\Gamma} \frac{\rho(y)}{\|x-y\|}\, ds(y) = u_0(x) \quad \text{for } x \in \Gamma. \qquad (73.36)$$

This is a *Fredholm integral equation of the first kind*, named after the Swedish mathematician Ivar Fredholm (1866–1927). In the beginning of the 20th century, Fredholm and Hilbert were competing to prove the existence of solutions of the basic boundary value problems of mechanics and physics using integral equation methods. The integral equation (73.36) is an alternative way of formulating the boundary value problem of finding u such that $\Delta u = 0$ in Ω, and $u = u_0$ on Γ. Integral equations may also be solved using Galerkin methods.

73.12   The Eigenvalue Problem for the Laplacian

The *eigenvalue problem* for the Laplace operator with Dirichlet boundary conditions on a domain Ω in $\mathbb{R}^d$ with boundary Γ takes the form: Find nonzero *eigen-functions* $\varphi(x)$ with corresponding *eigenvalues* λ such that

$$\begin{cases} -\Delta\varphi = \lambda\varphi & \text{in } \Omega, \\ \varphi = 0 & \text{on } \Gamma. \end{cases} \qquad (73.37)$$

In the one-dimensional case with $\Omega = (0, \pi)$, the eigenfunctions are (modulo normalization) $\varphi_n(x) = \sin(nx)$ with corresponding eigenvalues $\lambda_n = n^2$, $n = 1, 2, \ldots$. For a two-dimensional square $\Omega = (0, \pi) \times (0, \pi)$, the eigen-functions are $\varphi_{nm}(x_1, x_2) = \sin(nx_1)\sin(mx_2)$, $n, m = 1, 2, \ldots$, with eigen-values $\lambda_{nm} = n^2 + m^2$.

It follows by multiplication of (73.37) by φ and integration by parts, that all eigenvalues λ are positive. More precisely, there is an increasing se-quence of eigenvalues tending to infinity, and eigenfunctions corresponding to different eigenvalues are orthogonal with respect to the scalar product $(v, w) = \int_\Omega vw \, dx$.

If $\varphi(x)$ is an eigenfunction with corresponding eigenvalue λ, then the (real part of the) function $u(x, t) = \exp(it\sqrt{\lambda})\varphi(x)$ solves the homogeneous wave equation

$$\ddot{u} - \Delta u = 0 \text{ in } \Omega \times \mathbb{R}$$

corresponding to a vibrating elastic membrane (drum head) if $d = 2$ (string if $d = 1$). The smallest eigenvalue corresponds to the basic tone of the drum head.

In Fig. 73.11, we show contour plots for the first four eigenfunctions, corresponding to $\lambda_1 \approx 38.6$, $\lambda_2 \approx 83.2$, $\lambda_3 \approx 111.$, and $\lambda_4 \approx 122.$, in a case where Ω corresponds to the lid of a guitar with Dirichlet boundary conditions on the outer boundary, described as an ellipse, and Neumann boundary conditions at the hole in the lid,

Often the smaller eigenvalues are the most important in considerations of design. This is the case for example in designing suspension bridges, which must be built so that the lower eigenvalues of vibrations in the bridge are not close to possible wind-induced frequencies. This was not well under-stood in the early days of suspension bridges which caused the famous collapse of the Tacoma bridge in 1940.

The smallest eigenvalue is equal to the minimum value of the *Rayleigh quotient*

$$\frac{(\nabla\psi, \nabla\psi)}{(\psi, \psi)},$$

when varying over functions ψ satisfying the boundary conditions. More, generally, the eigenvalues corresponds to stationary values of the Rayleigh quotient.

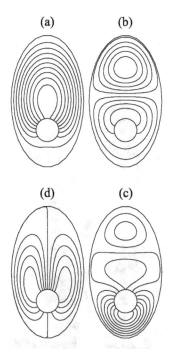

Fig. 73.11. Contour plots of the first four eigenfunctions of the guitar lid corresponding to (**a**) $\lambda_1 \approx 38.6$, (**b**) $\lambda_2 \approx 83.2$, (**c**) $\lambda_3 \approx 111$, and (**d**) $\lambda_4 \approx 122$. These were computed with Femlab with a fixed mesh size of diameter .02

73.13 Quantum Mechanics

The two most revolutionary achievements of physics during the 20th century was the development of of the *Theory of General Relativity* for Gravitation on astronomic scales by Einstein, and *Quantum Mechanics* for atomic scales by Schrödinger (1887–1961, Nobel Prize in Physics 1933), see Fig. 73.12. Einstein never fully accepted Quantum Mechanics, and the *Grand Unified Theory* connecting Gravitation and Quantum Mechanics is still missing, with *String Theory* being a recent attempt to fill the gap.

The basic equation of Quantum Mechanics is the *Schrödinger equation*, which for a system of N electrons (with the Born-Oppenheimer approximation) takes the following normalized form:

$$i\frac{\partial \varphi}{\partial t} = H\varphi = \left(-\frac{1}{2}\sum_j \Delta_j + V(r_1, \ldots, r_N) \right) \varphi, \qquad (73.38)$$

where $\varphi = \varphi(r_1, \ldots, r_N, t)$ is a *wave function* depending on the set of space coordinates $(r_1, \ldots, r_N)$ with each r_j varying over $\mathbb{R}^3$, together with time t, Δ_j denotes the Laplacian with respect to the coordinate $r_j \in \mathbb{R}^3$, and

Fig. 73.12. Schrödinger (1887–1961) at age 13: "I was a good student in all subjects, loved mathematics and physics, but also the strict logic of the ancient grammars, hated only memorizing incidental dates and facts. Of the German poets, I loved especially the dramatists, but hated the pedantic dissection of this works"

$V(r_1, \ldots, r_N)$ denotes a *potential* depending on $(r_1, \ldots, r_N)$ representing repulsive Coulomb forces between the electrons and attractive Coulomb forces between the electrons and the (fixed) nuclei of the system, $H = -\frac{1}{2} \sum_j \Delta_j + V$ is the *Hamiltonian* representing a sum of kinetic and potential energies, and i denotes the imaginary unit. The wave function is complex-valued and the square of its modulus represents an electron probability density.

The Schrödinger equation appears to give a very good description of phenomena on atomistic scales, but unfortunately it is not easy to deal with because of the large number of spatial dimensions involved: For a system with 100 electrons, which is still very small, the number of space dimensions is equal to 300, and standard techniques for either analytical or numerical solution fall very short. So, although the Schrödinger equation admittedly is a very beautiful equation which gives a surprisingly concise description of atomistic physics, it is certainly impossible to solve exactly analytically, and approximate solution becomes a key issue. The 1998 Nobel Prize in Chemistry was awarded Robert Kohn for his method for approximate solution of the Schrödinger equation based on using a single *electron density function* with the space dependence restricted to $\mathbb{R}^3$, independent of the number of electrons, and corresponding approximate potentials. Such simplified Schrödinger equations, referred to as *Kohn-Sham equations*, are today used extensively in computational chemistry.

The Hydrogen Atom

The Hydrogen atom consisting of one electron and one neutron is the only case in which analytical solution of the Schrödinger equation is feasible: In this case the Schrödinger equation takes the following (normalized) form assuming the neutron is positioned at the origin: Find the wave function $\varphi(x,t)$ with $x \in \mathbb{R}^3$, such that for $t > 0$

$$i\frac{\partial \varphi}{\partial t} = \left(-\frac{1}{2}\Delta + V\right)\varphi \quad \text{in } \mathbb{R}^3, \tag{73.39}$$

where Δ is the usual Laplacian with respect to x, $V(x) = -\frac{1}{|x|}$ is the Coulomb potential of the proton, with the normalization that

$$\int_{\mathbb{R}^3} |\varphi(x,t)|^2 \, dx = 1 \quad \text{for } t > 0.$$

For a domain $\Omega \in \mathbb{R}^3$, the integral

$$\int_{\Omega} |\varphi(x,t)|^2 \, dx$$

represents the probability to find the electron in the domain Ω at time t. Formally, $-\frac{1}{2}\Delta$ corresponds to the kinetic energy $\frac{p^2}{2m}$ with p the momentum and m the mass, replacing p by $-i\nabla$ and setting $m = 1$.

In the time-harmonic case with a time-dependence of the form $\exp(-i\omega t)$ with frequency ω, this leads to the eigenvalue problem: Find $\varphi(x) \neq 0$ and $\omega \in \mathbb{R}$ such that

$$H\varphi = \omega\varphi, \tag{73.40}$$

with $H = -\frac{1}{2}\Delta + V$ the Hamiltonian and the eigenvalue ω representing an energy level. The eigenvalues are real and the (real) eigenfunction corresponding to the smallest eigenvalue (smallest energy) is referred to as the *ground state* and the eigenfunctions corresponding to larger eigenvalues as *bound states*.

Assuming spherical symmetry (73.40) takes the following form in spherical coordinates with r the radius: Find $\varphi(r)$ such that

$$-\frac{1}{2}\frac{d^2\varphi}{dr^2} - \frac{1}{r}\frac{d\varphi}{dr} - \frac{1}{r}\varphi = \omega\varphi \quad \text{for } r > 0,$$

with the side condition that $\varphi(0)$ is finite and $\varphi(x)$ is square integrable over $\mathbb{R}^3$. The ground state is given by the eigenfunction $\varphi(r) = \exp(-r)$ corresponding to the eigenvalue $\omega = -\frac{1}{2}$.

Chapter 73 Problems

73.1. Interpret the fixed point iteration for Poisson's equation as an explicit time stepping scheme for the heat equation $\frac{du}{dt} - \Delta u = f$ with time step αh^2 with the starting value given by the initial approximation U^0. Explain why the convergence is slow if h is small.

73.2. Consider a horizontal elastic membrane spanned over a circular ring with constant tension H in all directions in unloaded configuration. Discuss under what conditions the membrane can carry a non-zero volume of water and try to compute the volume.

73.3. Prove that (73.32) is a fundamental solution of $-\Delta$ in $\mathbb{R}^2$.

73.4. Because the presented mathematical models of heat flow and gravitation, namely Poisson's equation, are the same, it opens the possibility of thinking of a gravitational potential as "temperature" and a gravitational field as "heat flux". Can you "understand" something about gravitation using this analogy?

73.5. Present the integral equation corresponding to (73.36) in the case $d = 2$.

73.6. What equation is obtained if $\partial D/\partial t$ is not neglected in the setting of time-dependent magnetics, but the x_3 independence is kept?

73.7. Derive the heat equation describing the heat conduction in a thin piece of wire of length one whose ends are kept at a fixed temperature (i.e., derive the heat equation in one dimension):

$$\begin{cases} \dot{u} - u'' = f & \text{in } (0,1) \times (0,T], \\ u(0,t) = u(1,t) = 0 & \text{for } t \in (0,T], \\ u(x,0) = u_0(x) & \text{for } x \in (0,1). \end{cases} \tag{73.41}$$

73.8. Let $F(x)$ be the gravitational field generated by a homogeneous ball of mass m occupying the volume $\{x \in \mathbb{R}^3 : \|x\| \le r\}$, satisfying $\nabla F(x) = \rho$ for $\|x\| < r$ and $\nabla F(x) = 0$ for $\|x\| > r$, where ρ is the density of the sphere. Argue that by symmetry $F(x) = f(\|x\|)\frac{-x}{\|x\|}$ for $\|x\| > r$ for some function $f : (0,\infty) \to \mathbb{R}$. Use the Divergence theorem to see that if $R > r$ then

$$\int_{S_R} F(x) \cdot n \, dS = 4\pi R^2 f(R) = \int_{B_R} \nabla F(x) \, dx = m,$$

where S_R is the boundary of the ball $B_R = \{x \in \mathbb{R}^3 : \|x\| \le R\}$. Conclude that $f(R) = \frac{m}{4\pi R^2}$, and thus that $F(x) = \frac{m}{4\pi} \frac{-x}{\|x\|^3}$ for $\|x\| > r$. This gives an alternative way of handling of Newton's nightmare. Note the change of normalization with the factor $1/4\pi$ appearing here.

73.9. To analyze the convergence of the fixed point iteration for the system of equations (73.15), we need to show that $\|I - \alpha A\| < 1$, where $A = (a_{ij})$ is the $(N-1) \times (N-1)$ matrix with $a_{ii} = 2$, $a_{i,i-1} = a_{i-1,i} = -1$ and $a_{ij} = 0$ if $|i - j| > 1$. Since A is symmetric, we have recalling the Chapter The Spectral Theorem:

$$\|I - \alpha A\| = \max_i |1 - \alpha \lambda_i|,$$

where λ_i, $i = 1, \ldots, N-1$, are the eigenvalues of A. To see this, diagonalize. Prove that for all nonzero $V \in \mathbb{R}^{N-1}$

$$AV \cdot V = \sum_{i,j=1}^{N-1} a_{ij} V_i V_j > 0,$$

and conclude that $\lambda_i > 0$ for all i (Hint: complete squares!). Show similarly that for all $V \in \mathbb{R}^{N-1}$

$$(I - \alpha A)V \cdot V \geq 0$$

if $\alpha \leq \frac{1}{4}$ (Hint: same as before!). Conclude that Fixed point iteration converges if $0 < \alpha \leq \frac{1}{4}$. Can you prove convergence if $\alpha < \frac{1}{2}$? What about convergence if $\alpha < 0$? Hint: Use that if A is a symmetric $m \times m$ matrix with eigenvalues $\lambda_1 \leq \lambda_2 \leq \ldots \leq \lambda_m$, then $\lambda_1 = \min_{V \in \mathbb{R}^m} (AV \cdot V)/(V \cdot V)$ and $\lambda_m = \max_{V \in \mathbb{R}^m} (AV \cdot V)/(V \cdot V)$, where $V \neq 0$.

73.10. Extend the above analysis to the 5-point scheme for the Laplacian and show that fixed point iteration converges if $0 < \alpha < \frac{1}{8}$ (or better $\alpha < \frac{1}{4}$).

73.11. Gather some friends and arrange them in a square regular grid, and ask them to keep updating their own value according to a Svensson's formula as the mean value of their neighbors (starting with zero), and assigning certain given values to the people at the boundary. Collect the values obtained after convergence. You have solved Laplace equation on a square with Dirichlet boundary values numerically. What value of α in fixed point iteration did you effectively use?

73.12. Prove Bernoulli's theorem stating that in stationary Euler flow satisfying $(u \cdot \nabla)u + \nabla p = 0$ the quantity $\frac{1}{2}\|u\|^2 + p$ is constant along streamlines.

73.13. Explain the *Magnus effect* causing a top-spin tennis ball curve downwards (see also Chapter *Analytic functions*).

73.14. Prove that the hydrogen atom is stable in the sense that the *Rayleigh quotient*

$$RQ(\psi) = \frac{\frac{1}{2}\int_\Omega |\nabla \psi|^2\, dx - \int_\Omega \psi^2/r\, dx}{\int_\Omega \psi^2\, dx},$$

satisfies

$$\min_{\psi \in V} RQ(\psi) \geq -2,$$

showing that the electron does not fall into the proton. Hint: estimate $\int_\Omega \psi \frac{\psi}{r}$ using Cauchy's inequality and the following Poincaré inequality for functions $\psi \in V$:

$$\int_\Omega \frac{\psi^2}{r^2} \, dx \leq 4 \int_\Omega |\nabla \psi|^2 \, dx. \tag{73.42}$$

This shows that the potential energy cannot outpower the kinetic energy in the Rayleigh quotient. To prove the last inequality, use the representation

$$\int_\Omega \frac{\psi^2}{r^2} \, dx = - \int_\Omega 2\psi \nabla \psi \cdot \nabla \ln(|x|) \, dx.$$

resulting from Green's formula, together with Cauchy's inequality.

73.15. (a) Show that the eigenvalue problem for the hydrogen atom for eigenfunctions with radial dependence only, may be formulated as the following one-dimensional problem

$$-\frac{1}{2}\varphi_{rr} - \frac{1}{r}\varphi_r - \frac{1}{r}\varphi = \lambda\varphi, \quad r > 0, \quad \varphi(0) \text{ finite}, \quad \int_{\mathbb{R}} \varphi^2 r^2 \, dr < \infty, \tag{73.43}$$

where $\varphi_r = \dfrac{d\varphi}{dr}$. (b) Show that $\psi(r) = \exp(-r)$ is an eigenfunction corresponding to the eigenvalue $\lambda = -\frac{1}{2}$. (b) Is this the smallest eigenvalue? (c) Determine λ_2 and the corresponding eigenfunction by using a change of variables of the form $\varphi(r) = v(r) \exp(-\frac{r}{2})$. (d) Solve (73.43) numerically.

> The idea of the continuum seems simple to us. We have somehow lost sight of the difficulties it implies ... We are told such a number as the square root of 2 worried Pythagoras and his school almost to exhaustion. Being used to such queer numbers from early childhood, we must be careful not to form a low idea of the mathematical intuition of these ancient sages; their worry was highly credible. (Schrödinger)

74

Chemical Reactions*

We already know the laws that govern the behavior of matter under all but the most extreme situations. In particular, we know the basic laws that underlie all of chemistry and biology. Yet we have certainly not reduced these objects to the status of solved problems; we have, as yet, had little success in predicting human behavior from mathematical equations. So even if we do find a complete set of basic laws, there will still be in the years ahead the intellectual challenging task of developing better approximation methods, so that we can make useful predictions of the probable outcomes in complicated and realistic situations. (S. Hawking in A Brief History of Time)

It is especially difficult to find exact solutions of the equations, as the equations (Einstein's equations) are non-linear. (Einstein)

Inasmuch as a propagating flame may be considered as a wave of chemical reactions sweeping across a flowing gas, it offers an excellent proving ground for the analytical skills of a fluid dynamicist, a heat and mass transfer specialist and a physical chemist, all put together into a well-rounded applied mathematician. (M. Kanury)

74.1 Constant Temperature

We consider N different chemical species $A_1, \ldots, A_N$, which participate in J reactions with stoichiometric (positive) integer coefficients $\nu_{n,j}$ for species n appearing as reactant in reaction j and $\lambda_{n,j}$ for species n appearing as product in reaction j (with the coefficients being zero if the species is not a reactant or product). This is commonly expressed as

$$\sum_{n=1}^{N} \nu_{n,j} A_n \rightarrow \sum_{n=1}^{N} \lambda_{n,j} A_n \quad \text{for } j = 1, \ldots, J. \tag{74.1}$$

We say that the *order* of reaction j is equal to $\sum_{n=1}^{N} \nu_{n,j}$. We denote the *molar concentration* (expressed in moles per unit volume) of species A_n by c_n. The *reaction rate* r_j of reaction j is supposed to be given by

$$r_j = k_j(T) \prod_{m=1}^{N} c_m^{\nu_{m,j}},$$

where the *reaction coefficient* or *Arrhenius factor* $k_j(T)$ is given by

$$k_j(T) = B_j T^{\alpha_j} \exp\left(-\frac{E_j}{RT}\right),$$

with $E_j > 0$ representing the *activation energy*, $B_j T^{\alpha_j}$ representing the *frequency factor*, B_j and α_j are positive constants, the absolute temperature T is assumed to be the same for all species, and R is the gas constant. The basic idea behind the product formula $\prod_{m=1}^{N} c_m^{\nu_{m,j}}$ is that the reaction rate is proportional to the molar concentrations of the reactants with each reactant A_m counted $\nu_{m,j}$ times. The Arrhenius factor is small if T is below some threshold value corresponding to the quotient $\frac{E_j}{RT}$ being moderately large.

The net production rate (moles per volume per unit time) of species A_n in reaction j is given by $\alpha_{n,j} r_j$, where

$$\alpha_{n,j} = \lambda_{n,j} - \nu_{n,j},$$

and the total net production rate s_n of species n is given by

$$s_n = \sum_{j=1}^{J} \alpha_{n,j} r_j.$$

We now assume that the temperature T is constant and is given, and we seek the vector of concentration $c(t) = (c_1(t), \ldots, c_N(t))$ as a function of time t describing the dynamics of the set of reactions for $t > 0$, assuming that $c(0) = c^0$, where $c^0 = (c_1^0, \ldots, c_N^0)$ is a given vector of initial concentrations. Using the balance equation $\dot{c}_n = s_n$ for each species $n = 1, 2, \ldots, N$, we obtain the following initial value problem for a system of ordinary differential equations: Find $c(t) = (c_1(t), \ldots, c_N(t))$ such that

$$\begin{cases} \dot{c}_n(t) = \sum_{j=1}^{J} \alpha_{n,j} k_j(T) \prod_{m=1}^{N} c_m(t)^{\nu_{m,j}} & \text{for } t > 0, \ n = 1, \ldots, N, \\ c(0) = c^0. \end{cases}$$

$$\tag{74.2}$$

This is an initial value problem of the form $\dot{u}(t) = f(u(t))$ for $t > 0$, $u(0) = u^0$, where $u(t) = c(t)$ and $f : \mathbb{R}^N \to \mathbb{R}^N$ is a given function.

An *Equilibrium* for a given temperature T corresponding to $\dot{c}_n(t) = 0$ for $t > 0$, $n = 1, \ldots, N$, is characterized by the algebraic system of equations

$$\sum_{j=1}^{J} \alpha_{n,j} k_j(T) \prod_{m=1}^{N} c_m^{\nu_{m,j}} = 0 \quad n = 1, \ldots, N, \tag{74.3}$$

corresponding to the equation $f(u) = 0$.

Example 74.1. The reaction

$$2NO + Cl_2 \to 2NOCl,$$

can be put in the form (74.1) with $A_1 = NO$, $A_2 = Cl_2$, $A_3 = NOCl$, $N = 3$, $J = 1$, $\nu_{1,1} = 2$, $\nu_{2,1} = 1$, $\nu_{3,1} = 0$, $\lambda_{1,1} = 0$, $\lambda_{2,1} = 0$, $\lambda_{3,1} = 2$, $\alpha_{1,1} = -2$, $\alpha_{2,1} = -1$, and $\alpha_{3,1} = 2$.

Example 74.2. The two reactions

$$2NO + Cl_2 \to^{k_1} 2NOCl,$$

$$2NOCl \to^{k_2} 2NO + Cl_2,$$

can be put in the form (74.1) with $A_1 = NO$, $A_2 = Cl_2$, $A_3 = NOCl$, $N = 3$, $J = 2$, $\nu_{1,1} = 2$, $\nu_{2,1} = 1$, $\nu_{3,1} = 0$, $\lambda_{1,1} = 0$, $\lambda_{2,1} = 0$, $\lambda_{3,1} = 2$, $\alpha_{1,1} = -2$, $\alpha_{2,1} = -1$, $\alpha_{3,1} = 2$, $\nu_{1,2} = 0$, $\nu_{2,2} = 0$, $\nu_{3,2} = 2$, $\lambda_{1,2} = 2$, $\lambda_{2,2} = 1$, $\lambda_{3,2} = 0$, $\alpha_{1,2} = 2$, $\alpha_{2,2} = 1$, and $\alpha_{3,2} = -2$. Equilibrium is characterized by

$$k_1 c_1^2 c_2 = k_2 c_3^2, \quad \text{or} \quad \frac{c_1^2 c_2}{c_3^2} = \frac{k_2}{k_1}.$$

Example 74.3. An *ideal first order tank reactor* is modeled by the equation

$$qc^0 - Vkc = qc,$$

where c^0 is the reactant concentration at inflow, c is the concentration in the reactor, q is the inflow (= outflow) rate, V is the volume of the reactor and k is a reaction coefficient. The equation expresses that the (rate of) reactant inflow minus the reactant consumed in the reaction is equal to the reactant outflow. Introducing $\tau = \frac{V}{q}$, which is the time the reactant stays in the reactor, we get

$$c = \frac{c^0}{1 + \tau k}.$$

The *efficiency* of the reactor is given by

$$\eta = \frac{c^0 - c}{c^0} = \frac{\tau k}{1 + \tau k} = \frac{1}{1 + \frac{1}{\tau k}}.$$

We see in particular that the efficiency decreases as τ decreases.

Example 74.4. An *ideal first order tube reactor* occupying the interval $(0,1)$, which may be viewed as a set of ideal first order tank reactors coupled in series, is modeled by

$$qc(x) - A\Delta xkc(x) = qc(x + \Delta x) \quad \text{for } 0 < x < 1,$$

where q is the (constant) flow rate, A the cross section of the tube, and Δx is a small increment in x. Dividing by Δx and letting Δx tend to zero leads to the initial value problem of finding the concentration $c(x)$ for $0 \le x \le 1$ such that

$$\frac{dc}{dx} = -\tau kc \quad \text{for } 0 < x \le 1, \quad c(0) = c^0,$$

where $\tau = \frac{A}{q}$. The solution is given by $c(x) = c^0 e^{-\tau kx}$, and the efficiency $\eta = \frac{c^0 - c(1)}{c^0} = 1 - e^{-\tau k}$. Using the fact that $\frac{x}{1+x} < 1 - e^{-x}$ for $x > 0$, it follows that the ideal tube reactor is more efficient than the ideal tank reactor.

74.2   Variable Temperature

Suppose now that the temperature $T(t)$ is variable with time t, and is unknown along with the concentrations $c_1(t), \ldots, c_N(t)$. The *heat of reaction* of reaction j is given by

$$\left(-\sum_{m=1}^{N} \alpha_{m,j} h_m \right) r_j,$$

where h_m is the *molar enthalpy* of species A_m. The heat of reaction is positive for an *exothermic reaction* and negative for an *endothermic reaction*.

The problem is now to find $c(t) = (c_1(t), \ldots, c_N(t))$ and $T(t)$ for $t > 0$ such that

$$\begin{cases} \dot{c}_n = \sum_{j=1}^{J} \alpha_{n,j} k_j(T) \prod_{m=1}^{N} c_m^{\nu_{m,j}}, & t > 0, \, n = 1, \ldots, N, \\ C_p \dot{T} = \sum_{j=1}^{J} (-\sum_{m=1}^{N} \alpha_{m,j} h_m) k_j(T) \prod_{m=1}^{N} c_j^{\nu_{m,j}}, \\ c(0) = c^0, \, T(0) = T^0, \end{cases} \tag{74.4}$$

where $c^0 = (c_1^0, \ldots, c_N^0)$ and T^0 are given initial concentrations and temperature, and C_p is the specific heat of the mixture of species.

74.3   Space Dependence

Adding spacial dependence in a domain Ω in $\mathbb{R}^3$, we are led to the following model: Find $c(x,t) = (c_1(x,t), \ldots, c_N(x,t))$ and $T(x,t)$ for $x \in \Omega$, $t > 0$, such that

$$
\begin{cases}
\dot{c}_n + \nabla \cdot (c_n \beta) - \nabla \cdot (\epsilon_n \nabla c_n) \\
\quad = \sum_{j=1}^{J} \alpha_{n,j} k_j(T) \prod_{m=1}^{N} c_m^{\nu_{m,j}} & \text{for } x \in \Omega,\, t > 0,\, n = 1, \dots, N, \\
C_p \dot{T} + \nabla \cdot (C_p T \beta) - \nabla \cdot (\epsilon_0 \nabla T) \\
\quad = \sum_{j=1}^{J} \left(-\sum_{m=1}^{N} \alpha_{m,j} h_m \right) k_j(T) \prod_{m=1}^{N} c_m^{\nu_{m,j}} & \text{for } x \in \Omega,\, t > 0, \\
c(x,0) = c^0, \quad T(x,0) = T^0 & \text{for } x \in \Omega,
\end{cases}
\tag{74.5}
$$

where $\beta(x,t)$ is a given convection velocity, and the ϵ_n are given diffusion coefficients. The system is complemented by boundary conditions of Dirichlet, Neumann or Robin type for each equation.

Example 74.5. A stationary one species constant temperature first order reaction with constant diffusion and zero convection is modeled in dimensionless form by the equation

$$
\Delta u = \varphi^2 u \quad \text{in } \Omega,
$$

together with Dirichlet, Neumann or Robin boundary conditions, where φ is the *Thiele modulus*, and Ω is a domain in $\mathbb{R}^d$, $d = 1, 2, 3$. A quantity of interest as a function of Ω, the reaction coefficient φ^2 and the boundary conditions, is the *total production* $\int_\Omega u(x)\, dx$.

Example 74.6. A simple model for *flame propagation* in a channel takes the form

$$
\begin{cases}
\dot{u}_1 - \Delta u_1 + \beta_1 \dfrac{\partial u_1}{\partial x_1} = u_2 f(u_1) & x \in \Omega,\, t > 0, \\
\dot{u}_2 - \Delta u_2 + \beta_1 \dfrac{\partial u_2}{\partial x_1} = -u_2 f(u_1) & x \in \Omega,\, t > 0,\, x \in \Omega,
\end{cases}
\tag{74.6}
$$

together with appropriate boundary conditions, where $\Omega = \mathbb{R} \times (0,1)$, u_1 represents temperature, u_2 represents a reactant concentration, β_1 is the velocity of the reactant in the x_1 direction, and $u_2 f(u_1)$ represents a reaction rate with $f : \mathbb{R}^+ \to \mathbb{R}^+$ given. With a proper choice of β_1 we may seek a stationary solution with $\dot{u} = 0$ corresponding to a propagating flame front.

Example 74.7. A basic model for *combustion* in a domain Ω in $\mathbb{R}^3$ takes the form: Find the concentration c and temperature T such that:

$$
\begin{cases}
\dot{c} - \epsilon_1 \Delta c = -B_1 e^{-\frac{E}{RT}} c, & x \in \Omega,\, t > 0, \\
\dot{T} - \epsilon_0 \Delta T = B_0 e^{-\frac{E}{RT}} c & x \in \Omega,\, t > 0,
\end{cases}
\tag{74.7}
$$

together with, say, homogeneous Neumann boundary conditions, and with B_0 and B_1 positive constants. Depending on the activation energy E and initial conditions, the process may be fast or slow locally in space and time.

Axiom 1: All bodies are either in motion or at rest.

Axiom 2: Each single body can move at varying speeds.

Lemma 1: Bodies are distinguished from one another in respect of motion and rest, quickness and slowness, and not in respect of substance.

Lemma 2: All bodies agree in certain respects.

Lemma 3: A body in motion or at rest must have been determined to motion or rest by another body, which likewise has been determined to motion or rest by another body, and that body by another, and so ad infinitum.

...

Lemma 6: If certain bodies composing an individual thing are made to change the existing direction of their motion, but in such a way that they can continue their motion and keep the same mutual relation as before, the individual thing will likewise preserve the same mutual relation as before, the individual thing will likewise preserve its own nature without change of form.

(Spinoza 1632–1677, Ethica II)

75
Calculus Tool Bag II

Timeo hominem unius libri. (St. Thomas of Aquino)

75.1 Introduction

We here collect the basic tools of Calculus of functions $f : \mathbb{R}^n \to \mathbb{R}^m$, that is Calculus of vector-valued functions of several real variables. The Euclidean norm of a vector $x = (x_1, \ldots, x_n) \in \mathbb{R}^n$ is denoted by $\|x\| = \sum_{i=1}^{n} x_i^2$.

75.2 Lipschitz Continuity

A function $f : A \to \mathbb{R}^m$ with a subset of $\mathbb{R}^n$ is Lipschitz continuous on A if there is a constant L such that

$$\|f(x) - f(y)\| \le L\|x - y\| \quad \text{for all } x, y \in A.$$

75.3 Differentiability

A function $f : A \to \mathbb{R}^m$ is *differentiable at* $\bar{x} \in A$, where A is an open subset of $\mathbb{R}^n$, if there is a $m \times n$ matrix $f'(\bar{x})$, called the *Jacobian* of the function $f(x)$ at $\bar{x}$, and a constant $K_f(\bar{x})$, such that for all $x \in A$ close to $\bar{x}$,

$$f(x) = f(\bar{x}) + f'(\bar{x})(x - \bar{x}) + E_f(x, \bar{x}),$$

where $E_f(x, \bar{x})$ is an m-vector satisfying $\|E_f(x, \bar{x})\| \leq K_f(\bar{x})\|x - \bar{x}\|^2$. We say that $f : A \to \mathbb{R}^m$ is *uniformly differentiable on* A if the constant $K_f(\bar{x}) = K_f$ can be chosen independently of $\bar{x} \in A$. We write $f' = \nabla f$ if $m = 1$ and call ∇f the gradient of f.

75.4   The Chain Rule

If $g : \mathbb{R}^n \to \mathbb{R}^m$ is differentiable at $\bar{x} \in \mathbb{R}^n$, and $f : \mathbb{R}^m \to \mathbb{R}^p$ is differentiable at $g(\bar{x}) \in \mathbb{R}^m$ and further $g : \mathbb{R}^n \to \mathbb{R}^m$ is Lipschitz continuous, then the composite function $f \circ g : \mathbb{R}^n \to \mathbb{R}^p$ is differentiable at $\bar{x} \in \mathbb{R}^n$ with Jacobian

$$(f \circ g)'(\bar{x}) = f'(g(\bar{x}))g'(\bar{x}).$$

75.5   Mean Value Theorem for $f : \mathbb{R}^n \to \mathbb{R}$

If $f : \mathbb{R}^n \to \mathbb{R}$ is differentiable on $\mathbb{R}^n$ with a Lipschitz continuous gradient ∇f, then for given x and $\bar{x}$ in $\mathbb{R}^n$, there is $y = x + \bar{t}(x - \bar{x})$ with $\bar{t} \in [0, 1]$, such that

$$f(x) - f(\bar{x}) = \nabla f(y) \cdot (x - \bar{x}).$$

75.6   A Minimum Point Is a Stationary Point

If $\bar{x} \in \mathbb{R}^n$ is a *local minimum point* of a differentiable function $f : \mathbb{R}^n \to \mathbb{R}$, that is, $f(\bar{x}) \leq f(x)$ for all x close to $\bar{x}$, then $\nabla f(\bar{x}) = 0$.

75.7   Taylor's Theorem

If $f : \mathbb{R}^n \to \mathbb{R}$ is twice differentiable with Lipschitz continuous Hessian $H = (h_{ij})$ with elements $h_{ij} = \frac{\partial^2 f}{\partial x_i \partial x_j}$, then, for given x and $\bar{x} \in \mathbb{R}^n$, there is $y = x + \bar{t}(x - \bar{x})$ with $\bar{t} \in [0, 1]$, such that

$$f(x) = f(\bar{x}) + \nabla f(\bar{x}) \cdot (x - \bar{x}) + \frac{1}{2} \sum_{i,j=1}^{n} \frac{\partial^2 f}{\partial x_i \partial x_j}(y)(x_i - \bar{x}_i)(x_j - \bar{x}_j)$$

$$= f(\bar{x}) + \nabla f(\bar{x}) \cdot (x - \bar{x}) + \frac{1}{2}(x - \bar{x})^\top H(y)(x - \bar{x}).$$

75.8 Contraction Mapping Theorem

If $g : \mathbb{R}^n \to \mathbb{R}^n$ is Lipschitz continuous with Lipschitz constant $L < 1$, then the equation $x = g(x)$ has a unique solution $\bar{x} = \lim_{i \to \infty} x^{(i)}$, where $\{x^{(i)}\}_{i=1}^{\infty}$ is a sequence in $\mathbb{R}^n$ generated by Fixed Point Iteration: $x^{(i)} = g(x^{(i-1)})$, starting with any initial value $x^{(0)}$.

75.9 Inverse Function Theorem

Let $f : \mathbb{R}^n \to \mathbb{R}^n$ and assume the coefficients of $f'(x)$ are Lipschitz continuous close to $\bar{x}$ and $f'(\bar{x})$ is non-singular. Then for y sufficiently close to $\bar{y} = f(\bar{x})$, the equation $f(x) = y$ has a unique solution x. This defines x as a function $x = f^{-1}(y)$ of y.

75.10 Implicit Function Theorem

If $f : \mathbb{R}^n \times \mathbb{R}^m \to \mathbb{R}^n$ with $f(x, y) \in \mathbb{R}^n$ and $x \in \mathbb{R}^n$ and $y \in \mathbb{R}^m$, $f(\bar{x}, \bar{y}) = 0$, and the Jacobian $f'_x(x, y)$ with respect to x is Lipschitz continuous for x close to $\bar{x}$ and y close to $\bar{y}$, and $f'_x(\bar{x}, \bar{y})$ is non-singular, then for y close to $\bar{y}$, the equation $f(x, y) = 0$ has a unique solution $x = g(y)$, which defines x as a function $g(y)$ of y.

75.11 Newton's Method

If $\bar{x}$ is a root of $f : \mathbb{R}^n \to \mathbb{R}^n$ such that $f(x)$ is uniformly differentiable with a Lipschitz continuous derivative close to $\bar{x}$ and $f'(\bar{x})$ is non-singular, then Newton's method $x^{(i+1)} = x^{(i)} - f'(x^{(i)})^{-1} f(x^{(i)})$ for solving $f(x) = 0$ converges quadratically if started sufficiently close to $\bar{x}$.

75.12 Differential Operators

Gradient of a function $u : \mathbb{R}^d \to \mathbb{R}$:

$$\operatorname{grad} u = \nabla u = \left(\frac{\partial u}{\partial x_1}, \frac{\partial u}{\partial x_2}, \ldots, \frac{\partial u}{\partial x_d} \right).$$

Divergence of a vector function $u : \mathbb{R}^d \to \mathbb{R}^d$:

$$\operatorname{div} u = \nabla \cdot u = \sum_{i=1}^{d} \frac{\partial u_i}{\partial x_i}.$$

Rotation of a vector function $u : \mathbb{R}^3 \to \mathbb{R}^3$:

$$\text{rot } u = \nabla \times u = \left(\frac{\partial u_3}{\partial x_2} - \frac{\partial u_2}{\partial x_3}, \frac{\partial u_1}{\partial x_3} - \frac{\partial u_3}{\partial x_1}, \frac{\partial u_2}{\partial x_1} - \frac{\partial u_1}{\partial x_2} \right).$$

Laplacian of a function $u : \mathbb{R}^d \to \mathbb{R}$:

$$\Delta u = \nabla \cdot (\nabla u) = \text{div (grad } u) = \sum_{i=1}^{d} \frac{\partial^2 u}{\partial x_i^2}.$$

Identities:

$$\nabla \cdot (\nabla \times u) = 0,$$
$$\nabla \times (\nabla u) = 0,$$
$$\nabla \times (\nabla \times u) = -\Delta u + \nabla(\nabla \cdot u).$$

Laplacian in $\mathbb{R}^2$ in polar coordinates $x = (x_1, x_2) = (r \cos(\theta), r \sin(\theta))$:

$$\Delta u = \frac{1}{r} \frac{\partial}{\partial r} \left(r \frac{\partial u}{\partial r} \right) + \frac{1}{r^2} \frac{\partial^2 u}{\partial \theta^2}.$$

Laplacian in spherical coordinates
$x = (r \sin(\varphi) \cos(\theta), r \sin(\varphi) \sin(\theta), r \cos(\varphi))$:

$$\Delta u = \frac{1}{r^2} \frac{\partial}{\partial r} \left(r^2 \frac{\partial u}{\partial r} \right) + \frac{1}{r^2 \sin(\theta)} \frac{\partial}{\partial \theta} \left(\sin(\theta) \frac{\partial u}{\partial \theta} \right) + \frac{1}{r^2 \sin^2(\theta)} \frac{\partial^2 u}{\partial \varphi^2}.$$

The Laplacian is invariant under orthogonal coordinate transformations in $\mathbb{R}^d$.

75.13   Curve Integrals

If $\Gamma = s([a, b])$ is a curve in $\mathbb{R}^n$ given by the function $s : [a, b] \to \mathbb{R}^n$, and $u : \Gamma \to \mathbb{R}$, then

$$\int_\Gamma u \, ds = \int_\Gamma u(x) \, ds(x) \equiv \int_a^b u(s(t)) \|s'(t)\| \, dt,$$
$$\int_\Gamma u \cdot ds = \int_a^b u(s(t)) \cdot s'(t) \, dt,$$
$$\int_\Gamma ds = \int_a^b \|s'(t)\| \, dt = \text{length of } \Gamma.$$

If $u = \nabla \varphi$, then

$$\int_\Gamma u \cdot ds = \varphi(s(b)) - \varphi(s(a)).$$

75.14   Multiple Integrals

Integral over the unit square: If $f : Q = [0,1] \times [0,1] \to \mathbb{R}$ is Lipschitz continuous, then

$$\int_Q f(x)\, dx = \int_0^1 \int_0^1 f(x_1, x_2)\, dx_1\, dx_2 = \lim_{n \to \infty} \sum_{i=1}^N \sum_{j=1}^N f(x_{1,i}^n, x_{2,j}^n) h_n h_n,$$

where $h_n = 2^{-n}$, $x_{j,i}^n = ih_n$, $N = 2^n$, and

$$\int_Q f(x)\, dx = \int_0^1 \left(\int_0^1 f(x_1, x_2)\, dx_2 \right) dx_1 = \int_0^1 \left(\int_0^1 f(x_1, x_2)\, dx_1 \right) dx_2.$$

Change of variables: If $y \to x = g(y)$ maps a domain $\tilde{\Omega}$ in $\mathbb{R}^d$ onto a domain Ω in $\mathbb{R}^d$, where the Jacobian of g is Lipschitz continuous and $f : \Omega \to \mathbb{R}$ be Lipschitz continuous, then

$$\int_\Omega f(x)\, dx = \int_{\tilde{\Omega}} f(g(y))|\det g'(y)|\, dy,$$

Polar coordinates:

$$\int_\Omega f(x_1, x_2)\, dx_1 dx_2 = \int_{\tilde{\Omega}} f(r\cos(\theta), r\sin(\theta))\, r dr\, d\theta,$$

where $(r, \theta) \to x$ is a one-to-one mapping of $\tilde{\Omega}$ onto Ω given by $x = (r\cos(\theta), r\sin(\theta))$.

Spherical coordinates:

$$\int_\Omega f(x)\, dx$$
$$= \int_{\tilde{\Omega}} f(r\sin(\varphi)\cos(\theta), r\sin(\varphi)\sin(\theta), r\cos(\varphi)) r^2 \sin(\varphi)\, dr\, d\theta\, d\varphi,$$

where $(r, \theta, \varphi) \to x$ is a one-to-one mapping of $\tilde{\Omega}$ onto Ω given by $x = (r\sin(\varphi)\cos(\theta), r\sin(\varphi)\sin(\theta), r\cos(\varphi))$.

75.15   Surface Integrals

If $S = s(\Omega)$ is a surface in $\mathbb{R}^3$ parameterized by the mapping $s : \Omega \to \mathbb{R}^3$, where Ω is a domain in $\mathbb{R}^2$, and $u : S \to \mathbb{R}$ is a real-valued function defined on S, then

$$\int_S u\, ds = \int_\Omega u(s(y)) \| s_{,1}'(y) \times s_{,2}'(y) \|\, dy,$$

where $s_{,i}' = (\frac{\partial s_1}{\partial y_i}, \frac{\partial s_2}{\partial y_i}, \frac{\partial s_3}{\partial y_i})$.

75.16   Green's and Gauss' Formulas

If Ω is a domain in $\mathbb{R}^3$ with boundary Γ with outward unit normal $n = (n_1, n_2, n_3)$, and $u : \Omega \to \mathbb{R}^3$ and $v, w : \Omega \to \mathbb{R}$, then

$$\int_\Omega \frac{\partial v}{\partial x_i} \, dx = \int_\Gamma v \, n_i \, ds, \quad i = 1, 2, 3.$$

$$\int_\Omega \frac{\partial v}{\partial x_i} w \, dx = \int_\Gamma v w \, n_i \, ds - \int_\Omega v \frac{\partial w}{\partial x_i} \, dx, \quad i = 1, 2, 3.$$

$$\int_\Omega \nabla \cdot u \, dx = \int_\Gamma u \cdot n \, ds, \quad \text{(Gauss' Divergence theorem)}$$

$$\int_\Omega \nabla \times u \, dx = \int_\Gamma n \times u \, ds,$$

$$\int_\Omega \nabla v \cdot \nabla w \, dx = \int_\Gamma v \partial_n w \, ds - \int_\Omega v \Delta w \, dx,$$

$$\int_\Omega v \Delta w \, dx - \int_\Omega \Delta v \, w \, dx = \int_\Gamma v \, \partial_n w \, ds - \int_\Gamma \partial_n v \, w \, ds.$$

75.17   Stokes' Theorem

If S is a surface in $\mathbb{R}^3$ bounded by a closed curve Γ, n is a unit normal to S, Γ is oriented in a clockwise direction following the positive direction of the normal n, and $u : \mathbb{R}^3 \to \mathbb{R}^3$ is differentiable, then

$$\int_S (\nabla \times u) \cdot n \, ds = \int_\Gamma u \cdot ds.$$

76

Piecewise Linear Polynomials in $\mathbb{R}^2$ and $\mathbb{R}^3$

...usually he sat in a comfortable attitude, looking down, slightly stooped, with hands folded above his lap. He spoke quite freely, very clearly, simply and plainly: but when he wanted to emphasize a new viewpoint ...then he lifted his head, turned to one of those sitting next to him, and gazed at him with his beautiful, penetrating blue eyes during the emphatic speech. ...If he proceeded from an explanation of principles to the development of mathematical formulas, then he got up, and in a stately very upright posture he wrote on a blackboard beside him in his peculiarly beautiful handwriting: he always succeeded through economy and deliberate arrangement in making do with a rather small space. For numerical examples, on whose careful completion he placed special value, he brought along the requisite data on little slips of paper. (Dedekind about Gauss)

76.1 Introduction

In this chapter, we prepare for the application of FEM to partial differential equations by discussing approximation of functions by piecewise linear functions in in $\mathbb{R}^2$ and $\mathbb{R}^3$. We consider three main topics: (i) the construction of a mesh, or *triangulation*, for a domain in $\mathbb{R}^2$ or $\mathbb{R}^3$, (ii) the construction piecewise linear functions on a triangulation, and (iii) estimation of interpolation errors.

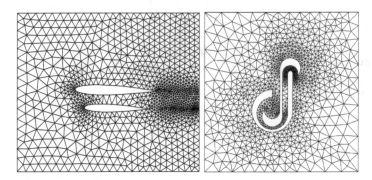

Fig. 76.1. The mesh on the *left* was used in a computation of the flow of air around two airfoils. The mesh on the *right* was used to discretize a piece of metal punched with a fancy character. In both cases, the meshes are adapted to allow accurate computation, taking into account both the behavior of the solution and the shape of the domain

76.2 Triangulation of a Domain in $\mathbb{R}^2$

We start by considering a two-dimensional domain Ω with a polygonal boundary Γ. A *triangulation* $\mathcal{T}_h = \{K\}$ is a sub-division of Ω into a non-overlapping set of triangles, or *elements*, K constructed so that no vertex of one triangle lies on the edge of another triangle, see Fig. 76.2. We use $\mathcal{N}_h = \{N\}$ to denote the set of *nodes* N or corners of the triangles, usually numbered $N_1, N_2, \ldots, N_M$, where M is the total number of nodes. A triangulation is specified by a list of the coordinates of the nodes, together with a list containing the numbers of the nodes of each triangle. We may also list the set of triangle sides or *edges* $\mathcal{S}_h = \{S\}$, with each edge S specified by the node numbers of its two end-points, and a list of the nodes and edges on the boundary Γ.

We measure the size of a triangle $K \in \mathcal{T}_h$, by the length h_K of its largest side, which is called the *diameter* of the triangle. The *mesh function* $h(x)$ associated to a triangulation $\mathcal{T}_h$ is the piecewise constant function defined so $h(x) = h_K$ for $x \in K$ for each $K \in \mathcal{T}_h$. We measure the degree of *isotropy* of an element $K \in \mathcal{T}_h$ by its smallest angle τ_K. If $\tau_K \approx \pi/3$ then K is almost isosceles, while if τ_K is small then K is thin, see Fig. 76.3. We use the smallest angle among the triangles in $\mathcal{T}_h$, i.e.

$$\tau = \min_{K \in \mathcal{T}_h} \tau_K$$

as a measure of the degree of anistropy of the triangulation $\mathcal{T}_h$. We shall see below that certain interpolation errors related to approximation with piecewise linear functions on a given triangulation get larger as τ tends to zero, corresponding to allowing the triangles to very thin.

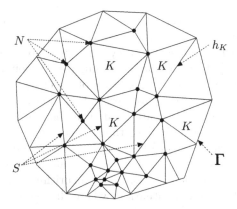

Fig. 76.2. A triangulation of a domain Ω

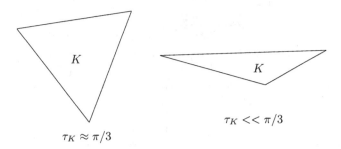

Fig. 76.3. Measuring the isotropy of a triangle

The basic problem of *mesh generation* is to generate a triangulation of a given domain with mesh size given by a prescribed mesh function $h(x)$. This problem arises in each step of an adaptive algorithm, where a new mesh function is computed from an approximate solution on a given mesh, and a new mesh is constructed with mesh size given by the new mesh function. The process is then repeated until a stopping criterion is satisfied. The new mesh may be constructed from scratch or by modification of the previous mesh including local refinement or coarsening.

In the *advancing front* strategy a mesh with given mesh size is constructed beginning at some point (often on the boundary) by successively adding one triangle after another, each with a mesh size determined by the mesh function. The curve dividing the domain into a part already triangulated and the remaining part is called the *front*. The front sweeps through the domain during the triangulation process. An alternative is to use a *h-refinement* strategy, where a mesh with a specified local mesh size is constructed by successively dividing elements of an initial coarse triangulation with the elements referred to as *parents*, into smaller elements, called the *children*. We illustrate the refinement and advancing front strategies in

Fig. 76.4. It is often useful to combine the two strategies using the advancing front strategy to construct an initial mesh that represents the geometry of the domain with adequate accuracy, and use adaptive h-refinement.

There are various strategies for performing the division in an h-refinement aimed at limiting the degree of anisotropy of the elements. After the refinements are completed, the resulting mesh is fixed up by the addition of edges aimed at avoiding nodes that are located in the middle of element sides. This causes a mild "spreading" of the adapted region. We illustrate one technique for h-refinement in Fig. 76.5. In general, refining a mesh tends to introduce elements with small angles, as can be seen in Fig. 76.5 and it is an interesting problem to construct algorithms for mesh refinement that avoid this tendency in situations where the degree of anisotropy has

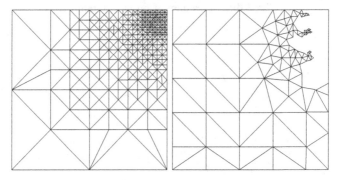

Fig. 76.4. The mesh on the *left* is being constructed by successive h refinement starting from the coarse parent mesh drawn with *thick lines.* The mesh on the *right* is being constructed by an advancing front strategy. In both cases, high resolution is required near the upper right-hand corner

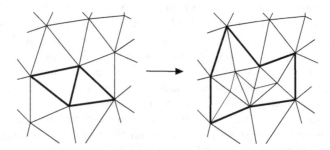

Fig. 76.5. On the *left*, two elements in the mesh have been marked for refinement. The refinement uses the Rivara algorithm in which an element is divided into two pieces by inserting a side connecting the node opposite the longest side to the midpoint of the longest side. Additional sides are added to avoid having a node of one element on the side of another element. The refinement is shown in the mesh on the *right* along with the boundary of all the elements that had to be refined in addition to those originally marked for refinement

to be limited. On the other hand, in certain circumstances, it is important to use "stretched" meshes that have regions of thin elements aligned together to give a high degree of refinement in one direction. In these cases, we also introduce mesh functions that give the local stretching, or degree of anisotropy, and the orientation of the elements. We discuss the construction and use of such meshes in the advanced companion volume.

76.3 Mesh Generation in $\mathbb{R}^3$

Mesh generation in three dimensions is analogous to that in two dimensions with the triangles being replaced by *tetrahedra*. In practice, the geometric constraints involved become more complicated and the number of elements also increases drastically. We show some examples in Fig. 76.6 and Fig. 76.7, and further examples in Fig. 76.14 and Fig. 76.13.

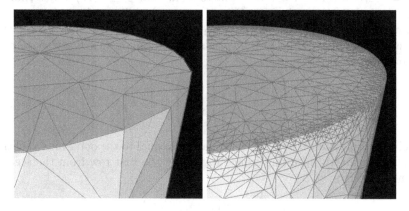

Fig. 76.6. Initial and refined mesh of cylinder

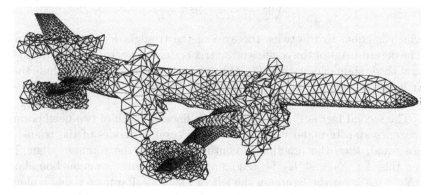

Fig. 76.7. The surface mesh on the body, and parts of a tetrahedral mesh around a Saab 2000

76.4   Piecewise Linear Functions

Let $T_h = \{K\}$ be a triangulation of a two-dimensional domain Ω with piece-wise polynomial boundary Γ, let $\mathcal{N}_h = \{N\}$ denote the nodes of T_h and introduce the corresponding the finite dimensional vector space V_h consisting of the continuous piecewise linear functions on T_h. In other words,

$$V_h = \{v : v \text{ is continuous on } \Omega, v|_K \in \mathcal{P}(K) \text{ for } K \in T_h\},$$

where $\mathcal{P}(K)$ denotes the set of linear functions on K, i.e., the set of functions $v(x) = v(x_1, x_2)$ of the form $v(x) = c_0 + c_1 x_1 + c_2 x_2$ for some constants c_i. We can describe a function $v(x)$ in V_h by the nodal values $v(N)$ with $N \in \mathcal{N}_h$ because of two facts. The first is that a linear function is uniquely determined by its values at three points, as long as they don't lie on a straight line. To prove this claim, let $K \in T_h$ have vertices $a^i = (a_1^i, a_2^i)$, $i = 1, 2, 3$, see Fig. 76.8. We want to show that $v \in \mathcal{P}(K)$ is determined uniquely by $\{v(a^1), v(a^2), v(a^3)\} = \{v_1, v_2, v_3\}$. A linear function v can be written $v(x_1, x_2) = c_0 + c_1 x_1 + c_2 x_2$ for some constants c_0, c_1, c_2. Substituting the nodal values of v into this expression yields a linear system of equations:

$$\begin{pmatrix} 1 & a_1^1 & a_2^1 \\ 1 & a_1^2 & a_2^2 \\ 1 & a_1^3 & a_1^3 \end{pmatrix} \begin{pmatrix} c_0 \\ c_1 \\ c_2 \end{pmatrix} = \begin{pmatrix} v_1 \\ v_2 \\ v_3 \end{pmatrix}.$$

The determinant of the coefficient matrix is equal to the determinant of the following matrix resulting from subtracting the first row from the second and third row:

$$\begin{pmatrix} 1 & a_1^1 & a_2^1 \\ 0 & a_1^2 - a_1^1 & a_2^2 - a_2^1 \\ 0 & a_1^3 - a_1^1 & a_2^2 - a_2^1 \end{pmatrix},$$

which is equal to the twice the area of the triangle K (up to the sign). The determinant of the coefficient matrix is thus non-zero, and we conclude that the system of equations (76.4) has a unique solution. We conclude that at linear function is uniquely specified by its values at three non-colinear points.

The second fact is that if a function is linear in each of two neighboring triangles and its nodal values on the two common nodes of the triangles are equal, then the function is continuous across the common edge. To see this, let K_1 and K_2 be adjoining triangles with common boundary $\partial K_1 = \partial K_2$; see the figure on the left in Fig. 76.9. Parameterizing v along this boundary, we see that v is a linear function of one variable there. Such functions are determined uniquely by the value at two points, and therefore since the values of v on K_1 and K_2 at the common nodes agree, the values

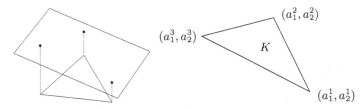

Fig. 76.8. On the *left*, we show that the three nodal values on a triangle determine a linear function. On the *right*, we show the notation used to describe the nodes of a typical triangle

of v on the common boundary between K_1 and K_2 agree, and v is indeed continuous across the boundary.

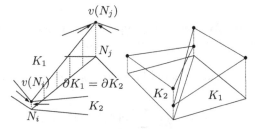

Fig. 76.9. On the *left*, we show that a function that is piecewise linear on triangles reduces to a linear function of one variable on triangle edges. On the *right*, we plot a function that is piecewise linear on triangles whose values at the common nodes on two neighboring triangles do not agree

To construct a set of basis functions for V_h, we begin by describing a set of *element basis functions* for triangles. Once again, assuming that a triangle K has nodes at $\{a^1, a^2, a^3\}$, the element nodal basis is the set of functions $\lambda_i \in \mathcal{P}(K)$, $i = 1, 2, 3$, such that

$$\lambda_i(a^j) = \begin{cases} 1, & i = j, \\ 0, & i \neq j. \end{cases}$$

We show these functions in Fig. 76.10.

We construct the *global* basis functions for V_h by piecing together the element basis functions on neighboring elements using the continuity requirement, i.e. by matching element basis functions on neighboring triangles that have the same nodal values on the common edge. The resulting set of basis functions $\{\varphi_j\}_{j=1}^M$, where $N_1, N_2, \ldots, N_M$ is an enumeration of the nodes $N \in \mathcal{N}_h$, is called the set of *tent* functions. The tent functions can also be defined by specifying that $\varphi_j \in V_h$ satisfy

$$\varphi_j(N_i) = \begin{cases} 1, & i = j, \\ 0, & i \neq j, \end{cases}$$

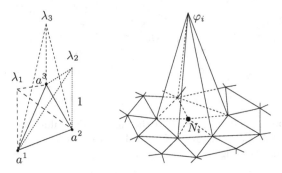

Fig. 76.10. On the *left*, we show the three element nodal basis functions for the linear functions on K. On the *right*, we show a typical global basis "tent" function

for $i, j = 1, \ldots, M$. We illustrate a typical tent function in Fig. 76.10. We see in particular that the *support* of φ_i is the set of triangles that share the common node N_i.

The tent functions are a nodal basis for V_h because if $v \in V_h$ then

$$v(x) = \sum_{i=1}^{M} v(N_i)\varphi_i(x).$$

76.5   Max-Norm Error Estimates

In this section we prove the basic pointwise maximum norm error estimate for linear interpolation on a triangle, which states that the interpolation error depends on the second order partial derivatives of the function being interpolated, i.e. on the "curvature" of the function, the mesh size and the shape of the triangle. Analogous results hold for other norms. The results also extend directly to more than two space dimensions.

Let K be a triangle with vertices $a^i, i = 1, 2, 3$. Given a continuous function v defined on K, let the linear interpolant $\pi_K v \in \mathcal{P}(K)$ be defined by

$$\pi_K v(a^i) = v(a^i), \quad i = 1, 2, 3.$$

We illustrate this in Fig. 76.11.

Theorem 76.1 *If v has continuous second derivatives, then*

$$\|v - \pi_K v\|_{L_\infty(K)} \leq \frac{1}{2} h_K^2 \|D^2 v\|_{L_\infty(K)}, \tag{76.1}$$

$$\|\nabla(v - \pi_K v)\|_{L_\infty(K)} \leq \frac{3}{\sin(\tau_K)} h_K \|D^2 v\|_{L_\infty(K)}, \tag{76.2}$$

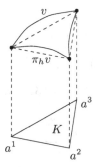

Fig. 76.11. The nodal interpolant $\pi_K v$ of v. (change from π_h to π_K)

where h_K is the largest side of K, τ_K is the smallest angle of K, and

$$D^2 v = \left(\sum_{i,j=1}^{2} \left(\frac{\partial^2 v}{\partial x_i \partial x_j} \right)^2 \right)^{1/2}.$$

If ∇v is continuous, then

$$\|v - \pi_K v\|_{L_\infty(K)} \leq h_K \|Dv\|_{L_\infty(K)}, \tag{76.3}$$

Note that the gradient estimate depends on the reciprocal of the sine of the smallest angle of K, and therefore this error bound deteriorates as the the triangle gets thinner.

The proof follows the same general outline as the proofs of corresponding results in the Chapter Piecewise linear approximation. Let λ_i, $i = 1, 2, 3$, be the element basis functions for $\mathcal{P}(K)$ defined by $\lambda_i(a^j) = 1$ if $i = j$, and $\lambda_i(a^j) = 0$ otherwise. A function $w \in \mathcal{P}(K)$ has the representation

$$w(x) = \sum_{i=1}^{3} w(a^i) \lambda_i(x) \quad \text{for } x \in K,$$

and thus

$$\pi_K v(x) = \sum_{i=1}^{3} v(a^i) \lambda_i(x) \quad \text{for } x \in K, \tag{76.4}$$

since $\pi_K v(a^i) = v(a^i)$. We shall derive representation formulas for the interpolation errors $v - \pi_K v$ and $\nabla(v - \pi_K v)$, using a Taylor expansion at $x \in K$:

$$v(y) = v(x) + \nabla v(x) \cdot (y - x) + R(x, y),$$

where

$$R(x, y) = \frac{1}{2} \sum_{i,j=1}^{2} \frac{\partial^2 v}{\partial x_i \partial x_j}(\xi)(y_i - x_i)(y_j - x_j),$$

is the remainder term of order 2 and ξ is a point on the line segment between x and y. In particular choosing $y = a^i = (a_1^i, a_2^i)$, we have

$$v(a^i) = v(x) + \nabla v(x) \cdot (a^i - x) + R_i(x), \tag{76.5}$$

where $R_i(x) = R(x, a^i)$. Inserting (76.5) into (76.4) gives for $x \in K$

$$\pi_K v(x) = v(x) \sum_{i=1}^{3} \lambda_i(x) + \nabla v(x) \cdot \sum_{i=1}^{3} (a^i - x) \lambda_i(x) + \sum_{i=1}^{3} R_i(x) \lambda_i(x). \tag{76.6}$$

We shall use the following identities that hold for $j, k = 1, 2$, and $x \in K$,

$$\sum_{i=1}^{3} \lambda_i(x) = 1, \quad \sum_{i=1}^{3} (a_j^i - x_j) \lambda_i(x) = 0, \tag{76.7}$$

$$\sum_{i=1}^{3} \frac{\partial}{\partial x_k} \lambda_i(x) = 0, \quad \sum_{i=1}^{3} (a_j^i - x_j) \frac{\partial \lambda_i}{\partial x_k} = \delta_{jk}, \tag{76.8}$$

where $\delta_{jk} = 1$ if $j = k$ and $\delta_{jk} = 0$ otherwise. The first of the identities in (76.7) follows by choosing $v(x) = 1$ in (76.6), and the second follows by choosing $v(x) = d_1 x_1 + d_2 x_2$ with $d_i \in \mathbb{R}$. Finally, (76.8) follows by differentiating (76.7).

Using (76.7), we obtain the following representation of the interpolation error,

$$v(x) - \pi_K v(x) = - \sum_{i=1}^{3} R_i(x) \lambda_i(x).$$

Since $|a^i - x| \le h_K$, we can estimate the remainder term $R_i(x)$ as

$$|R_i(x)| \le \frac{1}{2} h_K^2 \|D^2 v\|_{L_\infty(K)}, \quad i = 1, 2, 3.$$

where we used Cauchy's inequality twice to estimate an expression of the form $\sum_{ij} x_i c_{ij} x_j = \sum_i x_i \sum_j c_{ij} x_j$.

Now, using the fact that $0 \le \lambda_i(x) \le 1$ if $x \in K$, for $i = 1, 2, 3$, we obtain

$$|v(x) - \pi_K v(x)| \le \max_i |R_i(x)| \sum_{i=1}^{3} \lambda_i(x) \le \frac{1}{2} h_K^2 \|D^2 v\|_{L_\infty(K)} \quad \text{for } x \in K,$$

which proves (76.1).

To prove (76.2), we differentiate (76.4) with respect to x_k, $k = 1, 2$ to get

$$\nabla(\pi_K v)(x) = \sum_{i=1}^{3} v(a^i) \nabla \lambda_i(x),$$

which together with (76.5) and (76.8) gives the following error representation:

$$\nabla(v - \pi_K v)(x) = -\sum_{i=1}^{3} R_i(x)\nabla\lambda_i(x) \quad \text{for } x \in K.$$

We now note that

$$\max_{x \in K} |\nabla\lambda_i(x)| \leq \frac{2}{h_K \sin(\tau_K)},$$

which follows by an easy estimate of the shortest height (distance from a vertex to the opposite side) of K. We now obtain (76.2) and (76.3) finally follows using tha Mean Value theorem. The proof is now complete.

Let now $\mathcal{T}_h = \{K\}$ be a triangulation of a domain Ω with mesh function $h(x)$, and let π_h denote nodal interpolation into the corresponding space of continuous piecewise linear functions V_h on $\mathcal{T}_h$. The interpolation error estimates of Theorem 76.1 then take the form

$$\|v - \pi_h v\|_{L_\infty(\Omega)} \leq \frac{1}{2}\|h^2 D^2 v\|_{L_\infty(\Omega)}, \tag{76.9}$$

$$\|v - \pi_h v\|_{L_\infty(\Omega)} \leq \|hDv\|_{L_\infty(\Omega)}, \tag{76.10}$$

$$\|\nabla(v - \pi_h v)\|_{L_\infty(\Omega)} \leq \frac{3}{\sin(\tau)}\|hD^2 v\|_{L_\infty(\Omega)}, \tag{76.11}$$

where τ is the smallest of the τ_K. Below we shall use analogs of these estimates with the $L_\infty(\Omega)$ replaced by $L_2(\Omega)$.

76.6   Sobolev and his Spaces

Sergei Sobolev (1908–1989) played a leading role in the mathematical world of the former Soviet Union and made important contributions to the theory

Fig. 76.12. Sergei Lvovich Sobolev (1908–1989), creator of Functional Analysis and inventor of Sobolev spaces: "I wonder if my space of functions $H^1(\Omega)$ is large enough to contain the solution?"

and practice of partial differential equations, in particular on questions of existence, uniqueness, stability and regularity of solutions by developing tools of *Functional Analysis*. He also worked on numerical methods and gave important results on interpolation and quadrature of functions of several variables by developing techniques of *Sobolev spaces*. A basic Sobolev space is the space of real-valued functions defined on a domain Ω in $\mathbb{R}^d$, which are square integrable together with their first partial derivatives, denoted by $H^1(\Omega)$.

76.7   Quadrature in $\mathbb{R}^2$

To compute the stiffness matrix and load vector a FEM, we have to compute integrals of the form $\int_K g(x)\,dx$, where K is a triangle or tetrahedron and g a given function. Sometimes we may evaluate these integrals exactly, but usually this is either impossible or inefficient. Instead we usually evaluate the integrals approximately using quadrature formulas. We briefly present some quadrature formulas for integrals over triangles.

In general, we would like to use quadrature formulas that do not affect the accuracy of the underlying finite element method, which of course requires an estimate of the error due to quadrature. A quadrature formula for an integral over an element K has the form

$$\int_K g(x)\,dx \approx \sum_{i=1}^{q} g(y^i)\omega_i, \tag{76.12}$$

for a specified choice of *nodes* $\{y^i\}$ in K and *weights* $\{\omega_i\}$. We now list some possibilities using the notation a_K^i to denote the vertices of a triangle K, a_K^{ij} to denote the midpoint of the side connecting a_K^i to a_K^j, and a_K^{123} to denote the center of mass of K, and denote by $|K|$ the area of K:

$$\int_K g\,dx \approx g\big(a_K^{123}\big)|K|, \tag{76.13}$$

$$\int_K g(x)\,dx \approx \sum_{j=1}^{3} g(a_K^j)\frac{|K|}{3}, \tag{76.14}$$

$$\int_K g\,dx \approx \sum_{1\leq i<j\leq 3} g(a_K^{ij})\frac{|K|}{3}, \tag{76.15}$$

$$\int_K g\,dx \approx \sum_{j=1}^{3} g(a_K^i)\frac{|K|}{20} + \sum_{1\leq i<j\leq 3} g(a_K^{ij})\frac{2|K|}{15} + g\big(a_K^{123}\big)\frac{9|K|}{20}. \tag{76.16}$$

We refer to (76.13) as the center of gravity quadrature, to (76.14) as the vertex quadrature, and to (76.15) as the midpoint quadrature. Recall that the accuracy of a quadrature formula is related to the *precision* of the formula. A quadrature formula has precision r if the formula gives the exact value of the integral if the integrand is a polynomial of degree at most $r - 1$, but there is some polynomial of degree r such that the formula is not exact. The quadrature error for a quadrature rule of precision r is proportional to h^r, where h is the mesh size. More precisely, the error of a quadrature rule of the form (76.12) satisfies

$$\left| \int_K g\, dx - \sum_{i=1}^q g(y^i)\omega_i \right| \le Ch_K^r \sum_{|\alpha|=r} \int_K |D^\alpha g|\, dx,$$

where C is a constant. Vertex and center of gravity quadrature have precision 2, midpoint quadrature has precision 3, while (76.16) has precision 4.

In finite element methods based on continuous piecewise linear functions, we often use nodal or vertex quadrature, often also referred to as *lumped mass* quadrature, because the mass matrix computed this way becomes diagonal.

Example 76.1. In Fig. 76.13 and Fig. 76.14 we give two examples, one from fluid mechanics. and the other from electromagnetics.

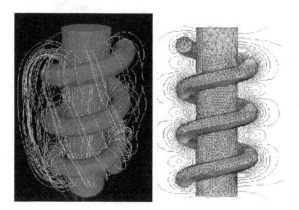

Fig. 76.13. Magnetic field around coil and mesh

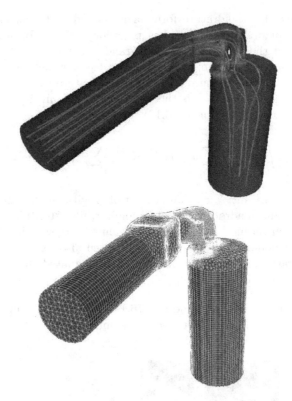

Fig. 76.14. Flow in diesel engine cylinder inlet and mesh

Chapter 76   Problems

76.1. For a given triangle K, determine the relation between the smallest angle τ_K, the triangle diameter h_K and the diameter ρ_K of the largest inscribed circle.

76.2. Draw the refined mesh that results from sub-dividing the smallest two triangles in the mesh on the right in Fig. 76.5.

76.3. Let K be a tetrahedron with vertices $\{a^i, i = 1, \ldots, 4\}$. Show that a linear polynomial $v(x) = c_0 + c_1 x_1 + c_2 x_2 + c_3 x_3$ on K is uniquely determined by the nodal values $\{v(a^i), i = 1, \ldots, 4\}$. Show that the corresponding finite element space V_h consists of continuous functions.

76.4. Prove that the quadrature formulas (76.13), (76.14), (76.15) and (76.16) have the indicated precision.

76.5. Prove that using nodal quadrature to compute a mass matrix for piecewise linears, gives a diagonal mass matrix where a diagonal term is the sum of the terms in the corresponding row in the exactly computed mass matrix. Motivate the term "lumped".

77

FEM for Boundary Value Problems in $\mathbb{R}^2$ and $\mathbb{R}^3$

> ...were very confused, skipping suddenly from one idea to another, from one formula to the next, with no attempt to give a connection between them. His presentations were obscure clouds, illuminated from time to time by flashes of pure genius. ...of the thirty who enrolled with me, I was the only one to see it through.
> (Menabrea about Cauchy 1832)

77.1 Introduction

In this chapter, we extend the cG(1) FEM for reaction-diffusion-convection problems in one space dimension to corresponding boundary value problems in $\mathbb{R}^2$ and $\mathbb{R}^3$ of the form: Find $u : \Omega \to \mathbb{R}$ such that

$$-\nabla \cdot (a\nabla u) + \nabla \cdot (ub) + cu = f \quad \text{in } \Omega, \tag{77.1}$$

together with boundary conditions of Dirichlet, Neumann or Robin type, where $a(x) > 0$, $b(x)$ and $c(x)$ are given variable coefficients, $f(x)$ is a given right hand side, and Ω is a bounded open domain in $\mathbb{R}^2$ or $\mathbb{R}^3$. Note that the coefficient b is a vector (typically corresponding to a given convection velocity), and that the equation (77.1) can alternatively be written

$$-\nabla \cdot (a\nabla u) + b \cdot \nabla u + \hat{c}u = f \quad \text{in } \Omega, \tag{77.2}$$

with $\hat{c} = c + \nabla \cdot b$. In general, problems of this form cannot be solved analytically and we have to rely on a numerical method such as FEM for computing the solution $u(x)$ for given data.

We consider below the extension to corresponding time dependent problems of the form

$$\dot{u} - \nabla \cdot (a\nabla u) + \nabla \cdot (ub) + cu = f, \qquad (77.3)$$

together with initial and boundary value problems, including extensions to systems of such equations, using the material in the Chapters *The General Initial Value Problem* and *Adaptive IVP-Solvers*.

The most fundamental example of the form (77.1) is Poisson's equation with homogeneous Dirichlet boundary conditions corresponding to setting $a = 1$, $b = 0$ and $c = 0$:

$$\begin{cases} -\Delta u(x) = f(x) & \text{for } x \in \Omega, \\ u(x) = 0 & \text{for } x \in \Gamma, \end{cases} \qquad (77.4)$$

where Ω is a bounded domain in $\mathbb{R}^2$ with polygonal boundary Γ. We recall that $\nabla \cdot (\nabla u) = \Delta u$. We shall now present the cG(1) method for (77.4) generalizing cG(1) for the two-point boundary value problem (53.9), and then extend to the general problem (77.1).

77.2   Richard Courant: Inventor of FEM

Richard Courant (1888–1972) was a student of Hilbert and published together with him the monumental work *Methoden der Mathematischen Physik*. In the mid 1930s he fled to New York away from the Nazis and created the Courant Institute of Mathematical Sciences, since 1964 occupying a 13 storey building close to Washington Square in Greenwich Village on

Fig. 77.1. Richard Courant (1888–1972), pioneer of finite elements: "In fact, already when writing my 1910 PhD thesis on using the Dirichlet minimum principle to prove the existence of solutions to Poisson's equation on a domain Ω, I had in mind of seeking approximate solutions in a subspace of the Sobolev space $H^1(\Omega)$ consisting of piecewise linear functions on a triangulation of Ω ..."

Manhattan. Courant presented in a famous paper from 1943 the basics of finite element approximation of differential equations, as an expansion of a foot-note in the 1924 *Methoden*. This foot-note must be one of the most productive remarks in the history of science generating hundreds of thousands of scientific articles and a flood of software from the mid 1960s and on.

77.3   Variational Formulation

We let $\mathcal{T}_h = \{K\}$ be a triangulation of Ω with mesh function $h(x)$ and internal nodes $N_1, \ldots N_M$, and we let V_h be the corresponding finite element space of continuous piecewise linear functions that vanish on the boundary Γ. We first give (77.4) the following preliminary variational formulation:

$$-\int_\Omega \Delta u\, v\, dx = \int_\Omega f\, v\, dx \tag{77.5}$$

for all suitable test functions v, which results from multiplying (77.4) by $v(x)$ and integrating over Ω. We now want to rewrite the left-hand side to move a derivative from Δu onto v. Assuming that the test function v is zero on Γ, Green's formula implies

$$-\int_\Omega \Delta u\, v\, dx = -\int_\Gamma \partial_n u v\, ds + \int_\Omega \nabla u \cdot \nabla v\, dx = \int_\Omega \nabla u \cdot \nabla v\, dx,$$

where $\partial_n = \frac{\partial}{\partial n}$ denotes the outward unit normal derivative on Γ. We find that a solution $u(x)$ of (77.4) satisfies

$$\int_\Omega \nabla u \cdot \nabla v\, dx = \int_\Omega f\, v\, dx, \tag{77.6}$$

for all test functions v with $v = 0$ on Γ.

77.4   The cG(1) FEM

We are thus led to the following formulation of the cG(1) FEM for (77.4): Find $U \in V_h$ such that

$$\int_\Omega \nabla U \cdot \nabla v\, dx = \int_\Omega f\, v\, dx \quad \text{for all } v \in V_h, \tag{77.7}$$

where V_h is the space of continuous piecewise linear functions on a triangulation $\mathcal{T}_h$ of Ω that vanish on the boundary Γ. Using the notation

$$(w, v) = \int_\Omega wv\, dx, \quad (\nabla w, \nabla v) = \int_\Omega \nabla w \cdot \nabla v\, dx,$$

we can write cG(1) in the form: Find $U \in V_h$ such that

$$(\nabla U, \nabla v) = (f, v) \quad \text{for all } v \in V_h. \tag{77.8}$$

We see that that the trial space and test spaces are equal ($= V_h$) and include the homogenous Dirichlet boundary condition. The Galerkin orthogonality is expressed by

$$(\nabla u - \nabla U, \nabla v) = 0 \quad \text{for all } v \in V_h, \tag{77.9}$$

which results upon subtracting (77.8) from (77.6) with $v \in V_h$.

We recall that the nodal basis functions $\{\varphi_i\}_{i=1}^M$ associated with the internal nodes $N_1, \ldots N_M$ of $\mathcal{T}_h$ is a basis for V_h. Expressing U in terms of this basis,

$$U(x) = \sum_{j=1}^M U(N_j)\varphi_j(x), \tag{77.10}$$

substituting into (77.8), and choosing $v = \varphi_i$ for $i = 1, \ldots, M$, gives

$$\sum_{j=1}^M (\nabla\varphi_j, \nabla\varphi_i)U(N_j) = (f, \varphi_i), \quad i = 1, \ldots, M.$$

This is equivalent to the linear system of equations

$$A\xi = b, \tag{77.11}$$

where $\xi = (\xi_j)$ is the vector of internal nodal values $\xi_j = U(N_j)$, $A = (a_{ij})$ is the *stiffness matrix* with elements $a_{ij} = (\nabla\varphi_j, \nabla\varphi_i)$ and $b = (b_i)$ with $b_i = (f, \varphi_i)$ is the *load vector*.

The stiffness matrix A is obviously symmetric and it is also positive-definite since for any $v = \sum_{i=1}^M \eta_i\varphi_i$ in V_h,

$$\sum_{i,j=1}^M \eta_i a_{ij}\eta_j = \sum_{i,j=1}^M \eta_i(\nabla\varphi_i, \nabla\varphi_j)\eta_j$$

$$= \left(\nabla\sum_{i=1}^M \eta_i\varphi_i, \nabla\sum_j^M \eta_j\varphi_j\right) = (\nabla v, \nabla v) > 0,$$

unless $\eta_i = 0$ for all i. This means in particular that (77.11) has a unique solution vector U and thus the cG(1) finite element problem (77.8) has a unique solution $U \in V_h$.

A triangle with associated linear approximation, i.e. the basic *finite element* of cG(1), is also called the *Courant element*, as a recognition of its inventor.

Uniform Triangulation of a Square

We compute the stiffness matrix A and load vector b in (77.11) explicitly on the uniform triangulation of the square $\Omega = [0,1] \times [0,1]$ pictured in Fig. 77.2. We choose an integer $m \geq 1$ and set $h = 1/(m+1)$, then construct the triangles as shown. The diameter of the triangles in $\mathcal{T}_h$ is $\sqrt{2}h$ and there are $M = m^2$ internal nodes. We number the nodes starting from the lower left and moving right, then working up across the rows.

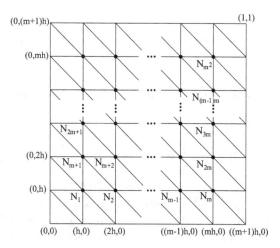

Fig. 77.2. The standard triangulation of the unit square

In Fig. 77.4, we show the support of the basis function corresponding to the node N_i along with parts of the basis functions for the neighboring nodes. As in one dimension, the basis functions are "almost" orthogonal in the sense that only basis functions φ_i and φ_j sharing a common triangle in their supports yield a non-zero value in $(\nabla \varphi_i, \nabla \varphi_j)$. We show the nodes neighboring N_i in Fig. 77.3. The support of any two neighboring basis

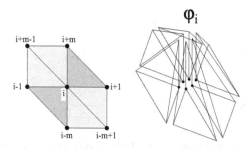

Fig. 77.3. The indices of the nodes neighboring N_i and an "exploded" view of φ_i

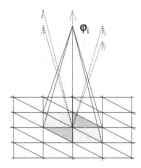

Fig. 77.4. The support of the basis function φ_i together with parts of the neighboring basis functions

functions overlap on just two triangles, while a basis function "overlaps itself" on six triangles.

We first compute

$$a_{ii} = (\nabla\varphi_i, \nabla\varphi_i) = \int_\Omega |\nabla\varphi_i|^2 \, dx = \int_{\text{support of } \varphi_i} |\nabla\varphi_i|^2 \, dx,$$

for $i = 1, \ldots, m^2$. As noted, we only have to consider the integral over the domain pictured in Fig. 77.3, which is written as a sum of integrals over the six triangles making up the domain. Examining φ_i on these triangles, see Fig. 77.3, we see that there are only two different integrals to be computed since φ_i looks the same, except for orientation, on two of the six triangles and similarly the same on the other four triangles. We shade the corresponding triangles in Fig. 77.4. The orientation affects the direction of $\nabla\varphi_i$ of course, but does not affect $|\nabla\varphi_i|^2$.

We compute $(\nabla\varphi_i, \nabla\varphi_i)$ on the triangle shown in Fig. 77.5. In this case, φ_i is one at the node located at the right angle in the triangle and zero at the other two nodes. We change coordinates to compute $(\nabla\varphi_i, \nabla\varphi_i)$ on the *reference triangle* shown in Fig. 77.5. Again, changing to these coordinates does not affect the value of $(\nabla\varphi_i, \nabla\varphi_i)$ since $\nabla\varphi_i$ is constant on the triangle.

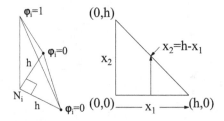

Fig. 77.5. First case showing φ_i on the *left* together with the variables used in the reference triangle

On the triangle, φ_i can be written $\varphi_i = ax_1 + bx_2 + c$ for some constants a, b, c. Since $\varphi_i(0, 0) = 1$, we get $c = 1$. Similarly, we compute a and b to find that $\varphi_i = 1 - x_1/h - x_2/h$ on this triangle. Therefore, $\nabla \varphi_i = \left(-h^{-1}, -h^{-1}\right)$ and the integral is

$$\int_{\triangleright} |\nabla \varphi_i|^2 \, dx = \int_0^h \int_0^{h-x_1} \frac{2}{h^2} \, dx_2 \, dx_1 = 1.$$

In the second case, φ_i is one at a node located at an acute angle of the triangle and is zero at the other nodes. We illustrate this in Fig. 77.6. We use the coordinate system shown in Fig. 77.6 to write $\varphi_i = 1 - x_1/h$. When we integrate over the triangle, we get $1/2$.

Summing the contributions from all the triangles gives

$$a_{ii} = (\nabla \varphi_i, \nabla \varphi_i) = 1 + 1 + \frac{1}{2} + \frac{1}{2} + \frac{1}{2} + \frac{1}{2} = 4.$$

Next, we compute $(\nabla \varphi_i, \nabla \varphi_j)$ for indices corresponding to neighboring nodes. For a general node N_i, there are two cases of inner products (see Fig. 77.3 and Fig. 77.4):

$$a_{i\,i-1} = (\nabla \varphi_i, \nabla \varphi_{i-1}) = (\nabla \varphi_i, \nabla \varphi_{i+1}) = (\nabla \varphi_i, \nabla \varphi_{i-m}) = (\nabla \varphi_i, \nabla \varphi_{i+m}),$$

and

$$a_{i\,i-m+1} = (\nabla \varphi_i, \nabla \varphi_{i-m+1}) = (\nabla \varphi_i, \nabla \varphi_{i+m-1}).$$

The orientation of the triangles in each of the two cases are different, but the inner product of the gradients of the respective basis functions is not affected by the orientation. Note that the the equations corresponding to nodes next to the boundary are special, because the nodal values on the boundary are zero, see Fig. 77.2. For example, the equation corresponding to N_1 only involves N_1, N_2 and N_{m+1}.

For the first case, we next compute $(\nabla \varphi_i, \nabla \varphi_{i+1})$. Plotting the intersection of the respective supports shown in Fig. 77.7, we conclude that there are equal contributions from each of the two triangles in the intersection. We choose one of the triangles and construct a reference triangle as above.

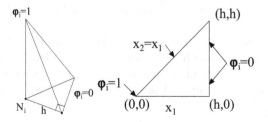

Fig. 77.6. Second case showing φ_i and the reference triangle

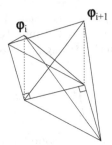

Fig. 77.7. The overlap of φ_i and φ_{i+1}

Choosing suitable variables, we find that

$$\nabla\varphi_i \cdot \nabla\varphi_{i+1} = \left(-\frac{1}{h}, -\frac{1}{h}\right) \cdot \left(\frac{1}{h}, 0\right) = -\frac{1}{h^2},$$

and integrating over the triangle gives $-1/2$. Similarly, we see that

$$(\nabla\varphi_i, \nabla\varphi_{i-m+1}) = (\nabla\varphi_i, \nabla\varphi_{i+m-1}) = 0.$$

We can now determine the stiffness matrix A using the information above. We start by considering the first row. The first entry is $(\nabla\varphi_1, \nabla\varphi_1) = 4$ since N_1 has no neighbors to the left or below. The next entry is $(\nabla\varphi_1, \nabla\varphi_2) = -1$. The next entry after that is zero, because the supports of φ_1 and φ_3 do not overlap. This is true in fact of all the entries up to and including φ_m. However, $(\nabla\varphi_1, \nabla\varphi_{m+1}) = -1$, since these neighboring basis functions do share two supporting triangles. Finally, all the rest of the entries in that row are zero because the supports of the corresponding basis functions do not overlap. We continue in this fashion working row by row. The result is pictured in Fig. 77.8. We see that A has a *block structure* consisting of banded $m \times m$ sub-matrices, most of which consist only of zeros. Note the pattern of entries around corners of the diagonal block matrices; it is a common mistake to program these values incorrectly.

The storage of a sparse matrix and the solution of a sparse system are both affected by the *structure* or *sparsity pattern* of the matrix. The sparsity pattern is affected in turn by the enumeration scheme used to mark the nodes.

There are several algorithms for reordering the coefficients of a sparse matrix to form a matrix with a smaller bandwidth. Reordering the coefficients is equivalent to computing a new basis for the vector space.

The load vector b is computed in the same fashion, separating each integral

$$\int_\Omega f\varphi_i \, dx = \int_{\text{support of } \varphi_i} f(x)\varphi_i(x) \, dx$$

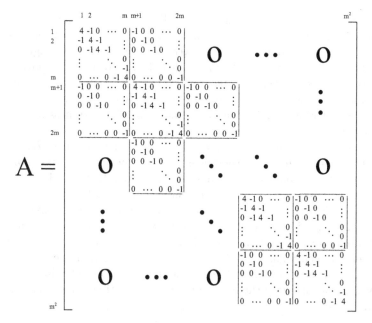

Fig. 77.8. The stiffness matrix

into integrals over the triangles making up the support of φ_i. To compute the elements (f, φ_i) of the load vector, we often use one of the quadrature formulas presented in Chapter 76.

77.5 Basic Data Structures

To compute the finite element approximation U, we have to compute the coefficients of the stiffness matrix A and load vector b and solve the linear system of equations (77.11). We just computed A and b for a uniform triangulation of the unit square, and we now discuss the case of a general triangulation of a general domain.

We have to compute the non-zero elements $a_{ij} = (\nabla \varphi_j, \nabla \varphi_i)$ of the stiffness matrix A. We know that $a_{ij} = 0$ unless both N_i and N_j are nodes of the same triangle K, because only then the supports of basis functions φ_i and φ_j overlap. The common support corresponding to a non-zero element a_{ij} is equal to the support of φ_i if $j = i$, and is equal to the two triangles with the common edge connecting N_j and N_i if $i \neq j$. In each case a_{ij} is the sum of contributions

$$a_{ij}^K = \int_K \nabla \varphi_j \cdot \nabla \varphi_i \, dx \qquad (77.12)$$

over the triangles K in the common support. The process of adding up the contributions a_{ij}^K from the relevant triangles K to get the element a_{ij}, is called *assembling* the stiffness matrix A. Arranging for a given triangle K the numbers a_{ij}^K, where N_i and N_j are nodes of K, into a 3×3 matrix, we obtain the *element stiffness matrix* for the triangle K. We refer to the assembled matrix A as the *global stiffness matrix*. Notice that we use *element* with two different meanings: as an element a_{ij} of the stiffness matrix A, and as a finite element or triangle of the triangulation.

To compute the element stiffness matrix a_{ij}^K for a given triangle K, we need the physical coordinates of the nodes of K. To perform the assembly where we loop over all elements and add the corresponding contributions to the global stiffness matrix, we need the node numbers of each triangle. Similar information is needed to compute the load vector. The required information is arranged in a *data structure*, or data base, containing (i) a list of the coordinates of the nodes numbered in some way, and (ii) a list of the node numbers of each triangle. A list of the numbers of the nodes on the boundary is also needed to handle the boundary conditions This information is typically the output of the mesh generator generating a triangulation of the domain.

77.6   Solving the Discrete System

Once we have assembled the stiffness matrix A and computed the load vector b, we have to solve the linear system $AU = b$ to obtain the finite element approximation $U(x)$. We now discuss this topic briefly based on the material presented in Chapter 44. The stiffness matrix resulting from discretizing the Laplacian is symmetric and positive-definite and therefore invertible. These properties also mean that there is a wide choice in the methods used to solve the linear system $AU = b$, which take advantage of the fact that A is sparse.

In the case of the standard uniform discretization of a square, we saw that A is a banded matrix with five non-zero diagonals and bandwidth $m + 1$, where m is the number of nodes on a side. The dimension of A is m^2 and the asymptotic operations count for using Gaussian elimination is $O(m^4) = O(h^{-4})$. Note that even though A has mostly zero diagonals inside the band, fill-in occurs as the elimination is performed, so we may as well treat A as if it has non-zero diagonals throughout the band. Clever rearrangement of A to reduce the amount of fill-in leads to a solution algorithm with an operations count on the order of $O(m^3) = O(h^{-3})$. In contrast, if we treat A as a full matrix, we get an asymptotic operations count of $O(h^{-6})$, which is considerably larger for a large number of elements.

In general, we get a sparse stiffness matrix, though there may not be a band structure. If we want to use a Gaussian elimination method efficiently in general, then it is necessary to first reorder the system to bring the matrix into banded form.

We can also apply both the Jacobi and Gauss-Seidel methods to solve the linear system arising from discretizing the Poisson equation. In the case of the uniform standard discretization of a square for example, the operations count is $O(M)$ per iteration for both methods if we make use of the sparsity of A. Therefore a single step of either method is much cheaper than a direct solve. The question is: How many iterations do we need to compute in order to obtain an accurate solution?

Typically the spectral radius of the iteration matrix of the Jacobi or Gauss-Seidel method is equal to $1 - Ch^2$ with C some moderate positive constant, which means that the convergence rate quickly gets slow as h decreases: to reduce the error a certain factor, we need of the order of $O(h^{-2})$ iterations, and since each iteration takes $O(h^{-2})$ operations, the total number of operations is $O(h^{-4})$, which is the same order as using a banded Gaussian elimination solver.

There has been a lot of activity in developing iterative methods that converge more quickly than Jacobi and Gauss-Seidel. In recent years, very efficient *multi-grid methods* have been developed and are now becoming a standard tool. A multi-grid method is based on a sequence of Gauss-Seidel or Jacobi steps performed on a hierarchy of successively coarser meshes and are optimal in the sense that the solution work is proportional to the total number of unknowns (that is h^{-2} in the model problem).

77.7   An Equivalent Minimization Problem

The variational problem (77.8) is equivalent to the following quadratic *minimization problem*: find $U \in V_h$ such that

$$F(u) \leq F(v) \quad \text{for all } v \in V_h, \tag{77.13}$$

where

$$F(v) = \frac{1}{2} \int_\Omega |\nabla v|^2 \, dx - \int_\Omega fv \, dx = \frac{1}{2}(\nabla v, \nabla v) - (f, v).$$

The quantity $F(v)$ may be interpreted as the *total energy* of the function $v(x)$ composed of the *internal energy* $\frac{1}{2}(\nabla v, \nabla v)$ and the *load potential* $-(f, v)$. Thus, the solution U minimizes the total energy $F(v)$ with v varying over V_h.

To see the equivalence of (77.8) and (77.13), we assume first that $U \in V_h$ satisfies (77.8). Let then $v \in V_h$ and write $v = U + (v - U) = U + w$ with

$w = v - U \in V_h$. Using $(\nabla U, \nabla w) = (\nabla w, \nabla U)$, we get

$$F(v) = F(U + w) =$$
$$\frac{1}{2}(\nabla U, \nabla U) + (\nabla U, \nabla w) + \frac{1}{2}(\nabla w, \nabla w) - (f, U) - (f, w)$$
$$= F(U) + \frac{1}{2}(\nabla w, \nabla w) \geq F(U),$$

with equality only if $w = 0$. We conclude that U satisfies (77.13).

Conversely, if U is the solution of (77.13), then we have for all $v \in V_h$

$$g_v(\epsilon) = F(U + \epsilon v) \geq g(0) = F(U) \quad \text{for all } \epsilon \in \mathbb{R},$$

and thus $\epsilon = 0$ is an interior minimum point of $g_v(\epsilon)$ with v fixed, and thus $g_v'(0) = 0$. Computing we get

$$0 = g_v'(0) = (\nabla U, \nabla v) - (f, v)$$

and thus U satisfies (77.8). We sum up in the following theorem:

Theorem 77.1 *The problems (77.8) and (77.13) are equivalent in the sense that they have the same unique solution.*

77.8 An Energy Norm a Priori Error Estimate

In this section, we derive a priori and a posteriori estimates of the error $u - U$ in the *energy norm* $\|\nabla(u - U)\|$ with

$$\|\nabla v\| = \left(\int_\Omega |\nabla v|^2 \, dx \right)^{1/2}, \tag{77.14}$$

where u is the exact solution and U a finite element solution of Poisson's equation with homogeneous Dirichlet boundary conditions. The energy norm, which is the L_2 norm of the gradient of a function in this problem, arises naturally in the error analysis of the finite element method because it is closely tied to the variational problem. The gradient of the solution, representing heat flow, electric field, flow velocity, or stress for example, can be a variable of physical interest as much as the solution itself, representing temperature, potential or displacement for example, and in this case, the energy norm is the relevant error measure.

We first prove that the Galerkin finite element approximation is the best approximation of the true solution in V_h with respect to the energy norm.

Theorem 77.2 *Assume that u satisfies the Poisson equation (77.4) and U is the Galerkin finite element approximation satisfying (77.8). Then*

$$\|\nabla(u - U)\| \leq \|\nabla(u - v)\| \quad \text{for all } v \in V_h. \tag{77.15}$$

The proof goes as follows: Using the Galerkin orthogonality (77.9) with v replaced by $U - v \in V_h$, we can write

$$\|\nabla e\|^2 = (\nabla e, \nabla(u - U)) = (\nabla e, \nabla(u - U)) + (\nabla e, \nabla(U - v)).$$

Adding the terms involving U on the right, whereby U drops out, and using Cauchy's inequality, we get

$$\|\nabla e\|^2 = (\nabla e, \nabla(u - v)) \le \|\nabla e\| \, \|\nabla(u - v)\|,$$

which proves the theorem after dividing by $\|\nabla e\|$.

Choosing $v = \pi_h u$ and using an $L_2(\Omega)$ analog of the interpolation estimate (76.11), we get the following quantitative a priori error estimate (with $\|v\| = \|v\|_{L_2(\Omega)}$):

Corollary 77.3 *There exists a constant C_i depending only on the minimal angle τ in $\mathcal{T}_h$, such that*

$$\|\nabla(u - U)\| \le C_i \|h D^2 u\|. \tag{77.16}$$

77.9 An Energy Norm a Posteriori Error Estimate

We now prove an a posteriori error estimate following the strategy used for the two-point boundary value problem in Chapter 53. A new feature occurring in higher dimensions is the appearance of integrals over the internal edges S in $\mathcal{S}_h$. We start by writing an equation for the error $e = u - U$ using (77.6) and (77.8) to get

$$\begin{aligned}\|\nabla e\|^2 &= (\nabla(u - U), \nabla e) = (\nabla u, \nabla e) - (\nabla U, \nabla e) \\ &= (f, e) - (\nabla U, \nabla e) = (f, e - \pi_h e) - (\nabla U, \nabla(e - \pi_h e)),\end{aligned}$$

where $\pi_h e \in V_h$ is an interpolant of e. We now break up the integrals over Ω into sums of integrals over the triangles K in $\mathcal{T}_h$ and integrate by parts over each triangle in the last term to get

$$\|\nabla e\|^2 = \sum_K \int_K (f + \Delta U)(e - \tilde\pi_h e) \, dx - \sum_K \int_{\partial K} \frac{\partial U}{\partial n_K}(e - \pi_h e) \, ds, \tag{77.17}$$

where $\partial U / \partial n_K$ denotes the derivative of U in the outward normal direction n_K of the boundary ∂K of K. In the boundary integral sum in (77.17), each internal edge $S \in \mathcal{S}_h$ occurs twice as a part of each of the boundaries ∂K of the two triangles K that have S as a common side. Of course the outward normals n_K from each of the two triangles K sharing S point in opposite directions. For each side S, we choose one of these normal directions and denote by $\partial_S v$ the derivative of a function v in that direction on S. We note that if $v \in V_h$, then in general $\partial_S v$ is different on the two triangles sharing S; see Fig. 76.9, which indicates the "kink" over S in the graph of v. We can express the sum of the boundary integrals in (77.17) as a sum of integrals over edges of the form

$$\int_S [\partial_S U](e - \pi_h e)\, ds,$$

where $[\partial_S U]$ is the difference, or jump, in the derivative $\partial_S U$ computed from the two triangles sharing S. The jump appears because the outward normal directions of the two triangles sharing S are opposite. We further note that $e - \tilde{\pi}_h e$ is continuous across S, but in general does not vanish on S even if π_h is the nodal interpolant. This is different than the one-dimensional case, where the corresponding sum over nodes does indeed vanish because $e - \pi_h e$ vanishes at the nodes. We may thus rewrite (77.17) as follows with the second sum replaced by a sum over internal edges S:

$$\|\nabla e\|^2 = \sum_K \int_K (f + \Delta U)(e - \pi_h e)\, dx + \sum_{S \in \mathcal{S}_h} \int_S [\partial_S U](e - \pi_h e)\, ds.$$

Next, we return to a sum over element edges ∂K by just distributing each jump equally to the two triangles sharing it, to obtain an *error representation* of the energy norm of the error in terms of the residual error:

$$\|\nabla e\|^2 = \sum_K \int_K (f + \Delta U)(e - \pi_h e)\, dx$$
$$+ \sum_K \frac{1}{2} \int_{\partial K} h_K^{-1}[\partial_S U](e - \pi_h e) h_K\, ds,$$

where we have prepared to estimate the second sum by inserting a factor h_K and compensating. In crude terms, the residual error results from substituting U into the differential equation $-\Delta u - f = 0$, but in reality straightforward substitution is not possible because U is not twice differentiable in Ω. The integral on the right over K is the remainder from substituting U into the differential equation inside each triangle K, while the integral over ∂K arises because $\partial_S U$ in general is different when computed from the two triangles sharing S.

We estimate the first term in the error representation by inserting a factor h, compensating and using the estimate $\|h^{-1}(e - \pi_h e)\| \leq C_i \|\nabla e\|$ analogous to (76.11), to obtain

$$\left| \sum_K \int_K h(f + \Delta U) h^{-1}(e - \pi_h e)\, dx \right|$$
$$\leq \|h R_1(U)\| \|h^{-1}(e - \pi_h e)\| \leq C_i \|h R_1(U)\| \|\nabla e\|,$$

where $R_1(U)$ is the function defined on Ω by setting $R_1(U) = |f + \Delta U|$ on each triangle $K \in \mathcal{T}_h$. We estimate the contribution from the jumps on the edges similarly. Formally, the estimate results from replacing $h_K \, ds$ by dx corresponding to replacing the integrals over element boundaries ∂K by integrals over elements K. Dividing by $\|\nabla e\|$, we obtain the following a posteriori error estimate:

Theorem 77.4 *There is an interpolation constant C_i only depending on the minimal angle τ such that the error of the Galerkin finite element approximation U of the solution u of the Poisson equation satisfies*

$$\|\nabla u - \nabla U\| \le C_i \|h R(U)\|, \tag{77.18}$$

where $R(U) = R_1(U) + R_2(U)$ with

$$R_1(U) = |f + \Delta U| \quad \text{on } K \in \mathcal{T}_h,$$

$$R_2(U) = \frac{1}{2} \max_{S \subset \partial K} h_K^{-1} |[\partial_S U]| \quad \text{on } K \in \mathcal{T}_h.$$

Note that $R_1(U)$ is the contribution to the total residual from the interior of the elements K. In the present case of piecewise linear approximation, $R_1(U) = |f|$. Further, $R_2(U)$ is the contribution to the residual from the jump of the normal derivative of U across edges. In the one dimensional problem considered in Chapter 53, this contribution does not appear because the interpolation error may be chosen to be zero at the node points.

77.10   Adaptive Error Control

The basic goal of adaptive error control is to find a triangulation $\mathcal{T}_h$ with a least number of nodes such that the corresponding finite element approximation U satisfies

$$\|\nabla u - \nabla U\| \le \text{TOL}. \tag{77.19}$$

Using the a posteriori error estimate we are thus led to find a triangulation $\mathcal{T}_h$ with a least number of nodes such that the corresponding finite element approximation U satisfies

$$C_i \|h R(U)\| \le \text{TOL}. \tag{77.20}$$

This is a nonlinear constrained minimization problem with U depending on $\mathcal{T}_h$. If (77.18) is a reasonably sharp estimate of the error, then a solution of this optimization problem will meet our original goal.

We cannot expect to be able to solve this minimization problem analytically. Instead, a solution has to be sought by an iterative process in which we start with a coarse initial mesh and then successively modify the mesh by seeking to satisfy the stopping criterion (77.20) with a minimal number of elements. More precisely, we follow the following *adaptive algorithm*:

1. Choose an initial triangulation $T_h^{(0)}$.

2. Given the jth triangulation $T_{h^{(j)}}$ with mesh function $h^{(j)}$, compute the corresponding finite element approximation $U^{(j)}$.

3. Compute the corresponding residuals $R_1(U^{(j)})$ and $R_2(U^{(j)})$ and check whether or not (77.20) holds. If it does, stop.

4. Find a new triangulation $T_{h^{(j+1)}}$ with mesh function $h^{(j+1)}$ and with a minimal number of nodes such that $C_i \| h^{(j+1)} R(U^{(j)}) \| \leq$ TOL, and then proceed to #2.

The success of this iteration hinges on the mesh modification strategy used to perform step #4. A natural strategy for error control based on the L_2 norm uses the *principle of equidistribution* of the error in which we try to equalize the contribution from each element to the integral defining the L_2 norm. The rationale is that refining an element with large contribution to the error norm gives a large pay-off in terms of error reduction per new degree of freedom.

In other words, the approximation computed on the optimal mesh T_h in terms of computational work satisfies

$$\| \nabla e \|_{L_2(K)}^2 \approx \frac{\text{TOL}^2}{M} \quad \text{for all } K \in T_h,$$

where M is the number of elements in T_h. Based on (77.18), we would therefore like to compute the triangulation at step #4 so that

$$C_i^2 \left(\| h^{(j+1)} R(U^{(j+1)}) \|_{L_2(K)}^2 \right) \approx \frac{\text{TOL}^2}{M^{(j+1)}} \quad \text{for all } K \in T_{h^{(j+1)}}, \quad (77.21)$$

where $M^{(j+1)}$ is the number of elements in $T_{h^{(j+1)}}$. However, (77.21) is a nonlinear equation, since we don't know $M^{(j+1)}$ and $U^{(j+1)}$ until we have chosen the triangulation. Hence, we replace (77.21) by

$$C_i^2 \left(\| h^{(j+1)} R(U^{(j)}) \|_{L_2(K)}^2 \right) \approx \frac{\text{TOL}^2}{M^{(j)}} \quad \text{for all } K \in T_h^{(j+1)}, \quad (77.22)$$

and use this formula to compute the new mesh size $h^{(j+1)}$.

There are several questions we may ask about the process described here: How much efficiency is lost by replacing (77.19) by (77.20)? Does the iterative process #1–#4 converge? Is the approximation (77.22) justified? We address such issues in the advanced companion volumes.

77.11   An Example

We want to compute the the solution

$$u(x) = \frac{a}{\pi} \exp\left(-a(x_1^2 + x_2^2)\right), \quad a = 400,$$

of Poisson's equation $-\Delta u = f$ on the square $(-.5, .5) \times (-.5, .5)$ with $f(x)$ being the following "approximate delta function":

$$f(x) = \frac{4}{\pi} a^2 \left(1 - ax_1^2 - ax_2^2\right) \exp\left(-a(x_1^2 + x_2^2)\right),$$

We plot f in Fig. 77.9 (note the vertical scale), together with the initial mesh with 224 elements. The adaptive algorithm took 5 steps to achieve an estimated .5% relative error. We plot the final mesh together with the associated finite element approximation in Fig. 77.10. The algorithm produced meshes with 224, 256, 336, 564, 992, and 3000 elements respectively.

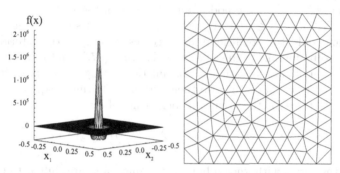

Fig. 77.9. The approximate delta forcing function f and the initial mesh used for the finite element approximation

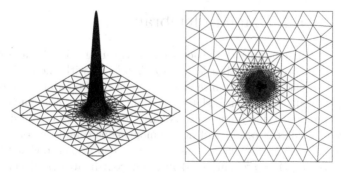

Fig. 77.10. The finite element approximation with a relative error of .5% and the final mesh used to compute the approximation. The approximation has a maximum height of roughly 5

77.12   Non-Homogeneous Dirichlet Boundary Conditions

We now consider Poisson's equation with non-homogeneous Dirichlet boundary conditions:

$$\begin{cases} -\Delta u = f & \text{in } \Omega, \\ u = g & \text{on } \Gamma, \end{cases} \tag{77.23}$$

where g is the given boundary data.

We compute a finite element approximation on a triangulation $\mathcal{T}_h$, where we now also include the nodes on the boundary, denoting the internal nodes by $\mathcal{N}_h$ as above and the set of nodes on the boundary by $\mathcal{N}_b$. We compute an approximation U of the form

$$U = \sum_{N_j \in \mathcal{N}_b} \xi_j \varphi_j + \sum_{N_j \in \mathcal{N}_h} \xi_j \varphi_j, \tag{77.24}$$

where φ_j denotes the basis function corresponding to node N_j in an enumeration $\{N_j\}$ of all the nodes, and, because of the boundary conditions, $\xi_j = g(N_j)$ for $N_j \in \mathcal{N}_b$. Thus the boundary values of U are given by g on Γ and only the coefficients of U corresponding to the interior nodes remain to be found. To this end, we substitute (77.24) into the variational formulation of (77.23) with the test functions being all the basis functions for the internal nodes, and we then get the following a square system of linear equations for the unknown coefficients of U:

$$\sum_{N_j \in \mathcal{N}_h} \xi_j (\nabla \varphi_j, \nabla \varphi_i) = (f, \varphi_i) - \sum_{N_j \in \mathcal{N}_b} g(N_j)(\nabla \varphi_j, \nabla \varphi_i), \quad N_i \in \mathcal{N}_h.$$

where the terms with known boundary values of U are shifted to the right hand side as data.

77.13   An L-shaped Membrane

We present an example that shows the performance of the adaptive algorithm on a problem with a *boundary singularity* with the derivatives of the exact solution being infinite at a corner of the boundary. We consider the Laplace equation in an L-shaped domain that has a non-convex corner at the origin satisfying homogeneous Dirichlet boundary conditions at the sides meeting at the origin and non-homogeneous conditions on the other sides, see Fig. 77.11. We choose the boundary conditions so that the exact solution is $u(r, \theta) = r^{2/3} \sin(2\theta/3)$ in polar coordinates (r, θ) centered at the origin, which has a typical singularity of a corner problem:

$$\frac{\partial u}{\partial r}(r, \theta) = \frac{2}{3} r^{-1/3} \sin(2\theta/3),$$

which tends to infinity as r tends to zero (unless $\theta = 0$ or $\theta = \frac{3\pi}{2}$).

We use the knowledge of the exact solution to evaluate the performance of the adaptive algorithm.

We compute using an adaptive FEM-solver with energy norm control based on (77.18) to achieve an error tolerance of TOL $= .005$ using h refinement mesh modification. In Fig. 77.11, we show the initial mesh $\mathcal{T}_{h^{(0)}}$ with 112 nodes and 182 elements. In Fig. 77.12, we show the level curves of the solution and the final mesh with 295 nodes and 538 elements that achieves the desired error bound. The interpolation constant was set to $C_i = 1/8$. The quotient between the estimated and true error on the final mesh was 1.5.

Since the exact solution is known in this example, we can also use the a priori error estimate to determine a mesh that gives the desired accuracy. We do this by combining the a priori error estimate (77.16) and the principle of equidistribution of error to determine $h(r)$ so that $C_i \| h D^2 u \| = $ TOL while keeping h as large as possible (and keeping the number of elements at a minimum). Since $D^2 u(r) \approx r^{-4/3}$, as long as $h \leq r$, that is up to the elements touching the corner, we determine that

$$\left(hr^{-4/3}\right)^2 h^2 \approx \frac{\text{TOL}^2}{M} \qquad \text{or} \qquad h^2 = \text{TOL}\, M^{-1/2} r^{4/3},$$

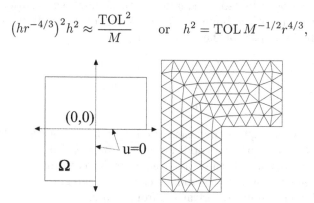

Fig. 77.11. The L-shaped domain and the initial mesh

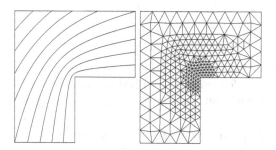

Fig. 77.12. Level curves of the solution and final adapted mesh on the L-shaped domain

where M is the number of elements and h^2 measures the element area. To compute M from this relation, we note that $M \approx \int_{\Omega} h^{-2} \, dx$, since the number of elements per unit area is $O(h^{-2})$, which gives

$$M \approx M^{1/2} \text{TOL}^{-1} \int_{\Omega} r^{-4/3} \, dx.$$

Since the integral is convergent (prove this), it follows that $M \propto \text{TOL}^{-2}$, which implies that $h(r) \propto r^{1/3} \text{TOL}$. Note that the total number of unknowns, up to a constant, is the same as that required for a smooth solution without a singularity, namely TOL^{-2}. This depends on the very local nature of the singularity in the present case. In general, of course solutions with singularities may require a much larger number of elements than smooth solutions do.

77.14 Robin and Neumann Boundary Conditions

Next, we consider Poisson's equation with homogeneous Dirichlet conditions on part Γ_1 of the boundary and non-homogeneous Robin conditions on the remaining part of the boundary Γ_2:

$$\begin{cases} -\Delta u = f & \text{in } \Omega, \\ u = 0 & \text{on } \Gamma_1, \\ \partial_n u + \kappa u = g & \text{on } \Gamma_2, \end{cases} \tag{77.25}$$

where $\kappa \geq 0$ is a given coefficient, and f and g are given data. Setting $\kappa = 0$ gives the Neumann condition $\partial_n u + \kappa u = g$. To find a variational formulation, we multiply the Poisson equation by a test function v satisfying the homogenous Dirichlet boundary condition, integrate over Ω, and use Green's formula to move derivatives from u to v:

$$(f, v) = -\int_{\Omega} \Delta u \, v \, dx = \int_{\Omega} \nabla u \cdot \nabla v \, dx - \int_{\Gamma} \partial_n u v \, ds$$

$$= \int_{\Omega} \nabla u \cdot \nabla v \, dx + \int_{\Gamma_2} \kappa u v \, ds - \int_{\Gamma_2} g v \, ds,$$

where we use the boundary conditions to rewrite the boundary integral. We are thus led to the following cG(1) FEM based on a space V_h of continuous piecewise linear functions vanishing on Γ_1: find $U \in V_h$ such that

$$(\nabla U, \nabla v) + \int_{\Gamma_2} \kappa U v \, ds = (f, v) + \int_{\Gamma_2} g v \, ds \quad \text{for all } v \in V_h. \tag{77.26}$$

We recall that boundary conditions like the Dirichlet condition that are enforced explicitly in the choice of the space V_h are called *essential boundary conditions*. Boundary conditions like the Robin condition that are implicitly contained in the weak formulation are called *natural boundary conditions*. (To remember that we must assume essential conditions: there are two "ss" in assume and essential.)

Note that the stiffness matrix and load vector related to (77.26) contain contributions from both integrals over Ω and Γ_2 related to the basis functions corresponding to the nodes on the boundary Γ_2.

To illustrate, we compute the solution of Laplace's equation with a combination of Dirichlet, Neumann and Robin boundary conditions on the domain shown in Fig. 77.13 using an adaptive FEM-solver. We show the boundary conditions in the illustration. The problem models e.g. stationary heat flow around a hot water pipe in the ground. We show the mesh used to compute the approximation so that the error in the L_2 norm is smaller than .0013 together with a contour plot of the approximation in Fig. 77.14. We notice that the level curves are parallel to a boundary with a homogeneous Dirichlet condition, and orthogonal to a boundary with a homogeneous Neumann condition.

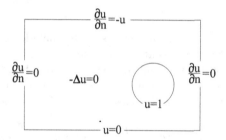

Fig. 77.13. A problem with Robin boundary conditions

Fig. 77.14. The adaptive mesh and contour lines of the approximate solution of the problem shown in Fig. 77.13 computed with error tolerance .0013

77.15   Stationary Convection-Diffusion-Reaction

We now consider the extension to a convection-diffusion-reaction problem of the form

$$-\nabla \cdot (a\nabla u) + \nabla \cdot (ub) + cu = f \quad \text{in } \Omega,$$
$$a\partial_n u + \kappa u = g \quad \text{on } \Gamma. \tag{77.27}$$

with Robin boundary conditions, where f and g are given data, and $a > 0$, b, c and $\kappa \geq 0$ are given coefficients, and Ω is a given domain in $\mathbb{R}^2$ with boundary Γ. The term cu models absorption if $c \geq 0$ and production $c < 0$.

Let V_h be the space of continuous piecewise linear functions on a triangulation of Ω with no restriction on the nodal values on the boundary. The cG(1) FEM for (77.27) takes the form: Find $U \in V_h$ such that

$$\int_\Omega a\nabla U \cdot \nabla v \, dx + \int_\Omega \nabla \cdot (Ub)v \, dx + \int_\Omega cUv \, dx + \int_\Gamma \kappa Uv \, ds$$
$$= \int_\Omega fv \, dx + \int_\Gamma gv \, ds, \tag{77.28}$$

for all $v \in V_h$. Note that the extension to include the terms $\nabla \cdot (Ub)$ and cu is very natural and that the corresponding terms in the variational formulation are obtained by multiplying by the test function v without any partial integration. For the term $-\nabla \cdot (a\nabla u)$ we note that multiplication by $v(x)$ and integration over Ω gives using the Divergence theorem

$$-\int_\Omega \nabla \cdot (a\nabla u)v \, dx = \int_\Omega a\nabla U \cdot \nabla v \, dx - \int_\Gamma a\partial_n u \, v \, ds$$

and the variational formulation results from replacing $-a\partial_n u \, ds$ by $ku - g$ using the Robin boundary condition.

The matrix equation corresponding to (77.28) has a banded and sparse stiffness matrix, but the symmetry is lost if $b \neq 0$, as is evident from the presence of the non-symmetric term $\int_\Omega \nabla \cdot (ub)v \, dx$. The non-symmetry of the convection term eliminates the best approximation property of FEM, but FEM still may give good results. If $c < 0$ then solutions may be non-unique corresponding to non-zero solutions (eigen-functions) of the homogeneous problem $-\nabla \cdot (a\nabla u) + \nabla \cdot (ub) + cu = 0$.

The Convection-Dominated Case: Streamline Diffusion

If $|b| > \frac{a}{h}$, where $h(x)$ is the mesh size, which we refer to as a *convection-dominated case*, and the exact solution is non-smooth with rapid variation, then the FEM-solution may exhibit spurious oscillations. In such cases the cG(1)-method (77.28) will have to be modified by changing the test functions from v to $v + \delta\nabla \cdot (vb)$ in all terms but the diffusion and boundary terms, where $\delta = \frac{h}{2|b|}$ acts as a parameter. The presence of the modification $\delta \nabla \cdot (vb)$ introduces the positive quadratic term $\int_\Omega \delta(\nabla \cdot (Ub))^2 \, dx$ upon choosing $v = U$, which gives enhanced stability and (almost) eliminates spurious oscillations. The fact that the modification is not made in the diffusion term does not destroy accuracy, because in the convection dominated case the diffusion coefficient is small. The modified method is referred to as *the streamline diffusion method* or *weighted least squares-stabilization*.

77.16 Time-Dependent Convection-Diffusion-Reaction

We now consider the time-dependent analog of (77.27), that is the problem

$$\dot{u} - \nabla \cdot (a\nabla u) + \nabla \cdot (ub) + cu = f \quad \text{in } \Omega \times (0, T],$$

$$a\partial_n u + \kappa u = g \quad \text{on } \Gamma \times (0, T], \qquad (77.29)$$

$$u(\cdot) = u^0 \quad \text{in } \Omega,$$

where $[0, T]$ is a given time interval, and u^0 a given initial value. For the time discretization we may use e.g. dG(0) or cG(1) on a subdivision $0 = t_0 < t_1 < \cdots < t_N = T$ into time intervals $I_n = (t_{n-1}, t_n]$ with time steps $k_n = t_n - t_{n-1}$. Using dG(0) we seek $U^n \in V_h$ for $n = 1, \ldots, N$, such that for $n = 1, \ldots, N$,

$$\int_\Omega U^n v \, dx + \int_{\Omega \times I_n} a\nabla U^n \cdot \nabla v \, dx \, dt$$

$$+ \int_{\Omega \times I_n} \nabla \cdot (U^n b) v \, dx \, dt + \int_{\Omega \times I_n} cU^n v \, dx \, dt + \int_{\Gamma \times I_n} \kappa U^n v \, ds \, dt \quad (77.30)$$

$$= \int_\Omega U^{n-1} v \, dx + \int_{\Omega \times I_n} fv \, dx \, dt + \int_{\Gamma \times I_n} gv \, ds \, dt,$$

for all $v \in V_h$, where $U^0 = u^0$. The corresponding discrete system for U^n takes the form

$$M\xi^n + k_n A_n \xi^n = M\xi^{n-1} + k_n b^n$$

where the vector ξ^n contains the nodal values of $U^n \in V_h$, M is the mass matrix related to V_h, A_n is the relevant stiffness matrix connected to the convection-diffusion-reaction terms, and b^n the relevant load vector.

In a convection-dominated case, the test functions v are again modified to $v + \delta \nabla \cdot (vb)$ in all terms with integration over $\Omega \times I_n$, but the diffusion term.

77.17 The Wave Equation

We now consider the extension to the wave equation with homogeneous Dirichlet boundary conditions:

$$\ddot{u} - \Delta u = f \quad \text{in } \Omega \times (0, T],$$
$$u = 0 \quad \text{on } \Gamma \times (0, T], \quad\quad (77.31)$$
$$u = u^0, \quad \dot{u} = \dot{u}^0 \quad \text{in } \Omega,$$

where u^0 and $\dot{u}^0$ are given initial conditions. As above we let V_h be the set of piecewise linears functions on a triangulation of Ω satisfying the homogeneous Dirichlet boundary conditions, and we let $0 = t_0 < t_1 < \cdots < t_N = T$ be a subdivision of $[0, T]$ into time intervals $I_n = (t_{n-1}, t_n]$ with time steps $k_n = t_n - t_{n-1}$. We apply cG(1) in space and cG(1) in time and seek a discrete solution U in the space of functions W_h spanned by the functions

$$v(x, t) = \sum_{n=0}^{N} \sum_{j=1}^{M} \eta_j^n \varphi_j(x) \psi_n(t),$$

where $\{\varphi_j(x)\}_{j=1}^{M}$ is a basis for V_h, and $\{\psi_n(t)\}_{n=0}^{N}$ is a basis for the space of continuous piecewise linear functions on the subdivision $0 = t_0 < t_1 < \cdots < t_N = T$. The corresponding discrete system takes the following explicit form if mass lumping is used in space as well as time and the time step is constant $k_n = k$:

$$\xi^{n+1} = 2\xi^n - \xi^{n-1} + k^2 A \xi^n \quad \text{for } n = 1, \dots, N - 1,$$

with appropriate starting values ξ^0 and ξ^1 computed from the initial conditions, and A the relevant stiffness matrix related to the Laplacian.

77.18 Examples

We present some examples of systems of nonlinear reaction-diffusion-convection equations (77.27) arising in physics, chemistry and biology of the form

$$\dot{u}_i - \nabla \cdot (a_i \nabla u_i) + \nabla \cdot (u_i b_i) + c_i u_i = f_i(u_1, \dots, u_d) \text{ in } \Omega \times I, \; i = 1, \dots, d,$$
$$(77.32)$$

where the $a_i > 0$, b_i and c_i are given coefficients, and the $f_i : \mathbb{R}^d \to \mathbb{R}$. These systems may be solved numerically by a direct extension of the cG(1) method in space and time presented above. We will return in detail to this issue below. In all the examples, a is a positive constant.

Example 77.1. The bistable equation for ferro-magnetism

$$\dot{u} - a\Delta u = u - u^3. \tag{77.33}$$

Example 77.2. Superconductivity of fluids

$$\dot{u}_1 - a\Delta u_1 = (1 - |u|^2)u_1,$$
$$\dot{u}_2 - a\Delta u_2 = (1 - |u|^2)u_2. \tag{77.34}$$

Example 77.3. Flame propagation

$$\dot{u}_1 - a\Delta u_1 = -u_1 e^{-\alpha_1/u_2},$$
$$\dot{u}_2 - a\Delta u_2 = \alpha_2 u_1 e^{-\alpha_1/u_2}, \tag{77.35}$$

where $\alpha_1, \alpha_2 > 0$ are constants.

Example 77.4. Interaction of two species

$$\dot{u}_1 - a\Delta u_1 = u_1 M(u_1, u_2),$$
$$\dot{u}_2 - a\Delta u_2 = u_2 N(u_1, u_2), \tag{77.36}$$

where $M(u_1, u_2)$ and $N(u_1, u_2)$ are given functions describing various situations such as (i) predator-prey ($M_{u_2} < 0$, $N_{u_1} > 0$) (ii) competing species ($M_{u_2} < 0$, $N_{u_1} < 0$) and (iii) symbiosis ($M_{u_2} > 0$, $N_{u_1} > 0$).

Example 77.5. Morphogenesis of patterns (zebra)

$$\dot{u}_1 - a\Delta u_1 = -u_1 u_2^2 + \alpha_1(1 - u_1),$$
$$\dot{u}_2 - a\Delta u_2 = u_1 u_2^2 - (\alpha_1 + \alpha_2)u_2. \tag{77.37}$$

Example 77.6. Belousov-Zhabotinski reaction in chemical kinetics

$$\dot{u}_1 - a\Delta u_1 = \alpha_1(u_2 - u_1 u_2 + u_1 - \alpha_2 u_2^2),$$
$$\dot{u}_2 - a\Delta u_2 = \alpha_1^{-1}(\alpha_3 u_3 - u_2 - u_1 u_2). \tag{77.38}$$
$$\dot{u}_3 - a\Delta u_3 = \alpha_4(u_1 - u_3),$$

where $\alpha \approx 10^2$, $\alpha_2 \approx 10^{-2}$, $\alpha_3 \approx 1$, $\alpha_4 \approx 10^{-1}$.

Chapter 77 Problems

77.1. Compute the coefficients of the mass matrix M on the standard triangulation of the square of mesh size h. Hint: it is possible to use quadrature based on the midpoints of the sides of the triangle because this is exact for quadratic functions. The diagonal terms are $h^2/2$ and the off-diagonal terms are all equal to $h^2/12$. The sum of the elements in a row is equal to h^2.

77.2. Compute the stiffness matrix for cG(1) for the problem $-\Delta u = 1$ in $\Omega = (0,1) \times (0,1)$ with $u = 0$ on the side with $x_2 = 0$ and $\partial_n u + u = 1$ on the other three sides of Ω using the standard triangulation. Note the contribution to the stiffness matrix from the nodes on the boundary.

77.3. Describe the sparsity pattern of the stiffness matrices A for the Poisson equation with homogeneous Dirichlet data on the unit square corresponding to the continuous piecewise linear finite element method on the standard triangulation using the three numbering schemes pictured in Fig. 77.15.

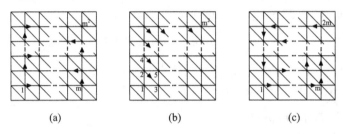

(a)   (b)   (c)

Fig. 77.15. Three node numbering schemes for the standard triangulation of the unit square

77.4. Compute the load vector for $f(x) = x_1 + x_2^2$ on the standard triangulation of the unit square using exact integration and the lumped mass (trapezoidal rule) quadrature.

77.5. Write a code to solve $A\xi = b$ using both the Jacobi and Gauss-Seidel iteration methods, making use of the sparsity of A in storage and operations. Compare the convergence rate of the two methods using the result from a direct solver as a reference value.

77.6. Write a code to solve the system $A\xi = b$ with A a band matrix.

77.7. Compute the stiffness matrix for the Poisson equation with homogeneous Dirichlet boundary conditions for (a) the *union jack* triangulation of a square shown in Fig. 77.16 and (b) the triangulation of triangular domain shown in Fig. 77.16.

77.8. Compute the discrete equations for the finite element approximation for $-\Delta u = 1$ on $\Omega = (0,1) \times (0,1)$ with boundary conditions $u = 0$ for $x_1 = 0$, $u = x_1$ for $x_2 = 0$, $u = 1$ for $x_1 = 1$ and $u = x_1$ for $x_2 = 1$ using the standard triangulation (Fig. 77.2).

77.9. (a) Show that the element stiffness matrix (77.12) for the linear polynomials on a triangle K with vertices at $(0,0)$, $(h,0)$, and $(0,h)$ numbered 1, 2 and 3, is given by

$$\begin{pmatrix} 1 & -1/2 & -1/2 \\ -1/2 & 1/2 & 0 \\ -1/2 & 0 & 1/2 \end{pmatrix}.$$

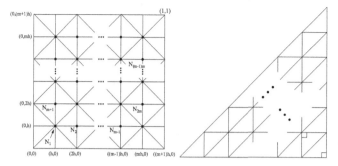

Fig. 77.16. The "union jack" triangulation of the unit square and a uniform triangulation of a right triangle

(b) Use this result to verify the formula computed for the stiffness matrix A for the continuous piecewise linear finite element method for the Poisson equation with homogeneous boundary conditions on the unit square using the standard triangulation. (c) Compute the element stiffness matrix for a triangle K with nodes $\{a^i\}$.

77.10. Compute the asymptotic operations count for the direct solution of the system $A\xi = b$ using the three A computed in Problem 77.3.

77.11. Apply the finite element method with piecewise linear approximation to the Poisson equation in three dimensions with a variety of boundary conditions. Compute the stiffness matrix and load vector in some simple cases.

77.12. Derive a priori error bound in the energy norm for cG(1) FEM for Poisson's equation with Robin boundary conditions. Generalize to problems of the form $-\nabla \cdot (a\nabla u) + cu = f$, where $a(x) > 0$ and $c \geq 0$.

77.13. Derive a posteriori error bound in the energy norm for cG(1) FEM for Poisson's equation with Robin boundary conditions. Generalize to problems of the form $-\nabla \cdot (a\nabla u) + cu = f$, where $a(x) > 0$ and $c \geq 0$.

77.14. Implement adaptive energy norm error control for cG(1) for Poisson's equation based on an a posteriori error estimate.

77.15. Find an exact solution of the L-shaped membrane problem with the Dirichlet condition replaced by a Neumann condition on one of the sides meeting at $\frac{3\pi}{2}$ corner. What is the nature of the singularity?

77.16. Let $\omega(x)$ be a positive weight function defined on the domain $\Omega \subset \mathbb{R}^2$. Assume that the mesh function $h(x)$ minimizes the integral $\int_\Omega h^2(x)\omega(x)\,dx$ under the constraint $\int_\Omega h^{-2}(x)\,dx = N$, where N is a given positive integer. Prove that $h^4(x)\omega(x)$ must be constant. Interpret the result as equidistribution in the context of error control. Hint: argue that $h^4(x)\omega(x)$ is the gain adding one more node.

78
Inverse Problems

I never guess. It is a capital mistake to theorize before one has data.
Insensibly one begins to twist facts to suit theories, instead of theories
to suit facts.

When you have eliminated the impossible, whatever remains, however improbable, must be the truth.
(Sherlock Holmes in the *The Sign of Four*, 1888)

78.1 Introduction

We have above in our study of Poisson's equation studied "forward" prob-
lems of the form: Given the function $f : \Omega \to \mathbb{R}$, find a function $u : \Omega \to \mathbb{R}$
such that

$$\begin{cases} -\Delta u = f & \text{in } \Omega, \\ \partial_n u + \kappa u = 0 & \text{on } \Gamma, \end{cases} \tag{78.1}$$

where $\kappa \geq 0$ is a given constant. A corresponding "inverse" problem would
be to assume knowledge of $u(x)$ and seek the corresponding function $f(x)$
so that (78.1) is satisfied! If we know $u(x)$ in the entire region Ω, this is
a problem of differentiation: we just compute $\Delta u(x)$ from $u(x)$. We then
have $-\Delta u = f$! We have studied this problem in Chapter *The derivative*,
and we recall that this problem is a bit delicate and that we have to balance
the step length h in a difference approximation of Δ to the precision in the
given data $u(x)$.

Suppose now that we know $u(x)$ only on the boundary Γ. Can we then determine $f(x)$ in Ω? This type of problem connects to a wealth of important applications of the following form: Suppose we can measure something on the *boundary* of an object. Can we then say something about what is *inside* the object? For example, suppose the object is a human body, and that we can measure something on the boundary (or outside) the body. Can we than get some information on what is inside the body? Or, suppose we can accumulate data on the surface of the Earth, can we then say something about what is in the interior of the Earth, such as the presence of layers of oil. These are all examples of *inverse problems*.

The nature of an inverse problem is to be "ill-posed" in the sense that solutions may be non-unique and/or that small changes in the data may cause large changes in the solution. To single out a unique solution which is not too sensitive to little errors in data, we may have to "regularize" the inverse problem e.g. by smoothing of the data and/or restricting the size of (derivatives of) the solution. Differentiation is such an ill-posed problem where we may need to "smooth" or regularize a given function before attempting to compute its derivative.

A typical forward problem is "well-posed" in the sense that small changes in data cause small changes in the solution. A basic example of a well-posed problem is integration corresponding to solving a differential equation. The corresponding inverse problem is differentiation which is ill-posed as we just noted. Solving a differential equation does not always correspond to a well-posed problem: in Chapter *Lorenz and the Essence of Chaos* we met a simple differential equation with solutions being highly sensitive to changes in data.

Example 78.1. An *electrocardiogram* ECG produces a curve reflecting the electrical activity of the heart from measurements of electric potentials on the chest, and the curve gives a specialist information on abnormal activities of the heart such as abnormal heart rhythm (arrhythmias). Similarly, an *electroencephalogram* EEG gives information on the electrical activity of the brain from measurements of electric potentials on the scalp. These techniques are however too imprecise for many diagnoses, and more recently techniques of *electrocardiographic imaging* have been developed, which build on solving inverse problems for Poisson-like equations. The geometry of the individual patient is then obtained from computer tomography, and a picture of the electrical activity inside the body is obtained from measurements of electric potentials on the boundary (e.g. the chest or scalp) by solving an inverse problem for a Poisson-like equation (using the finite element method). Electrocardiographic imaging may give more accurate information on e.g. abnormal cardiac or brain activity than ECG and EEG, and is now a part of practice in advanced neurological and radiological departments. Further development of in particular the computational process (adaptivity, geometric modeling) is needed to increase the accuracy.

Example 78.2. Another inverse problem of importance to mankind occurs in inverse seismic prospecting: Explosions on the surface of the Earth are set off and the reflections of the induced waves in the mantle of the Earth are recorded on the surface, and from this information one tries to determine subsurface structures such as layers of oil-bearing rock. To solve this reconstruction problem one uses computational methods based on solving the wave equation involving a wave speed coefficient characteristic of different materials, and through optimization one tries to find the local wave speed coefficient which gives best least squares fit to measured data on the surface of the Earth, and which then gives information on the unknown subsurface layering.

78.2   An Inverse Problem for One-Dimensional Convection

We start considering the simplest boundary value problem:

$$u'(x) = f(x), \text{ for } x \in (0,1], \ u(0) = 0, \qquad (78.2)$$

modeling convection with $u : [0,1] \to \mathbb{R}$ representing a concentration and $f : [0,1] \to \mathbb{R}$ a source. We seek to determine or *reconstruct* the function $f : [0,1] \to \mathbb{R}$ from the boundary value *observation* $u(1)$ of the corresponding solution $u(x)$ of (78.2). It is clear that we cannot hope to determine $f(x)$ for all $x \in (0,1)$ from this observation alone. This is because there are many functions $f(x)$ such that the corresponding function $u(x)$ satisfies (78.2) and $u(1) = 0$. To see this it is sufficient to choose a non-zero function $u(x)$ on $[0,1]$ satisfying $u(0) = u(1) = 0$ and define $f(x) = u'(x)$. Evidently, the reconstruction is undetermined up to such functions.

The indeterminancy of the reconstruction $f(x)$ reflects the *ill-posed* nature of the inverse problem; even if the measurements of the boundary value $u(1)$ is very precise, the corresponding source $f(x)$ is not well defined. We thus need some *extra condition* to single out a (hopefully) unique source $f(x)$. We may do this in many ways and depending on the extra condition imposed, we may get different reconstructions $f(x)$. We now indicate one possibility, where we reconstruct under the extra condition that $f(x)$ is *as small as possible* in a least squares sense, which is a common technique of regularization. We then reformulate the inverse problem as the following *least squares optimization problem*: Find the function $f : [0,1] \to \mathbb{R}$ which minimizes the total "cost"

$$J(f) = (u(1) - \bar{u}(1))^2 + \mu \int_0^1 f(x)^2 \, dx,$$

where $u(x)$ solves (78.2), $\bar{u}(1)$ is the observed boundary value at $x = 1$, and $\mu > 0$ is a constant. We may view this as a control problem where the objective is to find the *control* $f : [0, 1] \to \mathbb{R}$ which minimizes the total cost $J(f)$, where the μ-term measures the cost of the control f and the first term the cost of a boundary value misfit $u(1) - \bar{u}(1)$.

We can phrase this problem as finding the function $u(x)$ with $u(0) = 0$ which minimizes

$$(u(1) - \bar{u}(1))^2 + \mu \int_0^1 (u')^2 \, dx.$$

Recalling Chapter *FEM for two-point boundary value problems*, we understand that the solution $u(x)$ satisfies $u''(x) = 0$ in $(0, 1)$, $u(0) = 0$ and $u(1) - \bar{u}(1) + \mu u'(1) = 0$. We conclude that $u(x) = \frac{1}{1+\mu}\bar{u}(1)x$, and thus the reconstructed source $f(x) = u'(x)$ takes on a constant value and is given by

$$f(x) = \frac{1}{1 + \mu} \bar{u}(1) \quad \text{for } x \in [0, 1].$$

Evidently, we are led to choose the regularization parameter μ small; the smaller μ is the more accurately we will fit the boundary value observation $\bar{u}(1)$. Since the reconstructed function $f(x)$ is constant, we have effectively only one constant to determine and we may expect to be able to determine this single value from the single observation $\bar{u}(1)$.

We can also rephrase the optimization problem as follows introducing the integral operator B defined on functions on $[0, 1]$ by $Bf(x) = \int_0^x f(y) \, dy$ for $x \in [0, 1]$: Find the function $f : [0, 1] \to \mathbb{R}$ which minimizes

$$J(f) = (Bf(1) - \bar{u}(1))^2 + \mu \int_0^1 f(x)^2 \, dx.$$

The optimality condition obtained by setting $\frac{d}{d\epsilon} J(f + \epsilon g) = 0$ for $\epsilon = 0$, where $g : [0, 1] \to \mathbb{R}$ is an arbitrary function, takes the form:

$$(Bf(1) - \bar{u}(1))Bg(1) + \mu \int_0^1 f(x)g(x) \, dx = 0 \qquad (78.3)$$

for all functions $g : [0, 1] \to \mathbb{R}$. We shall now rewrite this condition by introducing the *adjoint operator* $B^\top$ defined on functions $w(x)$ as follows: for a given $w = w(x)$ we let $B^\top w$ be the function on $[0, 1]$ satisfying $(B^\top w)'(x) = 0$ for $x \in (0, 1)$ and $B^\top w(1) = w(1)$, that is, $B^\top w$ is the constant function on $[0, 1]$ taking the value $w(1)$ for all x. We can then rewrite (78.3) in the form

$$\int_0^1 B^\top (Bf(x) - \bar{u}(1))(x)g(x) \, dx + \mu \int_0^1 f(x)g(x) \, dx = 0 \qquad (78.4)$$

for all functions $g : [0, 1] \rightarrow \mathbb{R}$, because by partial integration

$$\int_0^1 B^\top (Bf(x) - \bar{u}(1))(x) \underbrace{(Bg(x))'}_{=g(x)} dx = (Bf(1) - \bar{u}(1))Bg(1),$$

where as indicated $(Bg(x))' = g(x)$ and $B^\top w(1) = w(1)$. We conclude that

$$\int_0^1 (B^\top Bf + \mu f)g \, dx = \int_0^1 B^\top \bar{u}(1)g \, dx$$

for all functions $g : [0, 1] \rightarrow \mathbb{R}$, and therefore

$$B^\top Bf + \mu f = B^\top \bar{u}(1) \quad \text{on } (0, 1), \tag{78.5}$$

or with I the identity operator:

$$(B^\top B + \mu I)f = B^\top \bar{u}(1). \tag{78.6}$$

We conclude that $f(x)$ is constant on $[0, 1]$ and takes the value

$$f(x) = \frac{1}{1 + \mu} \bar{u}(1) \quad \text{for } x \in [0, 1],$$

which is the same result as already derived. We note the form (78.6) of the optimality condition (78.6) with the operator $B^\top B + \mu I$ appearing. We shall meet the same equation below with different solution operators B and adjoints $B^\top$.

78.3 An Inverse Problem for One-Dimensional Diffusion

We continue with the boundary value problem

$$-u'' = f \quad \text{in } (0, 1), \quad u'(0) = 0, \quad u'(1) + u(1) = 0, \tag{78.7}$$

where we seek to determine the source $f(x)$ in $(0, 1)$ by observing the boundary values $u(0)$ and $u(1)$ of the corresponding solution $u(x)$ of (78.7). Again it is clear that we cannot hope to determine $f(x)$ for all $x \in (0, 1)$ from these two observations alone, because there are many functions $f(x)$ such that the corresponding function $u(x)$ satisfies (78.7) and $u(0) = u(1) = 0$. To see this it is sufficient to choose a non-zero function $u(x)$ on $[0, 1]$ satisfying $u(0) = u'(0) = u(1) = u'(1) = 0$ and set $f(x) = -u''(x)$.

As above we seek to reconstruct $f(x)$ under the extra condition that $f(x)$ is as small as possible and we therefore reformulate the inverse problem as the following least squares optimization problem: Find $f(x)$ in $(0,1)$ such that

$$J(f) = (u(0) - \bar{u}(0))^2 + (u(1) - \bar{u}(1))^2 + \mu \int_0^1 f^2(x)\, dx$$

is as small as possible, where $\mu > 0$ is a positive constant acting as a regularization, $\bar{u}(0)$ and $\bar{u}(1)$ are the boundary observations, and of course $u(x)$ solves (78.7). We thus seek $f(x)$ so that in a least squares sense we fit the boundary observations as well as the smallness of $f(x)$ as well as possible.

To state the optimality equations, we introduce the solution operator B corresponding to (78.7), that is, for a given function $f : [0,1] \to \mathbb{R}$ we let $Bf(x)$ be the function on $[0,1]$ satisfying

$$\int_0^1 (Bf)'v'\, dx + Bf(1)v(1) = \int_0^1 fv\, dx, \qquad (78.8)$$

for all functions $v(x)$ on $[0,1]$. This follows from Chapter *FEM for two-point boundary value problems*. Setting $\frac{d}{d\epsilon}J(f + \epsilon g) = 0$ for $\epsilon = 0$, we obtain the optimality condition in the form

$$(Bf(0) - \bar{u}(0), Bg(0)) + (Bf(1) - \bar{u}(1), Bg(1)) + \mu \int_0^1 f(x)g(x)\, dx = 0$$

$$(78.9)$$

for all functions $g(x)$ on $[0,1]$. Next we introduce the adjoint operator $B^\top$ defined as follows: given the values $w(0)$ and $w(1)$, we let $B^\top w$ be the function on $[0,1]$ which satisfies

$$\int_0^1 (B^\top w)'v'\, dx + B^\top w(1)v(1) = w(0)v(0) + w(1)v(1) \qquad (78.10)$$

for all $v(x)$. We see that $(B^\top w)'' = 0$ and $-(B^\top w)'(0) = w(0)$, $B^\top w(1) + (B^\top w)'(1) = w(1)$. In other words, $B^\top w$ is a linear function determined by the two boundary conditions. In particular, if $w(0) = 0$, then $B^\top w = w(1)$ is a constant. Now, setting $v = Bg$ in (78.10), we get

$$w(0)Bg(0) + w(1)Bg(1) = \int_0^1 (B^\top w)'(Bg)'\, dx + B^\top w(1)Bg(1)$$

$$= \int_0^1 B^\top wg\, dx,$$

where we used (78.8) with f replaced by g and v replaced by $B^\top w$, and thus we can write the optimality condition (78.9) in the same form as above:

$$(B^\top B + \mu I)f = B^\top \bar{u}. \qquad (78.11)$$

From this equation we can uniquely solve for the function $f(x)$, which will be a linear function defined by two constants, because $f = \frac{1}{\mu}B^\top(Bf - \bar{u})$. For example if $\bar{u}(0) = 0$ and $\bar{u}(1) = 1$, then we get choosing μ small, $f(x) \approx 10\bar{u}(1)x - 8\bar{u}(1)$ with corresponding solution $u(x) \approx -3x^3 + 4x^2$.

We now comment on the nature of the optimality equation (78.11). The operator B maps a space of sources, say F, into a space of observations, say O, and the adjoint operator $B^\top$ maps O into F. We may think of the dimension of F as large, and that of O as smaller. For the discussion we may assume that the dimension of F is n and the dimension of O is m and thus B corresponds to an $m \times n$ matrix and $B^\top$ to an $n \times m$ matrix with $m \ll n$. This will be the setting with computational approximations of the solution operators B and $B^\top$. In particular, the columns of B must be severely linearly independent since there are many more columns than rows, and thus the $n \times n$ matrix $B^\top B$ must be singular with many non-zero n-vectors f satisfying $B^\top Bf = 0$. On the other hand, the matrix $B^\top B + \mu I$ with $\mu > 0$ is nonsingular, because if $(B^\top B + \mu I)f = 0$, then scalar multiplication by the n-vector $f^\top$, we obtain $\|Bf\|^2 + \mu\|f\|^2 = 0$ and thus $f = 0$. The non-zero solutions f to $B^\top Bf = 0$ are eigenvectors corresponding to a zero eigenvalue, and by changing to the regularized operator $B^\top B + \mu I$ we shift the spectrum to the interval $[\mu, \infty)$ on the positive real axis.

78.4   An Inverse Problem for Poisson's Equation

We now pass to an inverse problem for Poisson's equation (78.1) assuming Ω, Γ and κ to be known: Given $u(x) = \hat{U}(x)$ for $x \in \Gamma$, find $f(x)$ for $x \in \Omega$.

We approach this problem directly in discrete form as the following least squares problem: Find $F \in V_h$ which minimizes

$$J(F) = \|U - \hat{U}\|_\Gamma^2 + \mu\|F\|_\Omega^2 \tag{78.12}$$

over V_h, where $U \in V_h$ satisfies

$$(\nabla U, \nabla v)_\Omega + (\kappa U, v)_\Gamma = (F, v)_\Omega \quad \text{for all } v \in V_h, \tag{78.13}$$

and V_h is the space of continuous piecewise linear functions on a given triangulation of Ω of mesh size $h(x)$. As above $\mu \geq 0$ acts as a *regularization parameter* which helps to cope with the ill-posed nature of the problem. Further, $\|\cdot\|_\Omega$ and $(\cdot,\cdot)_\Omega$ denote the $L_2(\Omega)$ norm and scalar product, and similarly, $\|\cdot\|_\Gamma$ and $(\cdot,\cdot)_\Gamma$ denote the $L_2(\Gamma)$ norm and scalar product.

We reformulate (78.12) by introducing the solution operator $B_h : V_h \to W_h$ defined by $B_h F = U_F$ on Γ, where $U_F \in V_h$ solves (78.13), and W_h is the restriction of the space V_h to the boundary Γ, that is a set of piecewise linear functions on Γ. By definition, $U_F \in V_h$ satisfies:

$$(\nabla U_F, \nabla v)_\Omega + (\kappa U_F, v)_\Gamma = (F, v)_\Omega \quad \text{for all } v \in V_h. \tag{78.14}$$

We can now formulate the minimization problem (78.12) as follows: Find $F \in V_h$ which minimizes

$$J(F) = \|B_h F - \hat{U}\|_\Gamma^2 + \mu \|F\|_\Omega^2, \tag{78.15}$$

over V_h. This is a quadratic minimization problem with unique solution $F \in V_h$ characterized by a least squares equation of the form

$$(B_h F, B_h G)_\Gamma + (\mu F, G)_\Omega = (\hat{U}, B_h G)_\Gamma \quad \text{for all } G \in V_h, \tag{78.16}$$

which expresses that $\frac{d}{d\epsilon} J(F + \epsilon G) = 0$ for $\epsilon = 0$ for all $G \in V_h$.

We can express (78.16) as

$$(B_h^\top B_h F, G)_\Omega + (\mu F, G)_\Omega = (B_h^\top \hat{U}, G)_\Omega \quad \text{for all } G \in V_h,$$

that is

$$(B_h^\top B_h + \mu I)F = B_h^\top \hat{U}, \tag{78.17}$$

where $B_h^\top : W_h \to V_h$ is the transpose of B_h defined as follows: Given $w \in W_h$, we let $B_h^\top w \in V_h$ satisfy

$$(\nabla v, \nabla B_h^\top w)_\Omega + (\kappa v, B_h^\top w)_\Gamma = (v, w)_\Gamma \quad \text{for all } v \in V_h. \tag{78.18}$$

In other words, $B_h^\top w$ is an approximation of the solution z of the Poisson-problem:

$$\begin{cases} -\Delta z = 0 & \text{in } \Omega, \\ \partial_n z + \kappa z = w & \text{on } \Gamma. \end{cases} \tag{78.19}$$

Choosing $v = B_h G$ in (78.18), we get using also (78.13) with $v = B_h^\top w$

$$(B_h G, w)_\Gamma = (\nabla B_h G, \nabla B_h^\top w)_\Omega + (\kappa B_h G, B_h^\top w)_\Gamma = (G, B_h^\top w)_\Omega$$

and thus as expected from a transpose

$$(B_h G, w)_\Gamma = (G, B_h^\top w)_\Omega,$$

that is, moving B_h from G onto w brings in the transpose $B_h^\top$.

Solving (78.17) gives an approximation $F(x)$ of the function $f(x)$ we are looking for. We may solve (78.17) by direct matrix inversion if the number of nodes is small, and by some iterative method such as the gradient or the conjugate gradient method for larger problems.

The gradient method takes the form:

$$\begin{aligned} F^{n+1} &= F^n - \alpha((B_h^\top B_h + \mu I)F^n - B_h^\top \hat{U}) \\ &= F^n - \alpha(B_h^\top (B_h F^n - \hat{U}) + \mu F^n). \end{aligned}$$

In each step we have to compute first $B_h F$ and then $B_h^\top (B_h F - \hat{U})$ corresponding to solving two Poisson problems.

Example 78.3. In our first application realized using Matlab we interpret (78.17) as a matrix equation explicitly formed by computing the inverses of the stiffness matrices for the problem (78.14) and the adjoint (78.18), and we then solve this matrix equation to get the nodal values of $F(x)$. One may handle a couple of hundreds of nodes this way. For simplicity, we have considered the case $\Omega = \{(x_1, x_2) : 0 < x_1, x_2 < 1\}$ with $\kappa = 1$ and $f = 0.5 + (x - y)(x + y - 1)$, observed the boundary values of the resulting solution u, and then solved for a reconstruction of the given data f using $\mu = 0.0001$. The result is shown in Fig. 78.1 with reconstruction error ~ 0.032 in f and ~ 0.000176 in the corresponding state (boundary values).

Example 78.4. We next take $\kappa = 50$ and show the resulting state u in Fig. 78.2. The reconstruction using $\mu = 0.0001$ is now rather poor, at least in terms of f with a reconstruction error of order ~ 0.4, while the corresponding state error is of order ~ 0.02. Taking $\mu = 0.00001$ brings the

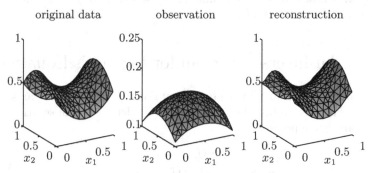

Fig. 78.1. Original data f (*left*), resulting state u (*middle*), and the reconstruction of f (*right*) with $\mu = 0.0001$ and reconstruction error ~ 0.032

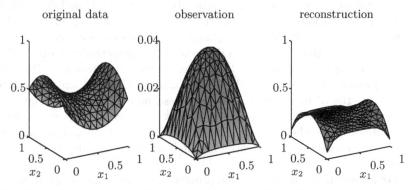

Fig. 78.2. Original data f (*left*), resulting state u (*middle*), and the reconstruction of f (*right*) with $\mu = 0.0001$, now with reconstruction error ~ 0.43

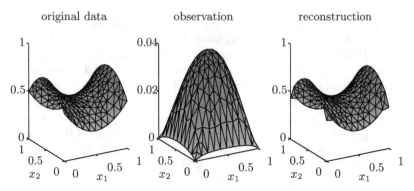

Fig. 78.3. Original data f (*left*), resulting state u (*middle*), and the reconstruction of f (*right*) after 10 steps of the conjugate gradient method with $\mu = 0.000001$, with a state error ~ 0.0003 in (the boundary values of) u and a reconstruction error ~ 0.07 in f

state error down to ~ 0.003, while a reconstruction of f with error ~ 0.04 requires taking $\mu = 0.0000005$.

78.5   An Inverse Problem for Laplace's Equation

Let Ω be a domain in $\mathbb{R}^2$ with boundary Γ composed of three parts Γ_0, Γ_1 and Γ_2. For a given function f defined on Γ_2, let u_f be the solution to the boundary value problem

$$\begin{cases} -\Delta u_f = 0 & \text{in } \Omega, \\ \quad u_f = 0 & \text{on } \Gamma_0 \cup \Gamma_1, \quad u_f = f \quad \text{on } \Gamma_2, \end{cases} \tag{78.20}$$

and define $Bf = \frac{\partial u_f}{\partial n}$ on Γ_1, where n is the unit outward normal to Γ_1. We may think of u_f as a stationary temperature defined in Ω satisfying given boundary conditions on Γ ($= 0$ on $\Gamma_0 \cup \Gamma_1$ and $= f$ on Γ_2) and with Bf representing the heat flux on Γ_1. Suppose now we can measure the heat flux on Γ_1 and that we want to determine the temperature f on Γ_2. We thus have a situation where we have access to the temperature ($= 0$) along $\Gamma_0 \cup \Gamma_1$ and may measure also the heat flux, say $\bar{q}$, along Γ_1, and we want to determine the temperature f on the inaccessible part of the boundary Γ_2. This problem arises in EKG with u being a potential and Γ_1 representing the surface of the chest and Γ_2 that of the heart, and the inverse problem being to reconstruct the potential on the heart from measurements on the chest.

We formulate the reconstruction problem as a least squares optimization problem of the form: Find f on Γ_2 which minimizes

$$J(f) = \|Bf - \bar{q}\|_{\Gamma_1}^2 + \mu \|f\|_{\Gamma_2}^2,$$

where we use the notation of the previous section, and $\mu > 0$. The optimality equation as usual takes the form

$$(B^\top B + \mu I) = B^\top \bar{q},$$

where $B^\top g = \frac{\partial u^g}{\partial n}$ on Γ_2 and u^g solves the problem

$$\begin{cases} -\Delta u^g = 0 & \text{in } \Omega, \\ \quad u^g = g & \text{on } \Gamma_1, \quad u^g = 0 \quad \text{on } \Gamma_0 \cup \Gamma_2. \end{cases} \tag{78.21}$$

This because by integrations by parts

$$(g, Bf)_{\Gamma_1} = (\nabla u^g, \nabla u_f)_\Omega = (B^\top g, f)_{\Gamma_2}$$

Example 78.5. We consider again the domain $\Omega = \{(x_1, x_2) : 0 < x_1, x_2 < 1\}$ now with $\Gamma_1 = \{(0, x_2) : 0 < x_2 < 1\}$, $\Gamma_2 = \{(1, x_2) : 0 < x_2 < 1\}$ and $\Gamma_0 = \Gamma \backslash \Gamma_1$ with an observed flow $\bar{q}$ along Γ_1 corresponding to $f = 6\,x_2^2\,(1 - x_2)$ along Γ_2. The figure shows the original (Dirichlet) boundary values to the left, the resulting state u and the associated observed flux q along Γ_1 in the middle, and the control/reconstruction f after a few conjugate gradient iterations to the right, with $\mu = 0.001$. The error in the (piecewise constant) reconstruction of the boundary values along Γ_2 is ~ 0.2 and the resulting error in flux through Γ_1 is ~ 0.006.

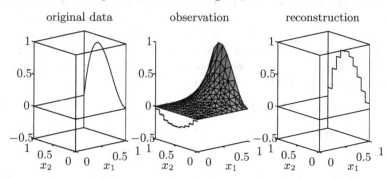

Fig. 78.4. Original Dirichlet boundary values (*left*), the corresponding state u with observed flow q along Γ_1 (*middle*), and the reconstruction of the boundary values along Γ_2 (to the *right*) using $\mu = 0.001$ and a few (5) conjugate gradient steps

78.6 The Backward Heat Equation

Another basic inverse problem is the *Backward heat equation*: Given the temperature at final time $t = T$, find the temperature at initial time $t = 0$.

We consider this problem in the following setting: let $f(x)$ be an initial temperature and let $u(x,t)$ be the corresponding solution of the heat equation:

$$\begin{cases} \dot{u} - \Delta u = 0 & \text{in } \Omega \times (0,T], \\ \partial_n u + \kappa u = 0 & \text{on } \Gamma \times (0,T], \\ u(x,0) = f(x) & \text{for } x \in \Omega, \end{cases}$$

where the domain $\Omega \in \mathbb{R}^d$ and the coefficient $\kappa \geq 0$ are given. We consider the following inverse problem: Given the final temperature $u(x,T)$, find the initial temperature $u(x,0) = f(x)$. This corresponds to solving the heat equation "backwards".

We consider the following discrete analog of (78.6) with discretization in space: Let V_h be the usual space of continuous piecewise linear functions on a triangulation of Ω with mesh size $h(x)$, and let $F \in V_h$ and let $U(t) \in V_h$ be the solution of the discrete heat equation

$$\begin{cases} (\dot{U},v)_\Omega + (\nabla U(t), \nabla v)_\Omega = 0 & \text{for } t \in (0,T], \ v \in V_h, \\ (U(0),v)_\Omega = (F,v)_\Omega & \text{for } v \in V_h. \end{cases} \tag{78.22}$$

The discrete inverse problem, corresponding to a discrete "backwards" heat equation, reads: Given the final temperature $U(T) = \hat{U} \in V_h$, find the initial temperature $U(0) = F \in V_h$.

To formulate this problem as a regularized least squares problem, we introduce the solution operator $B_h : V_h \rightarrow V_h$ defined as follows: $B_h F = U(T) \in V_h$, where U solves with $U(0) = F \in V_h$. The operator B_h thus takes an initial temperature F to a corresponding final temperature $B_h F$. The regularized least squares problem is now the same as that above, that is, we seek $F \in V_h$ which minimizes

$$J(F) = \|B_h F - \hat{U}\|_\Omega^2 + \mu \|F\|_\Omega^2, \tag{78.23}$$

over V_h. The unique solution $F \in V_h$ to this quadratic minimization problem is characterized by a least squares equation of the form

$$(B_h F, B_h G)_\Omega + (\mu F, G)_\Omega = (\hat{U}, B_h G)_\Omega \quad \text{for all } G \in V_h, \tag{78.24}$$

which we can express as

$$(B_h^\top B_h F, G)_\Omega + (\mu F, G)_\Omega = (B_h^\top \hat{U}, G)_\Omega \quad \text{for all } G \in V_h,$$

that is

$$(B_h^\top B_h + \mu I)F = B_h^\top \hat{U}, \tag{78.25}$$

where $B_h^\top : V_h \rightarrow V_h$ is defined as follows: $B_h^\top G = Z(0) \in V_h$, where $Z(t) \in V_h$ solves the discrete heat equation

$$\begin{cases} -(v,\dot{Z})_\Omega + (\nabla v, \nabla Z)_\Omega = 0 & \text{for } t \in (0,T], \ v \in V_h, \\ (v, Z(T))_\Omega = (G,v)_\Omega & \text{for } v \in V_h. \end{cases} \tag{78.26}$$

Note computing $B_h^\top$ corresponds to solving a heat equation: note the minus sign in the term $-(v, \dot{Z})$ and that we solve starting with $t = T$ and ending with $t = 0$. Changing variables introducing a new time variable $s = T - t$ brings this problem into the form of the usual heat equation. Note that (78.26) is a discrete analog of the problem:

$$\begin{cases} -\dot{z} - \Delta z = 0 & \text{in } \Omega \times (0, T], \\ \partial_n z + \kappa z = 0 & \text{on } \Gamma \times [0, T), \\ z(x, T) = g(x) & \text{for } x \in \Omega, \end{cases}$$

with $G \in V_h$ an approximation of $g(x)$ and $Z(t) \in V_h$ an approximation of $z(\cdot, t)$ for $t \in [0, T]$.

To solve the least squares equation by the gradient or conjugate gradient method, we have to compute $B_h F^n$ and $B_h^\top G$ for given vectors F^n and G in V_h, by using some time stepping method such as the dG(0) or cG(1) method.

Example 78.6. We consider the given problem with domain Ω as in the previous examples, $\kappa = 1000$ corresponding to boundary conditions $u \approx 0$, final time $T = 0.1$ and observed state at time T corresponding to initial values $u_0 = 16 \, x_1 \, (1 - x_1) \, x_2 \, (1 - x_2)$ and $u_0 = 2 \min(x_1, 1 - x_1, x_2, 1 - x_2)$, respectively. We then seek to reconstruct these initial data from the observations of the resulting solutions at time $T = 0.1$ with $\mu = 0.001$ and a few conjugate gradient iteration using the cG(1) (initiated by two dG(0) steps to filter out high frequency noise) with timesteps $k = 0.0025$. The results are shown in Fig. 78.5 and 78.6, respectively. The reconstruction error in the first case is ~ 0.057 and in the second case ~ 0.19. One would think that by decreasing μ it would be possible to better reconstruct the crisp details in the initial data u_0 in the second example. However, the observation we use here is a computed one modelling the fact that observations are imperfect or not considered in full detail in most cases, so that in this case the reconstruction does not get much better by decreasing μ. However, if we decrease T to say 0.02 we can reconstruct also the more detailed structure of u_0 also in the second example with $\mu = 0.00001$ and a resulting reconstruction error ~ 0.068:

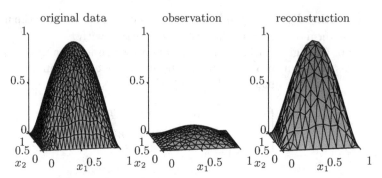

Fig. 78.5. Original initial data $u_0 = 16\,x_1\,(1-x_1)\,x_2\,(1-x_2)$ (*left*), corresponding observed state/solution at time $T = 0.1$ (*middle*), and reconstruction of u_0 (*right*) obtained with $\mu = 0.001$

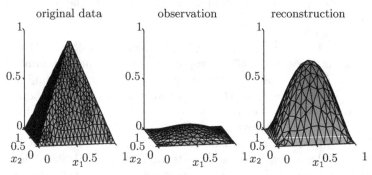

Fig. 78.6. Original initial data $u_0 = 2\max(\,x_1,\, 1 - x_1,\, x_2,\, 1 - x_2)$ (*left*), corresponding observed state/solution at time $T = 0.1$ (*middle*), and reconstruction of u_0 (*right*) obtained with $\mu = 0.001$

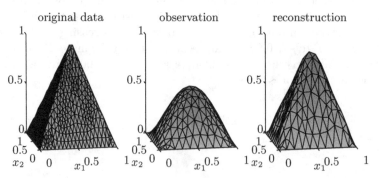

Fig. 78.7. Original initial data $u_0 = 2\max(\,x_1,\, 1 - x_1,\, x_2,\, 1 - x_2)$ (*left*), corresponding observed state/solution at time $T = 0.02$ (*middle*), and reconstruction of u_0 (*right*) obtained with $\mu = 0.00001$

79
Optimal Control

We're making the right decisions to bring the solution to an end.
(George W. Bush)

79.1 Introduction

In this chapter we continue with aspects of optimization connected to *optimal control* in the following setting: Consider an IVP of the form: Find the *state* $v : [0, T] \to \mathbb{R}^n$ satisfying the *state equation*

$$\dot{v}(t) + f(v(t), q(t)) = 0 \quad 0 < t \leq T, \quad v(0) = u^0, \tag{79.1}$$

where $f : \mathbb{R}^n \times \mathbb{R}^m \to \mathbb{R}^n$ is a given function, u^0 a given initial value, and $q : [0, T] \to \mathbb{R}^m$ is a *control*. We seek to determine an *optimal control* $p : [0, T] \to \mathbb{R}^m$ such that $J(p) \leq J(q)$ for all $q : [0, T] \to \mathbb{R}^m$, where

$$J(q) \equiv \frac{1}{2} \|v - \hat{u}\|^2 + \frac{\alpha}{2} \|q\|^2, \tag{79.2}$$

where v solves (79.1), $\hat{u} : [0, T] \to \mathbb{R}^n$ is a given function, and

$$\|w\|^2 = \int_0^T |w(t)|^2 \, dt$$

with $|\cdot|$ denoting the Euclidean norm, and α is a positive constant. We thus seek to choose the control q so that the corresponding state v is as close as possible to a given state $\hat{u}$ in the $\|\cdot\|$-norm and we also add a cost of the control measured by the factor $\alpha > 0$.

We reformulate this problem as the following *saddle point problem*:

$$\min_{v,q} \max_{\mu} L(v, q, \mu) \tag{79.3}$$

with the *Lagrangian* L defined by

$$L(v, q, \mu) = \frac{1}{2}\|v - \hat{u}\|^2 + \frac{\alpha}{2}\|q\|^2 + (\dot{v} + f(v, q), \mu) \tag{79.4}$$

with $(\cdot, \cdot)$ the scalar product corresponding to the norm $\|\cdot\|$, and (v, q, μ) varying freely (with $v(0) = u^0$ and $\mu(T) = 0$).

The condition for stationarity of $L(v, q, \mu)$ at (u, p, λ) is $L'(u, p, \lambda) = 0$, where L' is the Jacobian of $L : \mathbb{R}^n \times \mathbb{R}^m \times \mathbb{R}^n \to \mathbb{R}$, or in component form:

$$(\dot{u} + f(u, p), \mu) = 0 \quad \forall \mu, \tag{79.5}$$

$$(u - \hat{u}, v) + (\dot{v} + f'_v(u, p)v, \lambda) = 0 \quad \forall v, \tag{79.6}$$

$$(f'_q(u, p)q, \lambda) + \alpha(p, q) = 0 \quad \forall q, \tag{79.7}$$

where $f'_v(v, q)$ and $f'_q(v, q)$ denote the Jacobians of $f(v, q)$ with respect to v and q at (v, q), respectively, and we assume that $v(0) = 0$ and $\mu(T) = 0$. We can restate these equations in (u, p, λ) pointwise in time as follows:

$$\dot{u} + f(u, p) = 0 \quad \text{on } [0, T], \quad u(0) = u^0, \tag{79.8}$$

$$-\dot{\lambda} + f'_v(u, p)^\top \lambda = \hat{u} - u \quad \text{on } [0, T], \quad \lambda(T) = 0. \tag{79.9}$$

$$f'_q(u, p)^\top \lambda + \alpha p = 0 \quad \text{on } [0, T], \tag{79.10}$$

where $\top$ denotes transpose. Here (79.8) is the state equation, (79.9) is the *costate equation*, and (79.10) is the *feed back control* coupling the *optimal control* p to the *costate* λ.

To solve the stationarity equations we may consider the following *gradient method* in the control p:

$$p^{n+1} = p^n - \kappa(\alpha p^n + f'_q(u^n, p^n)^\top \lambda^n) \tag{79.11}$$

where u^n and λ^n solve the state and costate equations $\dot{u}^n + f(u^n, p^n) = 0$ and $-\dot{\lambda}^n + f'_v(u^n, p^n)^\top \lambda^n = \hat{u} - u^n$, respectively, and $\kappa > 0$ is a *step length*.

Example 79.1. If $f(v, p) = Av - Bq$ with A a $n \times n$ and B a $n \times m$ matrix, then the stationarity equations take the form:

$$\dot{u} + Au = Bp \quad \text{on } [0, T], \quad u(0) = u^0, \tag{79.12}$$

$$-\dot{\lambda} + A^\top \lambda = \hat{u} - u \quad \text{on } [0, T], \quad \lambda(T) = 0. \tag{79.13}$$

$$\alpha p = B^\top \lambda \quad \text{on } [0, T]. \tag{79.14}$$

79.2   The Connection Between $\frac{dJ}{dp}$ and $\frac{\partial L}{\partial p}$

We shall now prove that

$$J'(p) = \frac{dJ}{dp}(p) = \frac{\partial L}{\partial p}(u, p, \lambda), \qquad (79.15)$$

where the state $u = u(p)$ satisfies the state equation (79.8) with control p, and the costate λ satisfies the costate equation (79.9). We can thus express the gradient $J'(p) = \frac{dJ}{dp}(p)$ of the cost function $J(p)$ in terms of the corresponding state $u = u(p)$ and costate λ, while direct computation of $J'(p)$ requires computation of the derivative $u'(p)$ of the state $u(p)$ with respect to the control p: By the Chain rule we have

$$J'(p)q = \frac{\partial}{\partial \epsilon} J(p + \epsilon q)|_{\epsilon=0} = (u(p) - \hat{u}, u'(p)q) + \alpha(p, q),$$

where we thus want to eliminate $u'(p)$. To do so we differentiate the state equation in the form (assuming for simplicity that $u^0 = 0$),

$$0 = (u, -\dot{\mu}) + (f(u, p), \mu) \quad \forall \mu \text{ with } \mu(T) = 0,$$

with respect to p, to get $\forall \mu$ with $\mu(T) = 0$,

$$0 = \frac{d}{d\epsilon} \left((u(p + \epsilon q), -\dot{\mu}) + (f(u(p + \epsilon q), p + \epsilon q, \mu)) \right)|_{\epsilon=0},$$

that is,

$$0 = (u'(p)q, -\dot{\mu}) + (f'_u(u(p), p)u'(p)q + f'_p(u(p), p)q, \mu)$$

or

$$(u'(p)q, -\dot{\mu}) + (u'(p)q, f'_u(u(p), p)^\top \mu) = -(q, f'_p(u(p), p)^\top \mu).$$

Choosing now $\mu = \lambda$ and using that by the costate equation,

$$(u'(p)q, -\dot{\lambda}) + (u'(p)q, f'_u(u(p), p)^\top \lambda) = -(u(p) - \hat{u}, u'(p)q),$$

we can now express $J'(p)$ in the form

$$J'(p)q = (q, f'_p(u(p), p)^\top \lambda) + \alpha(p, q),$$

or

$$J'(p) = f'_p(u(p), p)^\top \lambda + \alpha p = \frac{\partial L}{\partial p}(u, p, \lambda)$$

as we set out to demonstrate.

Through the introduction of the costate λ we are thus able to express the gradient of the cost $J(p)$ with respect to the control p, and we may then apply a gradient method to search for the minimum of $J(p)$.

Example 79.2. We consider the problem of balancing an inverted pendulum on a fingertip, when the mass is subject to perturbations of horizontal force and initial condition. Assuming small displacements around the vertical position, the state equation takes the form $\dot{u}_2(t) - u_1(t) = f(t)$ and $\dot{u}_1(t) - u_2(t) = p(t)$ for $0 < t \leq T$, $u_1(0) = u_1^0$, $u_2(0) = u_2^0$, where $f(t)$ is the perturbation and $p(t)$ the control. The optimal control problem of keeping the pendulum in upright position with u_1 and u_2 close to zero, takes the form: Find $p : [0, T] \to \mathbb{R}$ which minimizes the cost

$$J(p) = \frac{1}{2} \int_0^T (a_1 u_1^2(t) + a_2 u_2^2(t))\, dt + \frac{\alpha}{2} \int_0^T p^2(t)\, dt,$$

where (u_1, u_2) solves the state equation with control p, and a_1, a_2 and α are positive constants. In Fig. 79.1 we show the result of applying the gradient method (79.11) for this problem with $f(t) = \sin(2t) + \sin(10t)$, $u_1^0 = 0.3$, $u_2^0 = 0$, $T = 2$, $a_1 = 100$, $a_2 = 1$, $\alpha = 0.0001$ and $\kappa = 0.005$. We note that the weighting with $a_1 \gg a_2$ gears the control towards keeping the position $u_1(t)$ close to zero, rather than the velocity $u_2(t)$.

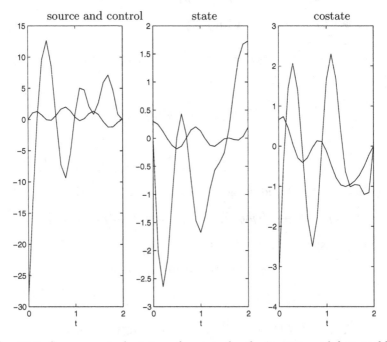

Fig. 79.1. Source, control, state and costate for the inverse pendulum problem

80

Differential Equations Tool Bag

It seems to me that there are at least four different viewpoints – or extremes of viewpoint – that one may reasonably hold:

1. All thinking is computation; in particular, feelings of conscious awareness are evoked merely by the carrying out of appropriate computations. (Hard AI)

2. Awareness is a feature of the brain's physiological action; and whereas any physical action can be simulated computationally, computational simulation cannot by itself evoke awareness. (Soft AI)

3. Appropriate physical action of the brain evokes awareness, but this physical action cannot even be properly simulated computationally. (Penrose's view)

4. Awareness cannot be explained by physical, computational, or any other scientific terms.

(R. Penrose in *Shadows of the Mind*)

80.1 Introduction

We here collect basic facts about solving differential equations analytically and numerically.

80.2 The Equation $u'(x) = \lambda(x)u(x)$

The solution to the scalar initial value problem

$$u'(x) = \lambda(x)u(x) \quad \text{for } x > a, \quad u(a) = u_a,$$

where $\lambda(x)$ is a given function of x, and u_a a given initial value, is

$$u(x) = \exp(\Lambda(x))u_a = e^{\Lambda(x)}u_a,$$

where $\Lambda(x)$ is a primitive function of $\lambda(x)$ such that $\Lambda(a) = 0$. In particular, if λ is a constant, then $u(x) = \exp(\lambda x)u_a$.

80.3 The Equation $u'(x) = \lambda(x)u(x) + f(x)$

The solution the scalar initial value problem

$$u'(x) = \lambda(x)u(x) + f(x) \quad \text{for } x > a, \quad u(a) = u_a,$$

where $\lambda(x)$ and $f(x)$ are given functions of x, and u_a a given initial value, can be expressed using Duhamel's principle in the form

$$u(x) = e^{\Lambda(x)}u_a + e^{\Lambda(x)} \int_a^x e^{-\Lambda(y)} f(y) \, dy.$$

where $\Lambda(x)$ is a primitive function of $\lambda(x)$ such that $\Lambda(a) = 0$.

80.4 The Differential Equation $\sum_{k=0}^n a_k D^k u(x) = 0$

A solution to the constant coefficient differential equation

$$p(D)u(x) = \sum_{k=0}^n a_k D^k u(x) = 0, \quad \text{for } x \in I,$$

where I is an interval of real numbers, has the form

$$u(x) = \alpha_1 \exp(\lambda_1) + \ldots + \alpha_n \exp(\lambda_n),$$

where the α_i are arbitrary constants and the λ_i are the roots of the polynomial equation $p(\lambda) = 0$ with $p(\lambda) = \sum_{k=0}^n a_k \lambda^k$, assuming there are n distinct roots. If $p(\lambda) = 0$ has a multiple root λ_i of multiplicity r, then the solution is the sum of terms of the form form $q(x) \exp(\lambda_i x)$, where $q(x)$ is a polynomial of degree at most $r - 1$. For example, if $p(D) = (D - 1)^2$, then a solution of $p(D)u = 0$ has the form $u(x) = (a_0 + a_1 x) \exp(x)$.

80.5 The Damped Linear Oscillator

A solution $u(t)$ to

$$\ddot{u} + \mu\dot{u} + ku = 0, \text{for } t > 0,$$

where μ and k are constants, has the form

$$u(t) = ae^{-\frac{1}{2}(\mu+\sqrt{\mu^2-4k})t} + be^{-\frac{1}{2}(\mu-\sqrt{\mu^2-4k})t},$$

if $\mu^2 - 4k > 0$, and

$$u(t) = ae^{-\frac{1}{2}\mu t} \cos\left(\frac{t}{2}\sqrt{4k-\mu^2}\right) + be^{-\frac{1}{2}\mu t} \sin\left(\frac{t}{2}\sqrt{4k-\mu^2}\right),$$

if $\mu^2 - 4k < 0$, and

$$u(t) = (a + bt)e^{-\frac{1}{2}\mu t},$$

if $\mu^2 - 4k = 0$, where a and b are arbitrary constants.

80.6 The Matrix Exponential

The solution to the initial value problem linear system

$$u'(x) = Au(x) \quad \text{for } 0 < x \leq T, \, u(0) = u_0,$$

where A is a *constant* $d \times d$ matrix, $u_0 \in \mathbb{R}^d$, $T > 0$, is given by

$$u(x) = \exp(xA)u_0 = e^{xA}u_0.$$

If A is diagonalizable so that $A = SDS^{-1}$, where S is nonsingular and D is diagonal with diagonal elements d_i (the eigenvalues of A), then

$$\exp(xA) = S\exp(xD)S^{-1}.$$

where $\exp(xD)$ be the diagonal matrix with diagonal elements equal to $\exp(xd_i)$.

The solution to the initial value problem

$$u'(x) = Au(x) + f(x) \quad \text{for } 0 < x \leq 1, \, u(0) = u_0,$$

where $f(x)$ is a given function, is given by Duhamel's principle:

$$u(x) = \exp(xA)u_0 + \int_0^x \exp((x-y)A)f(y)\, dy.$$

80.7   Fundamental Solutions of the Laplacian

The function $\Phi(x) = \frac{1}{4\pi} \frac{1}{\|x\|}$ for $x \in \mathbb{R}^3$ satisfies the differential equation $-\Delta\Phi = \delta_0$ in $\mathbb{R}^3$, where δ_0 represents a point mass at the origin. The function $\Phi(x) = \frac{1}{2\pi} \log(\frac{1}{\|x\|})$ for $x \in \mathbb{R}^2$ satisfies the differential equation $-\Delta\Phi = \delta_0$ in $\mathbb{R}^2$, where δ_0 represents a point mass at the origin.

80.8   The Wave Equation in 1d

The general solution to the one-dimensional wave equation

$$\ddot{u} - u'' = 0 \quad \text{for } x, t \in \mathbb{R},$$

is given by $u(x,t) = v(x - t) + w(x + t)$ where $v, w : \mathbb{R} \to \mathbb{R}$ are arbitrary functions.

80.9   Numerical Methods for IVPs

The dG(O), the discontinuous Galerkin method with piecewise constants, for the initial value problem $\dot{u}(t) = f(u(t), t)$ for $t > 0$, $u(0) = u^0$, with $f : \mathbb{R}^{d+1} \to \mathbb{R}^d$, takes the form

$$U^n = U^{n-1} + \int_{t_{n-1}}^{t_n} f(U^n, t)\, dt, \quad n = 1, 2, \ldots,$$

where $U(t)$ is piecewise constant on a partition $0 = t_0 < t_1 < \cdots < t_n < t_{n+1} < \cdots$, with $U(t) = U^n$ for $t \in (t_{n-1}, t_n]$ and $U(0) = u^0$. With right-end point quadrature we obtain the implicit backward-Euler method:

$$U^n = U^{n-1} + k_n f(U^n, t_n)\, dt, \quad n = 1, 2, \ldots,$$

where $k_n = t_n - t_{n-1}$. The explicit forward Euler method reads:

$$U^n = U^{n-1} + k_n f(U^{n-1}, t_{n-1})\, dt, \quad n = 1, 2, \ldots,$$

The cG(1), the continuous Galerkin method with continuous piecewise linear functions, takes the form

$$U(t_n) = U(t_{n-1}) + \int_{t_{n-1}}^{t_n} f(U(t), t)\, dt, \quad n = 1, 2, \ldots,$$

where $U(t)$ is continuous piecewise linear with nodal values $U(t_n) \in \mathbb{R}^d$ and $U(0) = u^0$.

80.10   cg(1) for Convection-Diffusion-Reaction

The cG(1) finite element method for the scalar convection-diffusion-reaction problem

$$-\nabla \cdot (a\nabla u) + \nabla \cdot (ub) + cu = f \quad \text{in } \Omega,$$

$$a\frac{\partial u}{\partial n} + \kappa u = g \quad \text{on } \Gamma,$$

with Robin boundary conditions, where f and g are given data, and $a > 0$, b, c and $\kappa \geq 0$ are given coefficients, and Ω is a given domain in $\mathbb{R}^2$ with boundary Γ, takes the form: Find $U \in V_h$ such that

$$\int_\Omega a\nabla U \cdot \nabla v \, dx + \int_\Omega \nabla \cdot (ub)v \, dx + \int_\Omega cuv \, dx$$

$$+ \int_\Gamma \kappa uv \, ds = \int_\Omega fv \, dx + \int_\Gamma gv \, ds,$$

where V_h is a space of continuous piecewise linear functions on a triangulation of Ω with no restriction on the nodal values on the boundary.

80.11   Svensson's Formula for Laplace's Equation

$$U_{i,j} = \frac{1}{4}(U_{i-1,j} + U_{i+1,j} + U_{i,j-1} + U_{i,j+1}), \quad \text{for } i, j \in \mathbb{Z},$$

where $U_{i,j}$ approximates $u(ih, jh)$ with $h > 0$ and $u : \mathbb{R}^2 \to \mathbb{R}$ solves $\Delta u = 0$.

80.12   Optimal Control

The stationary equations for the saddle point problem $\min_{v,q} \max_\mu L(v, q, \mu)$, with

$$L(v, q, \mu) = \frac{1}{2}\|v - \hat{u}\|^2 + \frac{\alpha}{2}\|q\|^2 + (\dot{v} + f(v, q), \mu)$$

with $(v, w) = \int_0^T v \cdot w \, dt$ and (v, q, μ) varying freely (with $v(0) = u^0$ and $\mu(T) = 0$), take the form:

$$\dot{u} + f(u, p) = 0 \quad \text{on } [0, T], \quad u(0) = u^0, \tag{80.1}$$

$$-\dot{\lambda} + f'_v(u, p)^\top\lambda = \hat{u} - u \quad \text{on } [0, T], \quad \lambda(T) = 0. \tag{80.2}$$

$$f'_q(u, p)^\top\lambda + \alpha p = 0 \quad \text{on } [0, T], \tag{80.3}$$

where $\top$ denotes transpose. Here (80.1) is the state equation, (80.2) is the *costate equation*, and (80.3) is the *feed back control* coupling the *optimal control* p to the *costate* λ.

81
Applications Tool Bag

81.1 Introduction

In this section we collect the basic models of engineering and science expressed as differential equations. For specification of boundary and initial values we refer to the text.

81.2 Malthus' Population Model

$$\dot{u} = \lambda u - \mu u,$$

where $u(t)$ is the population at time t, $\lambda \geq 0$ the birth rate and $\mu \geq 0$ the death rate.

81.3 The Logistics Equation

$$\dot{u} = u(1 - u)$$

81.4 Mass-Spring-Dashpot System

$$m\ddot{u} + \mu\dot{u} + ku = f, \quad \text{(force balance)},$$

where $u(t)$ is the displacement, m is the mass, μ the viscosity, and k the spring constant.

81.5   LCR-Circuit

$$L\ddot{u} + R\dot{u} + \frac{u}{C} = f, \quad \text{(balance of potentials)},$$

where $u(t)$ is a primitive function of the current, L is the inductance, R the resistance, C the capacitance, and f a potential.

81.6   Laplace's Equation for Gravitation

$$-\Delta u = \rho,$$

where $u : \mathbb{R}^3 \to \mathbb{R}$ is the gravitational potential and $\rho(x)$ the mass density.

81.7   The Heat Equation

$$\dot{u} - \nabla \cdot q = f, \quad q = k\nabla u \quad \text{(heat balance and Fourier's law)}$$

where $u(x,t)$ is a temperature, $q(x,t)$ a heat flux, $k(x,t) > 0$ a conduction coefficient and $f(x,t)$ a heat source. If $k = 1$, then we get the heat equation: $\dot{u} - \Delta u = f$.

81.8   The Wave Equation

$$\ddot{u} - \Delta u = f.$$

81.9   Convection-Diffusion-Reaction

$$\dot{u} + \nabla \cdot (\beta u) + \alpha u - \nabla \cdot (\epsilon \nabla u) = f.$$

where $u(x,t)$ a concentration, $\beta(x,t)$ is a convection velocity, $\alpha(x,t)$ a reaction coefficient, $\epsilon(x,t)$ a diffusion coefficient, and $f(x,t)$ a production rate.

81.10   Maxwell's Equations

$$
\begin{cases}
\dfrac{\partial B}{\partial t} + \nabla \times E = 0, & \text{(Faraday's law)} \\[2mm]
-\dfrac{\partial D}{\partial t} + \nabla \times H = J, & \text{(Ampère's law)} \\[2mm]
\nabla \cdot B = 0, \quad \nabla \cdot D = \rho, & \text{(Gauss' and Coulomb's laws)} \\[2mm]
B = \mu H, \quad D = \epsilon E, \quad J = \sigma E, & \text{(constitutive laws and Ohm's law)}
\end{cases}
$$

where E is the electric field, H is the magnetic field , D is the electric displacement, B is the em magnetic flux, J is the electric current, ρ is the charge, μ is the magnetic permeability , ϵ is the dielectric constant, and σ is the electric conductivity.

81.11   The Incompressible Navier-Stokes Equations

$$
\frac{\partial u}{\partial t} + (u \cdot \nabla)u + \nabla p - \nu \Delta u = f, \quad \nabla \cdot u = 0,
$$

where $u(x,t)$ is the fluid velocity, $p(x,t)$ the pressure, $f(x,t)$ a given force and $\nu > 0$ a constant viscosity.

81.12   Schrödinger's Equation

$$
i\frac{\partial \varphi}{\partial t} = \left(-\frac{1}{2}\sum_j \Delta_j + V(r_1, \ldots, r_N) \right) \varphi(r_1, \ldots, r_N), \quad r_j \in \mathbb{R}^3.
$$

$$
i\frac{\partial \varphi}{\partial t} = \left(-\frac{1}{2}\Delta + \frac{1}{|x|} \right) \varphi(x), \quad x \in \mathbb{R}^3, \quad \text{(Hydrogen atom)}.
$$

82
Analytic Functions

A mathematician of the first rank, Laplace quickly revealed himself as only a mediocre administrator, from his first work we saw we had been deceived. Laplace saw no question from its true point of view, he sought subtleties everywhere, had only doubtful ideas, and finally carried the spirit of the infinitely small into administration. (Napoleon)

We arrive at truth, not by reason only, but also by the heart. (Pascal)

In this chapter we give a short account of *analytic functions*, that is, differentiable functions $f : \mathbb{C} \to \mathbb{C}$, taking complex arguments and having complex values. We use heavily the material developed above on Calculus in $\mathbb{R}^d$, $d = 1, 2$, including, the definition of derivative of a function $f : \mathbb{R}^d \to \mathbb{R}^d$, and Green's formulas in $\mathbb{R}^2$.

82.1 The Definition of an Analytic Function

We recall that we can write each complex number $z \in \mathbb{C}$ in the form $z = x + iy$, with $x, y \in \mathbb{R}$ and i the imaginary unit, and we can identify $\mathbb{C}$ with $\mathbb{R}^2$ by identifying $x + iy \in \mathbb{C}$ with $(x, y) \in \mathbb{R}^2$. In particular, $i = (0, 1)$, and $|z| = (x^2 + y^2)^{1/2}$.

Let $f : \Omega \to \mathbb{C}$ be a complex-valued function of a complex variable $z = x + iy \in \Omega$, where $x, y \in \mathbb{R}$ and Ω is an open domain of the complex plane. Decomposing into real and imaginary parts, we can write

$$f(z) = f(x + iy) = u(x, y) + iv(x, y),$$

where $u : \mathbb{R}^2 \to \mathbb{R}$ and $v : \mathbb{R}^2 \to \mathbb{R}$ are the real and imaginary parts of $f(z)$, that is, $u(x, y) = \operatorname{Re} f(z)$ and $v(x, y) = \operatorname{Im} f(z)$, where we thus view $u(x, y)$ and $v(x, y)$ as functions of $(x, y) \in \mathbb{R}^2$ with values in $\mathbb{R}$.

We say that $f : \Omega \to \mathbb{C}$ is *differentiable* at $z_0 \in \Omega$ with derivative $f'(z_0) \in \mathbb{C}$, if for z close to z_0, we have

$$|f(z) - f(z_0) - f'(z_0)(z - z_0)| \le K_f(z_0)|z - z_0|^2, \qquad (82.1)$$

where $K_f(z_0)$ is a non-negative real constant depending on f and z_0. This is a direct extension of the corresponding definition of the derivative of a function $f : \mathbb{R} \to \mathbb{R}$ to a function $f : \mathbb{C} \to \mathbb{C}$, and the usual rules for differentiation of sums, products and quotients directly extend.

We recall that differentiability of a function $f : \mathbb{R} \to \mathbb{R}$ at a point x_0 means that $f(x)$ is well approximated (up to a quadratic term) by the linear function $c_0 + c_1(x - x_0) = f(x_0) + f'(x_0)(x - x_0)$ for x close to x_0, where $c_0 = f(x_0)$ and $c_1 = f'(x_0)$ are real constants. Similarly, differentiability of a function $f : \mathbb{C} \to \mathbb{C}$ at a point z_0 means that $f(z)$ for z close to z_0 is well approximated by the linear function $c_0 + c_1(z - z_0) = f(z_0) + f'(z_0)(z - z_0)$, involving a translation and multiplication by a complex constant. We conclude that differentiability of a function $f : \mathbb{C} \to \mathbb{C}$ at a point z_0 means that $f(z)$ in a neighborhood of z_0 acts like a combination of a translation, rotation and change of modulus, see Fig. 82.1.

We say that $f : \Omega \to \mathbb{C}$ is *analytic* in the open domain Ω of the complex plane if $f(z)$ is differentiable at all $z_0 \in \Omega$ with derivative $f'(z_0)$. The derivative f' of an analytic function $f : \Omega \to \mathbb{C}$ is again a function $f' : \Omega \to \mathbb{C}$. We shall shortly prove the surprising fact that if $f : \Omega \to \mathbb{C}$ is analytic, then also $f' : \Omega \to \mathbb{C}$ is analytic with derivative $f'' : \Omega \to \mathbb{C}$, which is also analytic, and so on. An analytic function $f : \Omega \to \mathbb{C}$ thus has derivatives of all orders $f^{(n)} : \Omega \to \mathbb{C}$, $n = 1, 2, \ldots$, which are all analytic. We recall that a differentiable function $f : \mathbb{R} \to \mathbb{R}$ need not have a differentiable derivative, and therefore does not have this very special property in general.

We can view a function $f : \mathbb{C} \to \mathbb{C}$ alternatively as a function $f : \mathbb{R}^2 \to \mathbb{R}^2$ if we identify $\mathbb{C}$ and $\mathbb{R}^2$ as indicated. The Jacobian $f'(x, y)$ of a function $f : \mathbb{R}^2 \to \mathbb{R}^2$ is a 2×2-matrix consisting of 4 real numbers, while the deriva-

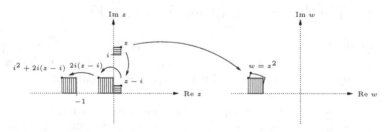

Fig. 82.1. Linear approximation of a function $f(z)$ near $z = z_0$, with $z_0 = i$ and $w = f(z) = z^2$ approximated by $f(z_0) + f'(z_0)(z - z_0) = i^2 + 2i(z - i)$

tive $f'(z) \in \mathbb{C}$ of a function $f : \mathbb{C} \to \mathbb{C}$ is supposed to be a complex number being represented by 2 real numbers. We conclude that differentiability of a complex-valued function $f : \mathbb{C} \to \mathbb{C}$, is a more stringent requirement than differentiability of the corresponding function $f : \mathbb{R}^2 \to \mathbb{R}^2$, which only requires the partial derivatives of the real and imaginary parts $u(x, y)$ and $v(x, y)$ of $f(z)$ to exist. In fact, we shall see that the partial derivatives of the real and imaginary parts of an analytic function must be coupled in a specific way, which is expressed through the *Cauchy-Riemann equations* stated below.

82.2 The Derivative as a Limit of Difference Quotients

Note that (82.1) implies that if $z \neq z_0$, then

$$|\frac{f(z) - f(z_0)}{z - z_0} - f'(z_0)| \leq K_f(z_0)|z - z_0|,$$

which we can write in the form

$$\lim_{z \to z_0} \frac{f(z) - f(z_0)}{z - z_0} = f'(z_0), \qquad (82.2)$$

by which we mean, of course, that

$$|\frac{f(z) - f(z_0)}{z - z_0} - f'(z_0)|$$

is as small as we please if we only choose $|z - z_0|$ small enough (respecting that $z \neq z_0$). In view of (82.2) we write as usual $\frac{df}{dz} = f'$.

82.3 Linear Functions Are Analytic

We consider the function $f : \mathbb{C} \to \mathbb{C}$ given by $f(z) = az + b$, where a and b are given complex numbers. We have for all z and $z_0 \in \mathbb{C}$ that

$$f(z) - f(z_0) - a(z - z_0) = 0,$$

and thus $f(z)$ is analytic in $\mathbb{C}$ with derivative $f'(z) = a$.

82.4 The Function $f(z) = z^2$ Is Analytic

If $f(z) = z^2$, then

$$f(z) - f(z_0) - 2z_0(z - z_0) = z^2 - z_0^2 - 2z_0 z + 2z_0^2 = (z - z_0)^2,$$

and thus $f'(z_0) = 2z_0$ for $z_0 \in \mathbb{C}$.

82.5   The Function $f(z) = z^n$ Is Analytic for $n = 1, 2, \ldots$

Consider the function $f : \mathbb{C} \to \mathbb{C}$, where $f(z) = z^n$ and $n = 1, 2, \ldots$, is a natural number, which may be viewed as an extension of the function $f : \mathbb{R} \to \mathbb{R}$ with $f(x) = x^n$. By a direct extension of the proof in the case $f(x) = x^n$, we find that

$$f'(z) = nz^{n-1}. \tag{82.3}$$

We conclude that z^n is differentiable in the whole of $\mathbb{C}$, with derivative nz^{n-1}. We just gave the proof in the case $n = 1, 2$.

82.6   Rules of Differentiation

As we said, the usual rules for differentiation of sums, products and quotients valid for functions $f : \mathbb{R} \to \mathbb{R}$ extend to functions $f : \mathbb{C} \to \mathbb{C}$. In particular we have if $f(z_0) \neq 0$ and $g(z) = \frac{1}{f(z)}$, that

$$g'(z_0) = -\frac{f'(z_0)}{f^2(z_0)}. \tag{82.4}$$

Further, the composition of two analytic functions is also analytic and the Chain rule for differentiation holds: if $g(z)$ is differentiable at z_0 and $f(z)$ is differentiable at $g(z_0)$, then the composite function $h(z) = f(g(z))$ is differentiable at z_0 with derivative $h'(z_0) = f'(g(z_0))g'(z_0)$. The proof is a direct extension of the corresponding proof for real-valued functions of a real variable.

82.7   The Function $f(z) = z^{-n}$

Applying the rule (82.4), we see that $f(z) = z^{-n}$ with $n = 1, 2, \ldots$, is differentiable for $z \neq 0$ and

$$f'(z) = -nz^{-n-1}, \quad \text{for } z \neq 0. \tag{82.5}$$

We can summarize by stating that if $f(z) = z^n$ with $n = \pm 1, \pm 2, \ldots$, then $f'(z) = nz^{n-1}$, where we assume that $z \neq 0$ if $n < 0$.

82.8   The Cauchy-Riemann Equations

We shall now derive the so-called *Cauchy-Riemann equations* connecting the partial derivatives of the real part $u(x, y)$ and the imaginary part $v(x, y)$

of a complex-valued function $f(z) = u(x,y) + iv(x,y)$ of a complex variable $z = x + iy$ at a point $z_0 = x_0 + iy_0$ with $x_0, y_0 \in \mathbb{R}$, such that (82.1) is satisfied. Writing $f'(z_0) = a + ib$, with $a, b \in \mathbb{R}$, we can express (82.1) as

$$|u(x,y) + iv(x,y) - u(x_0, y_0) - iv(x_0, y_0) - (a+ib)(x - x_0 + i(y - y_0))|$$
$$\leq K_f(z_0)|z - z_0|^2.$$

Separating into real and imaginary parts, we conclude (recalling (22.7)) that

$$|u(x,y) - u(x_0, y_0) - a(x - x_0) + b(y - y_0)| \leq K_f(z_0)|z - z_0|^2,$$
$$|v(x,y) - v(x_0, y_0) - a(y - y_0) - b(x - x_0)| \leq K_f(z_0)|z - z_0|^2. \tag{82.6}$$

Recalling the definition of the partial derivatives of $u(x,y)$ and $v(x,y)$ at (x_0, y_0) from Chapter *Vector-valued functions of several variables*, we conclude that

$$a = \frac{\partial u}{\partial x}(x_0, y_0), \quad b = -\frac{\partial u}{\partial y}(x_0, y_0),$$
$$a = \frac{\partial v}{\partial y}(x_0, y_0), \quad b = \frac{\partial v}{\partial x}(x_0, y_0),$$

and we thus find that

$$\frac{\partial u}{\partial x}(x_0, y_0) = \frac{\partial v}{\partial y}(x_0, y_0), \quad \frac{\partial u}{\partial y}(x_0, y_0) = -\frac{\partial v}{\partial x}(x_0, y_0). \tag{82.7}$$

These are the *Cauchy-Riemann equations* for $u(x,y)$ and $v(x,y)$ at the point (x_0, y_0).

We conclude that if $f(z) = u(x,y) + iv(x,y)$ is analytic in the open domain Ω of the complex plane, then the real and imaginary parts $u(x,y)$ and $v(x,y)$ satisfy the Cauchy-Riemann equations in Ω, that is,

$$\frac{\partial u}{\partial x} = \frac{\partial v}{\partial y}, \quad \frac{\partial u}{\partial y} = -\frac{\partial v}{\partial x}, \quad \text{in } \Omega. \tag{82.8}$$

Note that we can write the Cauchy-Riemann equations in the form $\nabla v = -\nabla \times u$, recalling that $\nabla \times u = (\frac{\partial u}{\partial y}, -\frac{\partial u}{\partial x})$.

Example 82.1. The analytic function $f(z) = z^2 = (x+iy)^2 = x^2 - y^2 + 2ixy$ with $u(x,y) = x^2 - y^2$ and $v(x,y) = 2xy$ satisfies $\frac{\partial u}{\partial x} = 2x = \frac{\partial v}{\partial y}$ and $\frac{\partial u}{\partial y} = -2y = -\frac{\partial v}{\partial x}$.

We have seen that the Cauchy-Riemann equations (82.8) follow from the analyticity of the complex valued function $f = u + iv$. In other words, the Cauchy-Riemann equations represents a *necessary* condition for the analyticity of $f = u + iv$. The Cauchy-Riemann equations also represent a *sufficient condition*: given a pair of functions $u(x, y)$ and $v(x, y)$ satisfying the Cauchy-Riemann equations, the function $f = u + iv$ is analytic. To see this, we note that if $u(x, y)$ and $v(x, y)$ are differentiable functions satisfying (82.8), then (82.6), and thus also (82.1) holds. That is, $f = u + iv$ is analytic.

Example 82.2. We consider the functions $u(x, y) = x + 2xy$ and $v(x, y) = y - x^2 + y^2$ and find that $\frac{\partial u}{\partial x} = 1 + 2y = \frac{\partial v}{\partial y}$ and $\frac{\partial u}{\partial y} = 2x = -\frac{\partial v}{\partial x}$, that is, u and v satisfy the Cauchy-Riemann equations and we thus conclude that the function $f(z) = u(x, y) + iv(x, y)$ must be analytic. In fact, $f(z) = u + iv = x + 2xy + i(y - x^2 + y^2) = x + iy - i(x + iy)^2 = z - iz^2$, and the analyticity is obvious.

We may summarize as follows:

Theorem 82.1 *The function $f(z) = u(x, y) + iv(x, y)$ is analytic if and only if the Cauchy-Riemann equations (82.8) are satisfied.*

82.9 The Cauchy-Riemann Equations and the Derivative

Using the limit definition (82.2) of the derivative $f'(z_0)$, we can write, varying first only x:

$$f'(z_0) = \lim_{x \to x_0} \frac{f(z) - f(z_0)}{x - x_0} = \frac{\partial u}{\partial x}(x_0, y_0) + i\frac{\partial v}{\partial x}(x_0, y_0), \qquad (82.9)$$

where $z = x + iy_0$, and then only y:

$$f'(z_0) = \lim_{y \to y_0} \frac{f(z) - f(z_0)}{i(y - y_0)} = \frac{1}{i}\frac{\partial u}{\partial y}(x_0, y_0) + \frac{\partial v}{\partial y}(x_0, y_0),$$

where $z = x_0 + iy$, from which the Cauchy-Riemann equations follow by equating the real and imaginary parts of $f'(z_0)$ using that $\frac{1}{i} = -i$.

Example 82.3. In the last example we found that $u(x, y) = x + 2xy$ and $v(x, y) = y - x^2 + y^2$ satisfy the Cauchy-Riemann equation. According to (82.9), the derivative of the corresponding analytic function $f = u + iv$ is given by $f' = \frac{\partial u}{\partial x} + i\frac{\partial v}{\partial x} = 1 + 2y + i(-2x)$ which agrees with our observation that $f(z) = z - iz^2$ with $f'(z) = 1 - 2iz = 1 - 2i(x + iy) = 1 + 2y - 2ix$.

Example 82.4. By direct verification using the Cauchy-Riemann equations one finds that $f(z) = e^z = e^x(\cos(y) + i\sin(y))$ is analytic in $\mathbb{C}$, and $\frac{d}{dz}e^z = e^z$. It follows that also $\sin(z) = \frac{1}{2i}(e^{iz} - e^{-iz})$ and $\cos(z) = \frac{1}{2}(e^{iz} + e^{-iz})$ are analytic in $\mathbb{C}$, and $\frac{d}{dz}\sin(z) = \cos(z)$ and $\frac{d}{dz}\cos(z) = -\sin(z)$. See Problem 82.1.

82.10   The Cauchy-Riemann Equations in Polar Coordinates

The Cauchy-Riemann equations take the following form in polar coordinates $z = re^{i\theta}$:

$$\frac{\partial u}{\partial r} = \frac{1}{r}\frac{\partial v}{\partial \theta}, \quad \frac{\partial v}{\partial r} = -\frac{1}{r}\frac{\partial u}{\partial \theta}. \tag{82.10}$$

Example 82.5. The function $\mathrm{Log}(z) = \log(|z|) + i\,\mathrm{Arg}\,z$ is analytic in $\{z \in \mathbb{C} : z \neq 0,\, 0 \leq \arg z < 2\pi\}$. This follows from the Cauchy-Riemann equations in polar coordinates. We recall that $\log(z) = \log(|z|) + i\arg z$ is multi-valued since $\arg z$ is multivalued. The function $\log(z)$ with $\arg z$ restricted to $0 \leq \arg z < 2\pi$, however, is single-valued analytic.

82.11   The Real and Imaginary Parts of an Analytic Function

We shall now prove that the Cauchy-Riemann equations (82.8) imply that both $u(x,y)$ and $v(x,y)$ are harmonic in Ω, that is,

$$\Delta u = 0 \quad \text{and} \quad \Delta v = 0 \quad \text{in } \Omega.$$

In fact, this follows directly by differentiating (82.8) with respect to x and y, if we assume that $u(x,y)$ and $v(x,y)$ are twice differentiable, since

$$\frac{\partial^2 u}{\partial x^2} = \frac{\partial^2 v}{\partial x \partial y} = \frac{\partial^2 v}{\partial y \partial x} = -\frac{\partial^2 u}{\partial y^2} \quad \text{in } \Omega \tag{82.11}$$

and thus $\Delta u = 0$ in Ω, and similarly $\Delta v = 0$ in Ω.

Now, one can show that solutions of the Cauchy-Riemann equations indeed must be twice differentiable, and thus the real and imaginary parts of an analytic function are harmonic. We sum up in the following theorem:

Theorem 82.2 *If $f : \Omega \to \mathbb{C}$ is analytic, where Ω is an open domain of the complex plane $\mathbb{C}$, then the real part $u(x,y) = \mathrm{Re}\,f(z)$ and the imaginary part $v(x,y) = \mathrm{Im}\,f(z)$ are harmonic in Ω.*

82.12   Conjugate Harmonic Functions

Suppose $u(x,y)$ is harmonic in a simply connected domain Ω in $\mathbb{R}^2$. We shall now prove that there exists a harmonic function $v(x,y)$, uniquely determined up to a constant, such that $f(z) = u(x,y) + iv(x,y)$ is analytic in Ω. We say that the function $v(x,y)$ is *conjugate* to $u(x,y)$. To prove this, we simply solve the Cauchy-Riemann equations $\nabla v = -\nabla \times u$ with u given using the basic result of the Chapter Potential fields, noting that $\nabla \times (-\nabla \times u) = \Delta u = 0$, that is, $-\nabla \times u$ is irrotational, and thus is the gradient of some function v. See also Problem 82.10.

Example 82.6. For the harmonic function $u(x_1, x_2) = x_1 x_2$, the conjugate $v(x_1, x_2)$ satisfies $\frac{\partial v}{\partial x_2} = \frac{\partial u}{\partial x_1} = x_2$, that is, $v = \frac{1}{2}x_2^2 + C(x_1)$ for some function C, and from $\frac{\partial v}{\partial x_1} = -\frac{\partial u}{\partial x_2} = -x_1$, that is $C'(x_1) = -x_1$, we conclude that $C(x_1) = -\frac{1}{2}x_1^2 + D$, for some arbitrary constant D. We note that also v is harmonic, and conclude that u and its conjugate v are the real and imaginary parts of the analytic function $f(z) = x_1 x_2 + i(\frac{1}{2}x_1^2 - \frac{1}{2}x_2^2) = -\frac{1}{2}z^2$.

82.13   The Derivative of an Analytic Function Is Analytic

Assume that $f(z)$ is *analytic* in the open domain Ω of the complex plane. This means that the derivative $f'(z)$ exists as a complex-valued function for $z \in \Omega$, and one may ask if $f'(z)$ itself has a derivative in Ω, that is, if $f'(z)$ is analytic in Ω. The plain answer is YES, which we prove below. Thus, if $f(z)$ is analytic in Ω, then also $f'(z)$ is analytic in Ω, and thus also the derivative of $f'(z)$, that is the second derivative $f''(z)$ is analytic, and so on. We conclude that an analytic function has derivatives of all orders. This is a remarkable property of an analytic function.

To answer the question posed, it is sufficient to notice that if $u(x, y)$ and $v(x, y)$ satisfy the Cauchy-Riemann equations, then so do all derivatives of $u(x, y)$ and $v(x, y)$, in particular $\frac{\partial u}{\partial x}$ and $\frac{\partial v}{\partial x}$, and thus $f' = \frac{\partial u}{\partial x} + i\frac{\partial v}{\partial x}$ is analytic in Ω. We state this important result as a theorem:

Theorem 82.3 *If $f : \Omega \to \mathbb{C}$ is analytic, where Ω is an open domain of the complex plane $\mathbb{C}$, then all the derivatives $f^{(n)}(z)$, $n = 1, 2, \ldots$, of $f(z)$ are analytic in Ω.*

We recall that if $f : \mathbb{R} \to \mathbb{R}$ is a real-valued function of a real variable, then the analogous statement may be wrong: even if $f'(x)$ exists, it is not clear that $f''(x)$ exists.

82.14   Curves in the Complex Plane

Let Ω be an open domain in the complex plane $\mathbb{C}$, and let $\gamma : I \to \Omega$, where $I = [a, b]$ is an interval of $\mathbb{R}$, be a Lipschitz continuous function. We say that $\Gamma = $ Range of $\gamma = \{\gamma(t) : t \in I\}$, is a *curve* in $\mathbb{C}$ parameterized by $\gamma(t)$. For example $\gamma(t) = \exp(it)$ where $0 \le t < 2\pi$ is a parametrization of the unit circle.

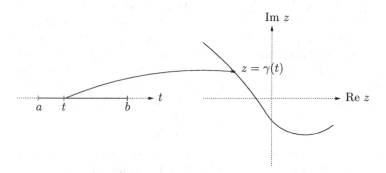

Fig. 82.2. A curve $z = \gamma(t)$

We say that Γ is a *differentiable curve* if the corresponding parametrization $\gamma : I \to \mathbb{C}$ is differentiable on I in the sense that the related function $\gamma : I \to \mathbb{R}^2$ is differentiable. In other words, decomposing $\gamma(t) = x(t) + iy(t)$ into real and imaginary parts $x : I \to \mathbb{R}$ and $y : I \to \mathbb{R}$, we have that $\gamma(t)$ is differentiable on I if $x(t)$ and $y(t)$ are differentiable on I. We also say that $\gamma : I \to \mathbb{C}$ is Lipschitz continuous on I if the corresponding function $\gamma : I \to \mathbb{R}^2$ is Lipschitz continuous. There are thus no surprises in this context.

A curve Γ with parametrization $\gamma : [a, b] \to \mathbb{C}$ is said to be *closed and simple* if $\gamma(s) \ne \gamma(t)$ for $s < t$, unless $s = a$ and $t = b$.

We say that a domain Ω in $\mathbb{C}$ which is bounded by a simple closed curve, is *simply connected*. A simply connected domain does not have any "holes".

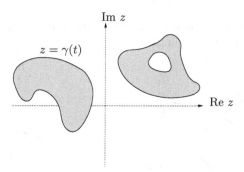

Fig. 82.3. A simply connected domain with boundary curve $z = \gamma(t)$, to the *left*, and a multiply connected domain with *one* hole to the *right* with the boundary consisting of *two* simple closed curves

82.15   Conformal Mappings

Let $f : \Omega \to \mathbb{C}$ be analytic where Ω is an open domain in $\mathbb{C}$. We shall now prove that the mapping $z \to w = f(z)$ is *conformal* in Ω in the sense that angles are preserved under the mapping $w = f(z)$. This is a direct consequence of the Chain rule and the analyticity of $f(z)$, as we now show. Let then $\gamma : I \to \mathbb{C}$, where $I = [-\delta, \delta]$ with $\delta > 0$, be a curve through $z_0 \in \Omega$ with $\gamma(0) = z_0$. Consider the curve $\kappa(t) = f(\gamma(t))$ which is the image of $\gamma(t)$ under the transformation $w = f(z)$. By the Chain rule we have

$$\frac{d\kappa}{dt} = \frac{df}{dz}\frac{d\gamma}{dt},$$

and we thus see, recalling that the argument of the product of two complex numbers is the sum of the arguments of the numbers, that

$$\arg \frac{d\kappa}{dt}(0) = \mathrm{Arg}\, f'(z_0) + \mathrm{Arg}\, \frac{d\gamma}{dt}(0),$$

where we assume that $f'(z_0) \neq 0$. Since $f(z)$ is analytic at z_0, we have that $f'(z_0)$ is independent of the curve γ, and thus the tangent direction $\frac{d\kappa}{dt}(0)$ differs from that of $\frac{d\gamma}{dt}(0)$ by the constant value $\mathrm{Arg}\, f'(z_0)$, independent of γ. We conclude that the angle between two curves passing through z_0 is the same as the angle between the corresponding transformed curves passing through $f(z_0)$. This means that the mapping $w = f(z)$ is *conformal at z_0*: angles are preserved locally.

Note that since

$$\lim_{z \to z_0} \left| \frac{f(z) - f(z_0)}{z - z_0} \right| = |f'(z_0)|,$$

the mapping $w = f(z)$ changes the length scale locally by the factor $|f'(z_0)| \neq 0$. Thus although the mapping $w = f(z)$ is locally conformal, the image of a large figure may be considerably distorted because of the change of scale.

We now present some basic analytic functions $w = f(z)$ and the corresponding conformal mappings of $f : \mathbb{C} \to \mathbb{C}$.

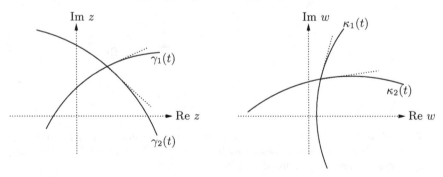

Fig. 82.4. An analytic mapping conforms angles

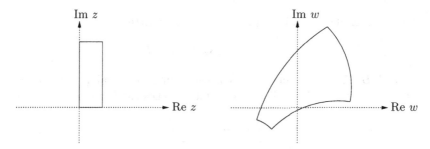

Fig. 82.5. A conformal mapping with large deformations

82.16   Translation-rotation-expansion/contraction

The linear transformation:

$$w = f(z) = az + b$$

where $a, b \in \mathbb{C}$, corresponds to a rotation with Arg a and an expansion/contraction with $|a|$, and a translation with b, see Fig. 82.6

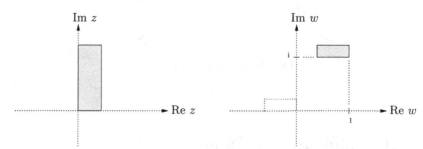

Fig. 82.6. The mapping $w = az + b$ with $a = \frac{1}{2}i$ and $b = 1 + i$

82.17   Inversion

The mapping

$$w = f(z) = \frac{1}{z},$$

is referred to as *inversion*. We now prove that an inversion maps every straight line or circle in the complex plane into a circle or straight line.

Indeed, a circle or straight line in $\mathbb{R}^2$ can be written

$$A(x^2 + y^2) + Bx + Cy + D = 0$$

with A, B, C, D real, and $A = 0$ corresponding to a straight line. In terms of $z = x + iy$ and $\bar{z} = x - iy$, the equation takes the form

$$Az\bar{z} + B\frac{z + \bar{z}}{2} + C\frac{z - \bar{z}}{2i} + D = 0,$$

and substitution of $z = \frac{1}{w}$ gives (after multiplication with $w\bar{w}$)

$$A + B\frac{\bar{w} + w}{2} + C\frac{\bar{w} - \bar{w}}{2i} + Dw\bar{w} = 0,$$

which represents a circle or straight line.

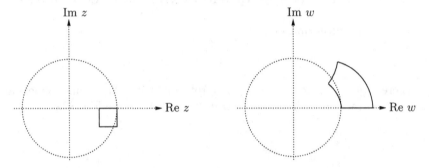

Fig. 82.7. The mapping $w = 1/z$

82.18 Möbius Transformations

A mapping of the form

$$w = f(z) = \frac{az + b}{cz + d},$$

where $a, b, c, d \in \mathbb{C}$, is said to be a *Möbius transformation*. We have

$$f'(z) = \frac{ad - bc}{(cz + d)^2},$$

and we are thus led to assume that $ad - bc \neq 0$ to guarantee conformity. Evidently, the inversion $w = \frac{1}{z}$ is a special case of a Möbius transformation. One can prove that a Möbius transformation maps every straight line or circle in the complex plane into a circle or straight line, see Problem 82.6.

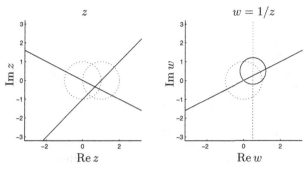

Fig. 82.8. Further illustration of the map $f(z) = 1/z$. Note that the unit circle is mapped onto itself, while a circle through the origin is mapped onto a straight line. Note also that the straight line $y = ax$ with $a \neq 0$ passing through the origin is mapped onto its conjugate line $y = -ax$, while other lines are mapped onto circles

Example 82.7. (**Disc onto disc**) The function

$$w = f(z) = e^{i\alpha} \frac{z - z_0}{1 - \bar{z}_0 z}$$

where $\alpha \in \mathbb{R}$ and $z_0 \in \mathbb{C}$ with $|z_0| < 1$, maps the closed unit disc $\{|z| \leq 1\}$ onto the closed unit disc $\{|w| \leq 1\}$ in a one-to-one fashion with $f(z_0) = 0$. For the verification it suffices to verify that the circle $\{|z| = 1\}$ is mapped onto the circle $\{|w| = 1\}$.

Example 82.8. (**Half-plane onto unit disc**) The function

$$w = f(z) = e^{i\alpha} \frac{z - z_0}{z - \bar{z}_0}$$

where $\alpha \in \mathbb{R}$ and $\operatorname{Im} z_0 > 0$, maps the upper half-plane $\{\operatorname{Im} z > 0\}$ onto the open unit disc $\{|w| < 1\}$ with $f(z_0) = 0$.

82.19 $w = z^{1/2}$, $w = e^z$, $w = \log(z)$ and $w = \sin(z)$

We describe in a couple of examples basic aspects of the mapping properties of some elementary functions.

Example 82.9. The function

$$w = f(z) = z^{1/2} = \sqrt{|z|}\, e^{\frac{i}{2}\operatorname{Arg} z},$$

maps the wedge $\{0 \leq \arg z < \theta\}$ where $0 \leq \theta < 2\pi$ onto the wedge $\{0 \leq \arg w < \frac{\theta}{2}\}$.

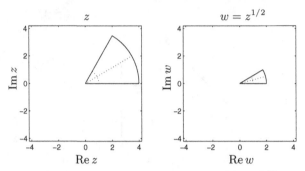

Fig. 82.9. Illustration of the map $f(z) = z^{1/2}$

Example 82.10. The function $w = e^z$ maps the strip $\{z = x + iy : x \in \mathbb{R}, 0 \le y < 2\pi\}$ onto the complex plane C minus the origin. The line $\{x + iy : x \in \mathbb{R}\}$ with y fixed is mapped onto the halfline $\{(r, \theta) : r > 0\}$ with $\theta = y$ using polar coordinates.

Example 82.11. The function $w = \text{Log}(z)$ maps $\mathbb{C}$ minus the origin onto the strip $\{w \in \mathbb{C} : 0 \le \text{Im}(w) < 2\pi\}$.

Example 82.12. The function

$$w = f(z) = \sin(z) = \frac{1}{2i}(e^{i(x+iy)} - e^{-i(x+iy)})$$
$$= \sin(x)\cosh(y) + i\,\cos(x)\sinh(y) = u(x, y) + iv(x, y),$$

maps the strip $\{z = x + iy : \frac{-\pi}{2} < x < \frac{\pi}{2}, y \in \mathbb{R}\}$ onto $\{w = u + iv : v \ne 0 \text{ if } |u| > 1\}$, which is the whole plane minus the two half-lines $\{u + iv : |u| > 1, v = 0\}$. The level curves of u and v are hyperbolas and ellipses, respectively. See Fig. 82.10.

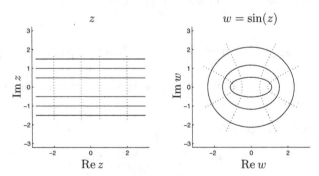

Fig. 82.10. Illustration of the map $f(z) = \sin(z)$

82.20   Complex Integrals: First Shot

We make a direct extension of the integral of a differentiable function $F :$ $\mathbb{R} \to \mathbb{R}$ to the integral of an analytic function $F : \mathbb{C} \to \mathbb{C}$, paralleling closely the presentation in Chapter *The Integral.*

Let $F(z)$ be analytic in the domain Ω of the complex plane, with Lipschitz continuous derivative $f(z) = F'(z)$. Let Γ be a differentiable curve in Ω parameterized by $\gamma : [a, b] \to \mathbb{C}$, connecting the point $z_a = \gamma(a)$ with the point $z_b = \gamma(b)$, and let $z_a = z_0, z_1, \ldots, z_n = z_b$ be a sequence of points on Γ connecting z_a and z_b, see Fig. 82.11. We assume that $z_k \neq z_{k-1}$ for $k = 1, \ldots, n$.

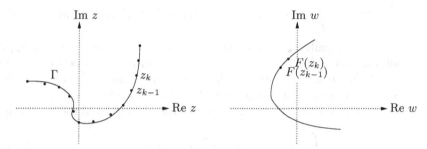

Fig. 82.11. A curve Γ with sample points z_k and corresponding function values $F(z_k)$

We can write

$$F(z_b) - F(z_a) = \sum_{k=1}^{n}(F(z_k) - F(z_{k-1})) = \sum_{k=1}^{n} \frac{F(z_k) - F(z_{k-1})}{z_k - z_{k-1}}(z_k - z_{k-1}).$$

(82.12)

Letting $\max_{k=1,\ldots,n} |z_k - z_{k-1}|$ tend to zero, we are led to write

$$F(z_b) - F(z_a) = \int_{\Gamma} f(z)\, dz,$$

(82.13)

where we replace $\frac{F(z_k)-F(z_{k-1})}{z_k-z_{k-1}}$ by the derivative $F'(z_{k-1}) = f(z_{k-1})$ and $z_k - z_{k-1}$ by dz.

We note that the integral $\int_{\Gamma} f(z)\, dz$, being equal to $F(z_b) - F(z_a)$, is thus independent of the choice of the curve Γ connecting z_a and z_b. As a special case we note that if Γ is closed, corresponding to choosing $z_b = z_a$, then

$$\int_{\Gamma} f(z)\, dz = 0.$$

(82.14)

Recalling that $f(z)$ is analytic if $F(z)$ is analytic, we have found a reason to believe in *Cauchy's theorem* stating that the integral of an analytic function

$f : \Omega \to \mathbb{C}$ around a simple closed curve in Ω enclosing a region contained in Ω, is zero. This is a corner-stone of the theory of analytic functions. Below we give a proof of Cauchy's theorem using a Green's formula.

82.21 Complex Integrals: General Case

Let Ω be an open domain in the complex plane and let Γ be a differentiable curve in $\mathbb{C}$ parameterized by $\gamma = (x, y) : [a, b] \to \mathbb{C}$. Let $f = u + iv : \Gamma \to \mathbb{C}$ be Lipschitz continuous and define

$$\int_\Gamma f(z)\, dz = \int_a^b \big(u(x(t), y(t)) + iv(x(t), y(t)) \big) (\dot{x}(t) + i\dot{y}(t))\, dt, \quad (82.15)$$

where thus formally $dz = dx + i\, dy = \dot{x}\, dt + i\dot{y}\, dt = (\dot{x} + i\dot{y})\, dt$. The integral is defined if $u(x, y)$ and $v(x, y)$ are Lipschitz continuous in (x, y) and $\dot{x}(t)$ and $\dot{y}(t)$ are Lipschitz continuous in t. As in the Chapter Curve Integrals, we see that the integral is independent of the parametrization.

We can express the integral as a limit of Riemann sums in the usual way:

$$\int_\Gamma f(z)\, dz = \lim_{n \to \infty} \sum_{k=1}^n f(z_{k-1})(z_k - z_{k-1}), \quad (82.16)$$

where $z_a = z_0, z_1, \ldots, z_n = z_b$ is a sequence of points along Γ with $\max_{k=1,\ldots,n} |z_k - z_{k-1}|$ tending to zero as n tends to infinity.

Below we use also ζ as a complex variable and thus write in particular

$$\int_\Gamma f(z)\, dz = \int_\Gamma f(\zeta)\, d\zeta.$$

Example 82.13. Let Γ be a circle around the origin of radius one oriented counter-clockwise and parameterized by $\gamma(t) = e^{it} = \cos(t) + i\sin(t) = (\cos(t), \sin(t))$ with $0 \le t < 2\pi$. Let $f(z) = z^n$ with n an integer. We have for $n \ne -1$, since $dz = (-\sin(t) + i\cos(t))\, dt = ie^{it}\, dt$,

$$\int_\Gamma f(z)\, dz = \int_\Gamma z^n\, dz = \int_0^{2\pi} e^{int}(-\sin(t) + i\cos(t))\, dt$$

$$= i \int_0^{2\pi} e^{int} e^{it}\, dt = i \int_0^{2\pi} e^{i(n+1)t}\, dt$$

$$= \frac{i}{n+1} [\sin((n+1)t) - i\cos((n+1)t)]_0^{2\pi} = 0.$$

This conforms with Cauchy's theorem for $n = 0, 1, 2, \ldots$, since then $f(z)$ is analytic in $\mathbb{C}$. For $n = -1$ with $f(z) = \frac{1}{z}$, we get

$$\int_\Gamma \frac{1}{z}\, dz = \int_\Gamma \frac{dz}{z} = \int_0^{2\pi} \frac{ie^{it}}{e^{it}}\, dt = 2\pi i. \quad (82.17)$$

Note the counter-clockwise orientation of Γ. The function $f(z) = \frac{1}{z}$ is not analytic in the domain enclosed by Γ, since $\frac{1}{z}$ is not differentiable for $z = 0$, and thus the integral $\int_\Gamma \frac{dz}{z}$ may be non-zero. We shall see below that the derivative of $\log(z)$ is equal to $\frac{1}{z}$, but $\log(z)$ is not uniquely defined for $z \neq 0$, and thus $\int_\Gamma \frac{dz}{z}$ may be non-zero.

The functions $f(z) = z^n$ with $n = -2, -3, \ldots$ are all derivatives of analytic functions and thus $\int_\Gamma f(z) \, dz = 0$ if Γ is a closed curve which does not pass through 0.

82.22 Basic Properties of the Complex Integral

The complex integral has properties analogous to those of the usual real integral such as linearity, additivity over subintervals and integration by parts. For example, we have if $|f(z)| \leq M$ for $z \in \Gamma$:

$$\left| \int_\Gamma f(z) \, dz \right| \leq M \int_\Gamma ds = ML(\Gamma), \qquad (82.18)$$

where $L(\Gamma)$ is the length of Γ:

$$L(\Gamma) = \int_a^b (\dot{x}^2(t) + \dot{y}^2(t))^{1/2} \, dt.$$

This follows by taking absolute values in (82.16) and then passing to the limit:

$$\left| \int_\Gamma f(z) \, dz \right| \leq \int_\Gamma |f(z)| \, |dz| = \int_\Gamma |f(z)| \, ds \leq ML(\Gamma),$$

where formally $|dz| = ds$, and thus the estimate may be viewed as a generalized triangle inequality.

82.23 Taylor's Formula: First Shot

If $f : \Omega \to \mathbb{C}$ is analytic and Γ is a straight line in Ω connecting z_0 and z, then we can write

$$f(z) = f(z_0) + \int_\Gamma f'(\zeta) \, d\zeta = f(z_0) + \int_\Gamma f'(\zeta) \frac{d}{d\zeta}(\zeta - z_0) \, d\zeta,$$

and thus by partial integration (the usual rules hold)

$$f(z) = f(z_0) + f'(z_0)(z - z_0) - \int_\Gamma f''(\zeta)(\zeta - z_0) \, d\zeta.$$

Continuing, writing $(\zeta - z_0) = \frac{1}{2}\frac{d}{d\zeta}(\zeta - z_0)^2$, we get

$$f(z) = f(z_0) + f'(z_0)(z - z_0) + \frac{f''(z_0)}{2}(z - z_0)^2 + \int_\Gamma f^{(3)}(\zeta)\frac{(\zeta - z_0)^2}{2}\,d\zeta.$$

We conclude that for z in a neighborhood of z_0, we have

$$f(z) = f(z_0) + f'(z_0)(z - z_0) + \frac{f''(z_0)}{2}(z - z_0)^2 + E_f(z, z_0), \qquad (82.19)$$

where

$$|E_f(z, z_0)| \le K \int_\Gamma \frac{|\zeta - z_0|^2}{2}\,|d\zeta| = \frac{K}{6}|z - z_0|^3,$$

and we assume that $|f^{(3)}(\zeta)| \le K$ for $\zeta \in \Gamma$. More generally, we have the following Taylor's formula:

Theorem 82.4 *If $f : \Omega \to \mathbb{C}$ is analytic in Ω with $|f^{(n+1)}(z)| \le K$ for $z \in \Omega$, then we have for $z, z_0 \in \Omega$ (with the straight line connecting z and z_0 contained in Ω):*

$$f(z) = f(z_0) + f'(z_0)(z - z_0) + \ldots + \frac{f^{(n)}(z_0)}{n!}(z - z_0)^n + R_n(z, z_0), \quad (82.20)$$

where $|R_n(z, z_0)| \le \frac{K}{(n+1)!}|z - z_0|^{n+1}$.

82.24   Cauchy's Theorem

We shall now prove that if $f(z)$ is analytic in Ω and Γ is a simple closed curve in Ω enclosing a domain Ω_Γ contained in Ω, then

$$\int_\Gamma f(z)\,dz = 0.$$

To see this we write

$$\int_\Gamma f(z)\,dz = \int_a^b \left(u\left(x(t), v(t)\right) + iv\left(x(t), y(t)\right)\right)\left(\dot{x}(t) + i\dot{y}(t)\right)\,dt,$$

where $\gamma(t) = (x(t), y(t))$ with $a \le t \le b$ a parametrization of Γ. Taking the real part, we get

$$\mathrm{Re}\left(\int_C f(z)\,dz\right) = \int_a^b \left(u(x(t), y(t))\dot{x}(t) - v(x(t), y(t))\dot{y}(t)\right)dt$$

$$= \int_a^b \left(u(x, y), -v(x, y)\right) \cdot (\dot{x}, \dot{y}))\,dt.$$

By the Cauchy-Riemann equations, we have

$$\nabla \times (u, -v) = \frac{\partial u}{\partial y} + \frac{\partial v}{\partial x} = 0 \quad \text{in } \Omega_\Gamma,$$

which proves, recalling Stokes' theorem (57.13), that

$$\int_a^b (u(x,y), -v(x,y)) \cdot (\dot{x}, \dot{y}) \, dt = \int_\Gamma (u(x,y), -v(x,y)) \cdot ds$$

$$= \int_{\Omega_\Gamma} \nabla \times (u, -v) dx dy = 0.$$

We conclude that $\text{Re}(\int_\Gamma f(z) \, dz) = 0$, and similarly we see that $\text{Im}(\int_\Gamma f(z) \, dz) = 0$ and we have thus proved *Cauchy's theorem*:

Theorem 82.5 (Cauchy's theorem) *If $f(z)$ is analytic in Ω and Γ is a simple closed curve in Ω enclosing a domain contained in Ω, then*

$$\int_\Gamma f(z) \, dz = 0.$$

Note that Γ is not allowed to enclose "holes" of Ω where $f(z)$ is not analytic. For example, we saw above that $\int_\Gamma \frac{1}{z} \, dz = 2\pi i \neq 0$, where Γ is a circle around the origin. This is because Γ encloses the point $z = 0$ where $\frac{1}{z}$ is not analytic.

82.25 Cauchy's Representation Formula

We prove that if $f(z)$ is analytic in an open domain Ω, and Γ is a simple closed curve in Ω oriented counter-clockwise and bounding the open domain Ω_Γ contained in Ω, then for $z_0 \in \Omega_\Gamma$,

$$f(z_0) = \frac{1}{2\pi i} \int_\Gamma \frac{f(z)}{z - z_0} \, dz, \qquad (82.21)$$

which is *Cauchy's representation formula*. Note the counter-clockwise orientation. Further, note that z_0 is not allowed to lie on the curve Γ; we assume that z_0 lies *inside* Γ. Cauchy's formula (82.21) shows that the values of $f(z)$ on Γ alone, determine the values of $f(z)$ in all of Ω_Γ. This shows that an analytic function is not allowed to bring surprises: if we know $f(z)$ on Γ, then we know $f(z)$ in the whole domain Ω_Γ bounded by Γ. The proof follows from realizing that the function

$$g(z) = \frac{f(z) - f(z_0)}{z - z_0} \quad \text{for } z \neq z_0, \quad g(z_0) = f'(z_0),$$

is analytic in Ω, because $g(z)$ is clearly differentiable if $z \neq z_0$ and using a Taylor expansion of $f(z)$, it follows that $g(z)$ is differentiable also at $z = z_0$ with derivative $g'(z_0) = \frac{f''(z_0)}{2}$. Indeed, recalling (82.19) we have

$$g(z) - g(z_0) = \frac{f(z) - f(z_0) - f'(z_0)(z - z_0)}{(z - z_0)} = \frac{f''(z_0)}{2}(z - z_0) + E_f(z, z_0)$$

with $|E_f(z, z_0)| \leq \frac{K}{6}|z - z_0|^2$ and K a bound for $|f^{(3)}(z)|$, which proves the desired result. We conclude that

$$\int_\Gamma \frac{f(z) - f(z_0)}{z - z_0} dz = 0,$$

and using that

$$\int_\Gamma \frac{f(z_0)}{z - z_0} dz = f(z_0) \int_\Gamma \frac{1}{z - z_0} dz = 2\pi i \, f(z_0),$$

we obtain the desired result (82.21). We summarize:

Theorem 82.6 (Cauchy's representation formula) *If $f(z)$ is analytic in an open domain Ω, and Γ is a simple closed curve in Ω oriented counter-clockwise and enclosing the open domain Ω_Γ contained in Ω, then for $z_0 \in \Omega_\Gamma$,*

$$f(z_0) = \frac{1}{2\pi i} \int_\Gamma \frac{f(z)}{z - z_0} dz. \tag{82.22}$$

Differentiating with respect to z_0 we obtain the following generalized representation formula:

Theorem 82.7 (Cauchy's generalized representation formula) *If $f(z)$ is analytic in an open domain Ω, and Γ is a simple closed curve in Ω oriented counter-clockwise and enclosing an open domain Ω_Γ in Ω, then for $z_0 \in \Omega_\Gamma$ and $n = 0, 1, 2, \ldots$,*

$$f^{(n)}(z_0) = \frac{n!}{2\pi i} \int_\Gamma \frac{f(z)}{(z - z_0)^{n+1}} dz. \tag{82.23}$$

We note that if z_0 lies outside the region bounded by Γ, then

$$\frac{1}{2\pi i} \int_\Gamma \frac{f(z)}{z - z_0} dz = 0,$$

simply because $\int_\Gamma \frac{1}{z - z_0} dz = 0$ in this case as a consequence of the fact that $\frac{1}{z - z_0}$ is analytic in a domain containing Γ. Choosing $z_0 \in \Gamma$ leads to a divergent integral because of the singularity of the factor $\frac{1}{z - z_0}$, and to define a proper value of the integral in this case leads to the so called *Cauchy principal value*, which we discuss below.

82.26   Taylor's Formula: Second Shot

By using Cauchy's formula we now give another version of Taylor's formula for a function $f(z)$ which is analytic in a neighborhood Ω of a point $z_0 \in \mathbb{C}$. We start writing Cauchy's formula in the form

$$f(z) = \frac{1}{2\pi i} \int_\Gamma \frac{f(\zeta)}{\zeta - z} d\zeta, \tag{82.24}$$

where for definiteness we choose Γ to be a counter-clockwise oriented circle around z_0 of radius r contained in Ω. Using the identity

$$\frac{1}{1-q} = 1 + q + q^2 + \ldots + q^n + \frac{q^{n+1}}{1-q},$$

where $q \in \mathbb{C}$ satisfies $|q| < 1$, setting $q = \frac{z-z_0}{\zeta-z_0}$ with $z \in \Omega$ and $\zeta \in \Gamma$, we can write

$$\frac{1}{\zeta - z} = \frac{1}{\zeta - z_0} \left[1 + \frac{z-z_0}{\zeta-z_0} + \ldots + \left(\frac{z-z_0}{\zeta-z_0} \right)^n \right] + \frac{1}{\zeta-z} \left(\frac{z-z_0}{\zeta-z_0} \right)^{n+1},$$

where we used that

$$\frac{1}{\zeta - z} = \frac{1}{\zeta - z_0 - (z - z_0)} = \frac{1}{\zeta - z_0} \frac{1}{1-q}.$$

Insertion into (82.24) now gives

$$f(z) = \frac{1}{2\pi i} \int_\Gamma \frac{f(\zeta)}{\zeta - z_0} d\zeta + \frac{z - z_0}{2\pi i} \int_\Gamma \frac{f(\zeta)}{(\zeta - z_0)^2} d\zeta + \ldots$$

$$+ \frac{(z - z_0)^n}{2\pi i} \int_\Gamma \frac{f(\zeta)}{(\zeta - z_0)^{n+1}} d\zeta + R_n(z),$$

where

$$R_n(z) = \frac{(z - z_0)^{n+1}}{2\pi i} \int_\Gamma \frac{f(\zeta)}{(\zeta - z_0)^{n+1}(\zeta - z)} d\zeta. \tag{82.25}$$

Using Cauchy's representation formulas we thus obtain the following Taylor formula:

Theorem 82.8 *If $f(z)$ is analytic in a neigborhood Ω of a $z_0 \in \mathbb{C}$, then*

$$f(z) = f(z_0) + f'(z_0)(z - z_0) + \ldots + \frac{f^{(n)}(z_0)}{n!}(z - z_0)^n + R_n(z), \tag{82.26}$$

where the remainder $R_n(z)$ is given by (82.25) with Γ a circle around z_0.

If $\lim_{n\to\infty} R_n(z) = 0$ for z in a neighborhood Ω of z_0, then we obtain the following *power series representation* of $f(z)$ for $z \in \Omega$:

$$f(z) = \sum_{n=0}^{\infty} \frac{f^{(n)}(z_0)}{n!}(z - z_0)^n. \tag{82.27}$$

We conclude by proving that indeed $\lim_{n\to\infty} R_n(z) = 0$ for z in a neighborhood of z_0. We then assume that the disc $D_r(z_0) = \{z \in \mathbb{C} : |z - z_0| \le r\}$ is contained in the domain Ω of analyticity of $f(z)$ and we assume that $|f(z)| \le M$ for $z \in D_r(z_0)$. Assuming that $|z - z_0| < \frac{r}{2}$, we obtain by inserting absolute values in (82.25) using that $|\zeta - z| \ge \frac{r}{2}$ and $L(\Gamma) = 2\pi r$:

$$|R_n(z)| \le \left(\frac{|z - z_0|}{r}\right)^{n+1} 2M,$$

which proves that $\lim_{n\to\infty} R_n(z) = 0$ for $|z - z_0| \le \frac{r}{2}$. We can extend the argument to z satisfying $|z - z_0| < r$, and we summarize as follows:

Theorem 82.9 (Taylor's formula) *If $f : \Omega \to \mathbb{C}$ is analytic and $D_r(z_0) = \{z \in \mathbb{C} : |z - z_0| \le r\}$ is contained in Ω and f is bounded on $D_r(z_0)$, then $f(z)$ can be represented as the convergent power series (82.27) for $|z - z_0| < r$.*

Power series representations of analytic functions of the form (82.27) play an important role and we devote the next section to this topic starting with the case $z_0 = 0$.

82.27 Power Series Representation of Analytic Functions

Consider a series of the form

$$\sum_{m=0}^{\infty} a_m z^m \tag{82.28}$$

where the coefficients $a_m \in \mathbb{C}$ and we assume $z \in \mathbb{C}$. The concepts of convergence and absolute convergence for (82.28) are direct analogs of the corresponding concepts for series with a_m and z being real, see Chapter *Series*. In particular we say that $\sum_{m=0}^{\infty} a_m z^m$ is absolutely convergent if $\sum_{m=0}^{\infty} |a_m z^m|$ is convergent, and note that an absolutely convergent series is convergent.

Each term of the series (82.28) is analytic in $\mathbb{C}$ and each partial sum

$$\sum_{m=0}^{n} a_m z^m$$

is thus analytic in $\mathbb{C}$. Suppose now that the series (82.28) is convergent for a particular $z = \hat{z}$ with $|\hat{z}| = r$. Since the terms b_m of a convergent series $\sum_{m=0}^{\infty} b_m$ must tend to zero, there is a constant C such that

$$|a_n \hat{z}^n| = |a_n| r^n \le C \quad n = 0, 1, 2, \dots$$

Suppose now that $|z| < r$. We then have

$$\sum_{n=0}^{\infty} |a_n z^n| = \sum_n |a_n r^n| \left| \frac{z}{r} \right|^n \le C \sum_{n=0}^{\infty} \left| \frac{z}{r} \right|^n < \infty,$$

because $\left| \frac{z}{r} \right| < 1$. This proves that $\sum_{n=0}^{\infty} a_n z^n$ is absolutely convergent for $|z| < r$ and is thus convergent for $|z| < r$.

We say that the *radius of convergence* of $\sum_{n=0}^{\infty} a_n z^n$ is equal to r, if $\sum_{n=0}^{\infty} a_n z^n$ is convergent for $|z| < r$ but not convergent for some z with $|z| \ge r$.

One can (easily) show that inside its radius of convergence r a series $\sum_{n=0}^{\infty} a_n z^n$ is differentiable with

$$\left(\sum_{n=0}^{\infty} a_n z^n \right)' = \sum_{n=1}^{\infty} n a_n z^{n-1},$$

where the termwise differentiated series $\sum_{n=1}^{\infty} n a_n z^{n-1}$ is also convergent for $|z| < r$.

More generally, we consider power series of the form

$$\sum_{m=0}^{\infty} a_m (z - z_0)^m, \qquad (82.29)$$

where we made a shift of variable from z to $z - z_0$ with $z_0 \in \mathbb{C}$ given. The notion of convergence and radius of convergence extend in the obvious way. Of course, (82.29) connects to the Taylor series of $f(z)$ at z_0 with $a_m = \frac{f^{(m)}(z_0)}{m!}$.

Example 82.14. The series

$$\sum_{n=0}^{\infty} \frac{z^n}{n!}$$

is convergent for any fixed $z \in \mathbb{C}$, since $n! = 1 \cdot 2 \cdot 3 \cdots n$ grows much quicker than r^n for any $r > 0$. We can thus differentiate termwise and we get

$$\left(\sum_{n=0}^{\infty} \frac{z^n}{n!} \right)' = \sum_{n=1}^{\infty} \frac{z^{n-1}}{(n-1)!} = \sum_{n=0}^{\infty} \frac{z^n}{n!},$$

which shows that $\sum_{n=0}^{\infty} \frac{z^n}{n!}$ satisfies the differential equation $u'(z) = u(z)$ with the "initial" condition $u(0) = 1$. It follows that

$$\exp(z) = \sum_{n=0}^{\infty} \frac{z^n}{n!}. \qquad (82.30)$$

Alternatively, this follows by noting that this is the Taylor series representation of $f(z) = \exp(z)$ around $z_0 = 0$, noting that $f^{(n)}(z) = \exp(z)$ for $n = 1, 2, \ldots,$.

Using that $\cos(z) = \frac{1}{2}(\exp(iz) + \exp(-iz))$ and $\sin(z) = \frac{1}{2i}(\exp(iz) - \exp(-iz))$, we obtain the following Taylor series representations valid for $z \in \mathbb{C}$:

$$\cos(z) = \sum_{n=0}^{\infty}(-1)^n \frac{z^{2n}}{(2n)!}, \qquad \sin(z) = \sum_{n=0}^{\infty}(-1)^n \frac{z^{2n+1}}{(2n+1)!}.$$

Example 82.15. Another basic example is given by

$$\log(1+z) = \sum_{n=1}^{\infty} \frac{(-1)^{n-1}}{n} z^n \quad \text{for } |z| < 1,$$

which is readily obtained differentiating $\log(1+z)$.

82.28   Laurent Series

Consider a series of the form

$$\sum_{m=1}^{\infty} b_m z^{-m}, \qquad (82.31)$$

obtained by replacing z by $\frac{1}{z}$ in a power series $\sum_{m=1}^{\infty} b_m z^m$ with radius of convergence r. The series (82.31) will thus converge for $|z| > r$. More generally we may consider a *Laurent series* of the form

$$f(z) = \sum_{m=0}^{\infty} a_m z^m + \sum_{m=1}^{\infty} b_m z^{-m}, \qquad (82.32)$$

which we assume to be convergent in an annulus $\{r_1 < |z| < r_2\}$. The function $f(z)$ defined by (82.32) is analytic in the annulus $\{r_1 < |z| < r_2\}$. Conversely, if $f(z)$ is analytic in the annulus $\{r_1 < |z| < r_2\}$, then $f(z)$ admits the Laurent series expansion (82.32) with the coefficients a_m and b_m being given by

$$a_m = \frac{1}{2\pi i} \int_\Gamma \frac{f(\zeta)}{\zeta^{m+1}} d\zeta, \quad b_m = \frac{1}{2\pi i} \int_\Gamma f(\zeta)\zeta^{m-1} d\zeta, \qquad (82.33)$$

where Γ is a simple closed counter-clockwise oriented curve in the annulus encircling the origin. The formula for the coefficients is obtained by multiplying by a proper power of z and integrating around Γ.

We may generalize to shifts of the origin to a given point z_0 replacing z by $z - z_0$.

Example 82.16. We have

$$\frac{1}{1-z} = \sum_{m=0}^{\infty} z^m \quad \text{for } |z| < 1,$$

$$\frac{1}{1-z} = \frac{-1}{z(1-z^{-1})} = \sum_{m=1}^{\infty} z^{-m} \quad \text{for } |z| > 1.$$

82.29 Residue Calculus: Simple Poles

Let $f(z)$ be analytic in a simply connected open domain Ω, except at an isolated point $z_0 \in \Omega$, and let Γ be a simple closed curve in Ω oriented counter-clockwise with z_0 contained in the open domain Ω_Γ bounded by Γ. We say that the simple closed curve Γ *surrounds* z_0 counter clockwise. In general the integral

$$\int_\Gamma f(z) \, dz$$

will then not be zero, but the integral will have the same value for any such simple closed curve Γ surrounding z_0 clockwise. To see this we consider two such curves Γ_1 and Γ_2 and introduce the two coinciding curves $\Gamma_3^{\pm}$ with opposite orientation joining Γ_1 and Γ_2 according to Fig. 82.12, and by joining the curves Γ_1, Γ_3^+, $-\Gamma_2$ (Γ_2 backwards) and Γ_3^- we obtain a single closed curve enclosing a domain where $f(z)$ is analytic (that is, not containing z_0 in its interior) over which the integral of $f(z)$ vanishes because of Cauchy's theorem. Thus, noting that the integrals over Γ_3^+ and Γ_3^- cancel, we have

$$0 = \int_{\Gamma_1} f(z) \, dz + \int_{-\Gamma_2} f(z) \, dz = \int_{\Gamma_1} f(z) \, dz - \int_{\Gamma_2} f(z) \, dz$$

where we used that the orientation of $-\Gamma_2$ and Γ_2 are reversed. It follows that the integral over Γ_1 is equal to the integral over Γ_2.

Suppose now that $f(z)$ has the form

$$f(z) = \frac{g(z)}{z - z_0},$$

where $g(z)$ is analytic in Ω and $z_0 \in \Omega$. We then say that $f(z)$ has a *simple pole* at $z = z_0$. We have by Cauchy's representation formula with Γ a simple

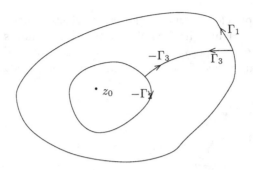

Fig. 82.12. Two simple curves Γ_1 and $-\Gamma_2$ (Γ_2 backwards), surrounding z_0, adjoined by curves $\Gamma_3^{\pm}$ to form one simply connected curve *not* surrounding z_0

closed curve surrounding z_0 counter-clockwise,

$$\int_\Gamma f(z)\,dz = \int_\Gamma \frac{g(z)}{z - z_0}\,dz = 2\pi i g(z_0).$$

The value $g(z_0)$ is called the *residue* of $f(z)$ at z_0, which we denote by Res $f(z_0)$, and thus

$$\text{Res } f(z_0) = g(z_0) = \lim_{z \to z_0} (z - z_0) f(z).$$

Example 82.17. Let $f(z) = \frac{z}{z-1}$ and let Γ be the circle $\gamma(t) = (\cos(t) - 1,\ \sin(t))$ with $0 \le t \le 2\pi$, surrounding $(1, 0)$ counter-clockwise. By the Residue Theorem, we have since obviously Res $f(1) = 1$

$$\int_\Gamma \frac{z}{z - 1}\,dz = 2\pi i.$$

Example 82.18. To evaluate $\int_\Gamma f(z)\,dz$, where $f(z) = \frac{1}{e^z - 1}$ and Γ is a circle centered at the origin and oriented counter-clockwise, we note that

$$f(z) = \frac{z}{e^z - 1}\frac{1}{z} = \frac{g(z)}{z}$$

with

$$\frac{1}{g(z)} = \frac{e^z - 1}{z} = h(z).$$

Since $\lim_{z \to 0} h(z) = 1$, we have Res $f(0) = g(0) = 1$, and thus $\int_\Gamma f(z)\,dz = 2\pi i$.

82.30   Residue Calculus: Poles of Any Order

Suppose now $f(z)$ has a (multiple) *pole of order* $n = 2, 3, \ldots$, at z_0, that is, $f(z)$ is of the form

$$f(z) = \frac{g(z)}{(z - z_0)^n},$$

with $g(z)$ analytic in a neighborhood of z_0. By Cauchy's generalized representation formula we have if Γ is a simple closed curve surrounding z_0 counter-clockwise:

$$\int_\Gamma f(z)\,dz = \int_\Gamma \frac{g(z)}{(z - z_0)^n}\,dz = \frac{2\pi i}{(n-1)!}\, g^{(n-1)}(z_0).$$

We now extend the definition of the residue $\operatorname{Res} f(z_0)$ to a pole of order $n = 1, 2, \ldots$, by setting

$$\operatorname{Res} f(z_0) = \frac{g^{(n-1)}(z_0)}{(n-1)!},$$

and thus we have again

$$\int_\Gamma f(z)\,dz = 2\pi i\, \operatorname{Res} f(z_0).$$

Example 82.19. The function

$$f(z) = \frac{1}{(z-1)^2(z-3)}$$

has a pole of order 2 at $z = 1$ and order 1 at $z = 3$. We compute $\operatorname{Res} f(3) = \frac{1}{4}$, and further $\operatorname{Res} f(1) = -\frac{1}{4}$ since $\frac{d}{dz}\frac{1}{z-3} = -\frac{1}{(z-3)^2} = -\frac{1}{4}$ if $z = 1$.

82.31   The Residue Theorem

We now prove the following basic result of residue calculus:

Theorem 82.10 (The Residue Theorem) *Let $f(z)$ be analytic in a simply connected open domain Ω, except at finitely many isolated points z_1, $z_2, \ldots, z_n$ in Ω where $f(z)$ has simple or multiple poles, and let Γ be a simple closed curve in Ω surrounding all the z_m counter-clockwise. Then*

$$\int_\Gamma f(z)\,dz = \sum_{m=1}^{n} 2\pi i \operatorname{Res} f(z_m).$$

The result follows by surrounding each of the z_m with a little circle $= \Gamma_m$ inside Γ oriented counter-clockwise. By Cauchy's theorem we then have

$$\int_\Gamma f(z)\,dz + \sum_{m=1}^n \int_{-\Gamma_m} f(z)\,dz = 0,$$

arguing as in the Section on Residue Calculus: simple poles, from which follows that

$$\int_\Gamma f(z)\,dz = \sum_{m=1}^n \int_{\Gamma_m} f(z)\,dz = 2\pi i \sum_{m=1}^n \operatorname{Res} f(z_m),$$

which proves the desired result.

Example 82.20. We compute

$$I = \int_\Gamma \frac{4-3z}{z^2-z}\,dz = \int_\Gamma \frac{4-3z}{z(z-1)}\,dz$$

where Γ is a simple closed curve surrounding counter-clockwise the two simple poles $z=1$ and $z=0$ of $\frac{4-3z}{z^2-z}$ and get $I = 2\pi i(-4+1) = -6\pi i$.

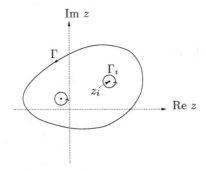

Fig. 82.13. A curve Γ and curves Γ_i, surrounding poles of $f(z)$

82.32   Computation of $\int_0^{2\pi} R(\sin(t), \cos(t))\,dt$

Consider an integral of the form

$$\int_0^{2\pi} R(\sin(t), \cos(t))\,dt$$

where $R(x, y)$ is a rational function of $x, y \in \mathbb{R}$. By the substitution

$$z = e^{it}, \quad dz = ie^{it} dt = iz \, dt$$

$$\cos(t) = \frac{e^{it} + e^{-it}}{2} = \frac{1}{2}\left(z + \frac{1}{z}\right)$$

$$i\sin(t) = \frac{e^{it} - e^{-it}}{2} = \frac{1}{2}\left(z - \frac{1}{z}\right),$$

the integral is converted into

$$\int_{|z|=1} R\left(\frac{z^2 - 1}{2iz}, \frac{z^2 + 1}{2z}\right) \frac{dz}{iz},$$

which can be evaluated using residue calculus, provided the integrand has no poles on $|z| = 1$.

Example 82.21. We compute

$$I = \int_0^{2\pi} \frac{dt}{a + \cos(t)},$$

where $a > 1$ is a constant. Using the transformation just indicated we get

$$I = -2i \int_{|z|=1} \frac{dz}{z^2 + 2az + 1} = -2i \int_{|z|=1} \frac{dz}{(z - \alpha)(z - \beta)},$$

where $\alpha = -a + \sqrt{a^2 - 1}$ and $\beta = -a - \sqrt{a^2 - 1}$. Since $|\alpha| < 1$ and $|\beta| > 1$, the residue at α is $\frac{1}{\alpha - \beta}$ and thus $I = \frac{2\pi}{\sqrt{a^2 - 1}}$.

82.33   Computation of $\int_{-\infty}^{\infty} \frac{p(x)}{q(x)} \, dx$

Integrals of the form

$$\int_{-\infty}^{\infty} f(x) \, dx \tag{82.34}$$

can be evaluated using residue calculus under the assumption that the extended function $f(z)$ with $z \in \mathbb{C}$ has no poles on the real axis and that $|f(z)| \leq M|z|^{-2}$ for $|z|$ large. We start out showing how to use residue calculus to compute the integral

$$I = \int_{-\infty}^{\infty} \frac{1}{1 + x^2} \, dx,$$

(thus without using that $\arctan(x)$ is a primitive function of $\frac{1}{1+x^2}$). We write

$$I = \lim_{R \to \infty} \int_{-R}^{R} \frac{1}{1 + x^2} \, dx = \lim_{R \to \infty} \int_{\Gamma_R} f(z) \, dz,$$

where $f(z) = \frac{1}{1+z^2}$ and Γ_R is the boundary of the semi-disc $|z| \le R$ with $\mathrm{Re}\, z = x \ge 0$. This follows from the fact that

$$\lim_{R\to\infty} \int_{\Gamma_R^+} f(z)\,dz = 0,$$

where Γ_R^+ is the upper part of the semi-circle with $\mathrm{Re}\, z = x > 0$. By the Residue theorem we have

$$\int_{\Gamma_R} f(z)\,dz = 2\pi i \frac{1}{2i} = \pi$$

since the residue of $f(z) = \frac{1}{(z-i)(z+i)}$ inside Γ_R is equal to $\frac{1}{z+i}$ with $z = i$. We conclude that

$$\int_{-\infty}^{\infty} f(x)\,dx = \pi,$$

which of course conforms with the result obtained using that $\frac{d}{dx}\arctan(x) = \frac{1}{1+x^2}$.

The same technique can be used if $f(x) = \frac{p(x)}{q(x)}$ is a rational function with the degree of $q(x)$ two units (or more) higher than that of the numerator $p(x)$. The same technique can be used to evaluate the Fourier transform (cf below) of $\frac{p(x)}{q(x)}$:

$$\frac{1}{2\pi}\int_{-\infty}^{\infty} \frac{p(x)e^{i\xi x}}{q(x)}\,dx.$$

82.34 Applications to Potential Theory in $\mathbb{R}^2$

There is a strong coupling between analytic functions and potential theory in $\mathbb{R}^2$, because if $f(z) = u(x,y) + iv(x,y)$ is analytic in Ω, then the real and imaginary parts $u(x,y)$ and $v(x,y)$ are harmonic in Ω, that is, $\Delta u = \Delta v = 0$ in Ω. Conversely, as we saw above, if $u(x,y)$ is harmonic in a simply connected domain Ω in $\mathbb{R}^2$, then there exists a *harmonic conjugate* function $v(x,y)$ uniquely determined up to a constant such that $f(z) = u(x,y) + iv(x,y)$ is analytic in Ω, see Problem 82.10. The Cauchy-Riemann equations state that $\nabla u = \nabla \times v$:

$$\nabla u = \left(\frac{\partial u}{\partial x}, \frac{\partial u}{\partial y}\right) = \nabla \times v = \left(\frac{\partial v}{\partial y}, -\frac{\partial v}{\partial x}\right).$$

from which follows that ∇u and ∇v are orthogonal:

$$\nabla u \cdot \nabla v = \frac{\partial u}{\partial x}\frac{\partial v}{\partial x} + \frac{\partial u}{\partial y}\frac{\partial v}{\partial y} = \frac{\partial v}{\partial y}\frac{\partial v}{\partial x} - \frac{\partial v}{\partial x}\frac{\partial v}{\partial y} = 0.$$

We conclude that the level curves of $u(x, y)$ and its conjugate $v(x, y)$ are orthogonal. We note that level curves of $u(x, y)$ and $v(x, y)$ in the $z = (x, y)$-plane correspond to the level lines $u = $ constant and $v = $ constant of the analytic function $w = u + iv$ in the $w = (u, v)$-plane.

In fact, much of the interest in analytic functions comes from the connection to potential theory in $\mathbb{R}^2$. Today, computational methods capable of solving also problems in $\mathbb{R}^3$, have changed this picture and analytic functions now play a less important role in areas of applications such as fluid and solid mechanics.

Applications to fluid mechanics typically concern incompressible irrotational flow in 2d with $u(x, y)$ representing a velocity potential and $v(x, y)$ an associated so called *stream function*. The velocity U of the flow is then given by $U = \nabla u = \nabla \times v$

$$U = \nabla u = \left(\frac{\partial u}{\partial x}, \frac{\partial u}{\partial y} \right) = \nabla \times v = \left(\frac{\partial v}{\partial y}, -\frac{\partial v}{\partial x} \right).$$

We have $\nabla \cdot U = \Delta u = 0$ and $\nabla \times U = -\Delta v = 0$ and thus U is incompressible and irrotational. The level curves of $u(x, y)$ with normal ∇u correspond to equi-potential curves, and the level curves of v with normal $\nabla v = -\nabla \times u$ will correspond to the *streamlines* followed by a fluid particle moving with the velocity U. We conclude that each analytic function $f(z) = u(x, y) + iv(x, y)$ may be associated to a particular stationary incompressible and irrotational fluid flow, and the level curves of u and v form a mutually orthogonal set of curves with the level curves of v describing the streamlines of the flow.

In applications to electromagnetics, $u(x, y)$ represents an electric potential with ∇u an electric field, and the level curves of $v(x, y)$ represent the curves traced by electrically charged particles in the electric field.

In applications to heat flow $u(x, y)$ may represent temperature and the level curves for u thus become isolines for temperature and ∇u is proportional to the heat flow.

Example 82.22. (**Flow in a corner**) The function $w = u + iv = z^2$ describes a certain flow in the quarter-plane $\{z = x + iy : x, y \geq 0\}$, with corresponding potential $u(x, y) = x^2 - y^2$ and stream-function $v(x, y) = 2xy$, see Fig. 82.14.

The equi-potential lines $u(x, y) = $ constant and streamlines $v(x, y) = $ constant in the (x, y)-plane are the images of the lines $u = $ constant and $v = $ constant under the mapping $z = w^{1/2}$ from the halfplane $\{w = u + iv : v \geq 0\}$ onto the quarter-plane $\{z = x + iy : x, y \geq 0\}$.

Example 82.23. (**The spinning tennis ball**) We consider two types of rotation-free incompressible flow around the unit disc $\{(x_1, x_2) : x_1^2 + x_2^2 < 1\}$ in two dimensions. The first flow is given by the function $f(z) = z + \frac{1}{z}$,

level curves for Im(w) and Re(w)

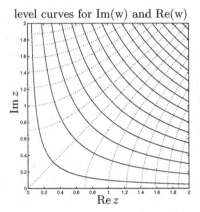

Fig. 82.14. Level curves of $\text{Im}(w) = 2xy$ (*solid*) and $\text{Re}(w) = x^2 - y^2$ (*dotted*) for $w = z^2$

which in polar coordinates with $z = re^{i\theta}$ takes the form

$$f(z) = u(r, \theta) + iv(r, \theta) = \left(r + \frac{1}{r} \right) \cos(\theta) + i \left(r - \frac{1}{r} \right) \sin(\theta). \quad (82.35)$$

This represents a symmetric flow with the velocity $\approx (1, 0)$ (far) away from the disc, and the level curves of $v(r, \theta)$ give the streamlines of the flow around the disc, see Fig. 82.15.

level curves for Im(w) for $w = z + 1/z$

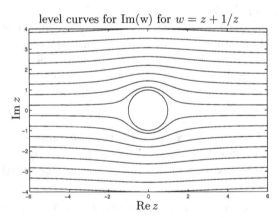

Fig. 82.15. Level curves of $\text{Im}(w)$ for $w = z + 1/z$

The second flow is a flow circulating around the disc given by

$$g(r, \theta) = -\frac{iK}{2\pi} \log(z) = \frac{K}{2\pi} \theta + i \left(-\frac{K}{2\pi} \log(r) \right) \quad (82.36)$$

Consider now the flow given by $f(z) + g(z)$ with stream-function $(r - \frac{1}{r})\sin(\theta) - \frac{K}{2\pi}\log(r)$. We may consider this to be the flow around a spinning tennis ball in a horisontal stream of air, see Fig. 82.16.

level curves for Im(w) for $w = z + 1/z - i\log(z)$

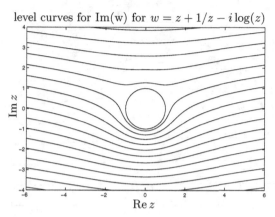

Fig. 82.16. Level curves of $\text{Im}(w)$ for $w = z + 1/z - i\log(z)$

We now recall *Bernouilli's law* stating that for steady inviscid irrotational flow, the quantity

$$p + \frac{|U|^2}{2}$$

is constant because $\nabla(p + \frac{|U|^2}{2}) = 0$, which follows by direct computation, see Problem 82.13. We conclude that high velocity implies low pressure. Now inspecting Fig. 82.16 we see that the velocity is high below the ball (dense level curves of the stream function), and thus the pressure is low below the ball and thus there will be a resulting force downward, which is referred to as *lift*. This is the reason a top-spin in tennis is so efficient in bringing down the ball inside the lines. The more top spin the more curved path of the ball! Björn Borg was one of the first to really exploit this law of mechanics. One can show that the *lift* is proportional to the *circulation* given by

$$\int_\Gamma u \cdot ds,$$

where Γ is the unit circle oriented counter-clockwise. The circulation of the flow given by (82.35) is equal to zero because of symmetry, while the circulation of the flow given by (82.36) is equal to K. The lift of the spinning tennis ball gives a hint to the mechanism behind flying. In fact, the design of an airplane wing with non-symmetric cross section and a sharp trailing edge creates a circulation around the wing which causes a lift, see Fig. 82.17.

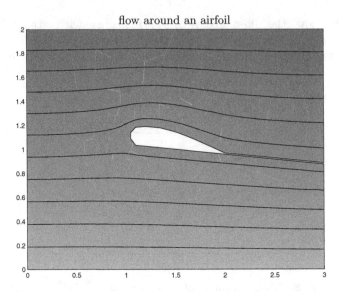

Fig. 82.17. Potential flow around an airfoil

The lifting clockwise circulation around the wing may be considered compensated for by a counter-clockwise vortex in the turbulent layer behind the wing, here localized to the line of discontinuity behind the wing, because the total rotation of the flow must be zero.

Example 82.24. (**Flow through an aperture**) The function

$$z = \sin(w) = \frac{1}{2i}\left(e^{i(u+iv)} - e^{-i(u+iv)}\right)$$
$$= \sin(u)\cosh(v) + i(\cos(u)\sinh(v)),$$

maps the strip $\{w = u + iv : \frac{-\pi}{2} < u < \frac{\pi}{2}, v \in \mathbb{R}\}$ onto $\{z = x + iy : y \neq 0 \text{ if } |x| > 1\}$, that is, the whole plane minus the two half-lines $\{x + iy : |x| > 1, y = 0\}$. The corresponding inverse function $w = f(z) = \arcsin(z) = \sin^{-1}(z)$ may be viewed as the potential for flow through an aperture, see Fig. 82.19.

The streamlines are hyperbolas and the equipotential lines are ellipses.

Example 82.25. (**Discontinuous electric potential**) Consider the function $f(z) = u(x, y) + iv(x, y) = i\log(z) = i\log(|z|) - \text{Arg } z$ in the right half-plane $\{z \in \mathbb{C} : \text{Re } z \geq 0\}$. We have $u(x, y) = \arctan(\frac{y}{x})$ and $v(x, y) = \log(r)$ with $r = (x^2 + y^2)^{1/2}$ and we plot the equi-potential and level curves of the curves in Fig. 82.18.

Note that the potential $u(x, y)$ approaches the value $\frac{\pi}{2}$ for x tending to zero if $y > 0$ and the value $-\frac{\pi}{2}$ for x tending to zero if $y < 0$, corresponding to discontinuous boundary values for $x = 0$.

level curves for Im(w) and Re(w) for $w = i \log(z)$

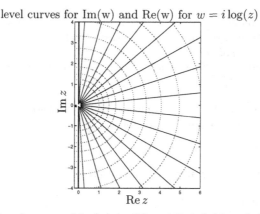

Fig. 82.18. Level curves of $\text{Im}(w)$ (*solid*) and $\text{Re}(w)$ (*dotted*) for $w = i\log(z)$

level curves for Im(w) and Re(w) for $w = \sin^{-1}(z)$

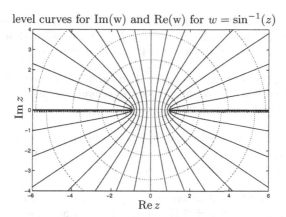

Fig. 82.19. Level curves of $\text{Im}(w)$ (*solid*) and $\text{Re}(w)$ (*dotted*) for $w = \arcsin(z)$

Chapter 82 Problems

82.1. (a) Prove that $f(z) = e^z$ is analytic and that $f'(z) = e^z$. (b) Prove that $\sin(z)$ and $\cos(z)$ are analytic with derivatives $\cos(z)$ and $-\sin(z)$, respectively.

82.2. It is possible to view an analytic function $f : \mathbb{C} \to \mathbb{C}$ as a function $F : \mathbb{R}^2 \to \mathbb{R}^2$ if we set $f(z) = u(x,y) + iv(x,y)$, $z = x + iy$ and $F(x,y) = (u(x,y), v(x,y))$. Explain the connection between the Jacobian F' of $F(x,y)$ and the derivative f', and motivate the Cauchy-Riemann equations this way.

82.3. What happens if we try to choose $z_0 \in \Gamma$ in Cauchy's representation formula?

82.4. Prove Liouville's theorem stating that if $f(z)$ is analytic in the whole complex plane and bounded, then $f(z)$ is constant. Hint: Use the representation formula for $f'(z)$ with Γ a circle with large radius.

82.5. Prove Morera's theorem stating that if $f : \Omega \to \mathbb{C}$ satisfies $\int : \Gamma f(z)\, dz = 0$ for all sinmple closed curves in Ω, then $f(z)$ is analytic in Ω. Hint: Define $F(z) = \int_{\Gamma_z} f(\zeta)\, d\zeta$, where Γ_z is a courve joining a fixed point z_0 with the variable point $z \in \Omega$. Show independence of the specific choice of Γ_z and then that $F'(z) = f(z)$.

82.6. Prove that a Möbius transformation maps every straight line or circle in the complex plane into a circle or straight line. Hint: write $w = \frac{az+b}{cz+d}$ in the form $w = -\frac{ad-bc}{c}\frac{1}{cz+d} + \frac{a}{c}$.

82.7. Compute (a) $\int_0^{2\pi} \frac{d\theta}{5-3\sin(\theta)}$, (b) $\int_{-\infty}^{\infty} \frac{x}{1+x^4}\, dx$.

82.8. Prove that $\int_{-\infty}^{\infty} \frac{\sin\theta}{\theta} = 2\pi$.

82.9. Prove (82.10).

82.10. Prove that if $u(x,y)$ is harmonic in a simply connected domain Ω, then there exists a function v such that $u + iv$ satisfies the Cauchy-Riemann equations in Ω. Hint use the central result of Chapter *Potential fields*.

82.11. Construct your own examples of 2d irrotational potential flow, electrostatics, and heat flow, by combining elementary functions such as z^α, e^z, $\log(z)$, $\sin(z)$, $\sinh(z)$ and Möbius transformations.

82.12. Give a different proof of Cauchy's representation theorem using that $\frac{g(z)}{z-z_0}$ is a analytic in the domain $\Omega_\epsilon = \{z \in \Omega : |z - z_0| > \epsilon\}$, so that $\int_{\Gamma_\epsilon} \frac{g(z)}{z-z_0}\, dz = 0$, where Γ_ϵ is the boundary of Ω_ϵ. Then let ϵ tend to zero.

82.13. Show that if the fluid velocity $u = (u_1, u_2)$ defined in a domain Ω in $\mathbb{R}^2$ satisfies $\nabla \cdot u = \nabla \times u = 0$ and solves the stationary momentum equation $(u \cdot \nabla)u + \nabla p - \Delta u = 0$ in Ω, then $\nabla(p + \frac{|u|^2}{2}) = 0$ in Ω. This proves Bernouilli's Law stating that $p + \frac{|u|^2}{2}$ is constant so that high velocity corresponds to low pressure.

82.14. Determine the images the circle $|z| = 1$ and the unit disc $|z| < 1$ under the mapping $w = \frac{i(1-z)}{1+z}$. Use the result to determine the elektrostatic potential $\varphi(x,y)$, $(z = x + iy)$ in the unit disc $|z| < 1$ with boundary values

$$\varphi(x,y) = \begin{cases} P, & \text{om } |z| = 1, \quad x > 0,\ y > 0, \\ 0, & \text{om } |z| = 1, \quad x < 0,\ \text{eller } y < 0. \end{cases}$$

82.15. Let T be a triangle with corners at 0, 1 and $1 + i$. Determine the image of T under the mapping $w = \frac{z}{1-z}$.

82.16. Determine a harmonic function $\varphi(x,y)$ in the domain between the hyoerbolas $x^2 - y^2 = 1$ and $x^2 - y^2 = 4$ with boundary values $\varphi(x,y) = 2xy$ on $x^2 - y^2 = 1$ and $\varphi(x,y) = 4xy$ on $x^2 - y^2 = 4$.

83
Fourier Series

Yesterday was my 21st birthday, at that age Newton and Pascal had already acquired many claims to immortality. (Fourier 1789, age 21)

83.1 Introduction

We give in the following two chapters a short account of *Fourier analysis* starting with *Fourier series* in this chapter and continuing in the next chapter to *Fourier transforms*. The basic idea is to represent (or approximate) given functions as linear combinations of trigonometric functions. We have met the same general idea in the Chapter Piecewise linear approximation, where we studied approximation of given functions as a linear combination of piecewise polynomials. Fourier representations have particular properties which are useful in for example signal/image processing with important applications to e.g. computer tomography. In recent years variants of Fourier techniques referred to as *Wavelets* have been developed with applications to for example compression of images. We touch this topic at the end of the Chapter *Fourier transforms*.

Fourier (1768–1830), see Fig. 83.1 used trigonometric series in his famous *Théorie analytique de la chaleur* (1822) to study properties of solutions of the heat equation. The idea of expressing a general function as a Fourier series (or as a power series) has influenced the development of mathematical analysis profoundly with the driving force being the formidable success of these techniques for certain classes of problems, for example linear constant coefficient differential equations. However, as any highly specialized tool or

Fig. 83.1. Fourier, Inventor of Fourier series: "Mathematics compares the most diverse of phenomena and discovers the secret analogies between them"

organism, these techniques have not been able to adapt to the needs of a changing world with computational methods for nonlinear differential equations taking over as work-horse in applications. Nevertheless, Fourier analysis still plays a fundamental role for the basic understanding of many phenomena.

We start with Fourier series in complex form and then pass to the real form as a special case. Fourier series concern functions $f : \mathbb{R} \rightarrow \mathbb{C}$ which are *periodic* with a certain period $a > 0$, that is $f(x + a) = f(x)$ for $x \in \mathbb{R}$. We often normalize to $a = 2\pi$ and thus consider 2π-periodic functions $f : \mathbb{R} \rightarrow \mathbb{C}$ satisfying $f(x + 2\pi) = f(x)$ for $x \in \mathbb{R}$. Usually we restrict attention to real-valued functions $f : \mathbb{R} \rightarrow \mathbb{R}$. Fourier transforms concern non-periodic functions $f : \mathbb{R} \rightarrow \mathbb{C}$.

We shall see that representing a given 2π-periodic function $f(x)$ as a Fourier series corresponds to expressing $f(x)$ as a linear combination of a certain set a trigonometric functions $\{e_m(x)\}$:

$$f(x) = \sum_m c_m e_m(x) \tag{83.1}$$

with certain coefficients $c_m \in \mathbb{C}$. We thus view the functions $e_m(x)$ as *basis functions* and express a general function $f(x)$ as a certain *linear combination* of basis functions. For example $f(x) = 0.5 \sin(2x) - 0.8 \sin(7x)$ is a linear combination of the two basis functions $\sin(2x)$ and $\sin(7x)$ with coefficients 0.5 and 0.8, see Fig. 83.2.

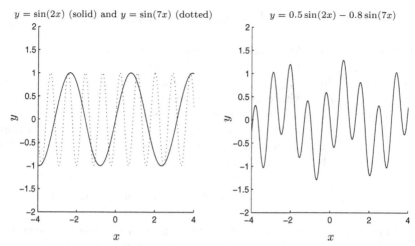

Fig. 83.2. The functions $\sin(2x)$ and $\sin(7x)$, and the linear combination $0.5\sin(2x) + 0.8\sin(7x)$ of the two

The trigonometric basis functions $e_m(x)$ used in Fourier series are of the form

$$\sin(mx), \quad \cos(mx), \quad m = 0, 1, 2, \ldots, \quad \text{(real Fourier series)} \quad (83.2)$$

or

$$e^{imx} = \cos(mx) + i\sin(mx), \quad m = 0, \pm 1, \pm 2, \ldots, \quad \text{(complex Fourier series)} . \quad (83.3)$$

Each basis function or "harmonic" $\sin(mx)$, $\cos(mx)$ or e^{imx}, is periodic with period $\frac{2\pi}{|m|}$ and (angular) *frequency* or *wave number* $|m|$. The larger $|m|$ is the higher is the frequency and the quicker do the basis functions $\sin(mx)$, $\cos(mx)$ and e^{imx} "oscillate". The series (83.1) expresses $f(x)$ as a linear combination of basis functions of increasing frequencies. Since the basis functions are all periodic with period 2π, so is their linear combination $f(x)$.

The basis functions (83.2) and (83.3) are *orthogonal* with respect to the $L_2(-\pi, \pi)$ scalar product

$$(v, w) = \int_{-\pi}^{\pi} v(x)\overline{w(x)} \, dx \quad (83.4)$$

with $\overline{w(x)}$ the complex conjugate of $w(x)$, with corresponding norm $\|v\| = (v, v)^{1/2}$. The orthogonality makes the coefficients c_m directly computable upon taking the $L_2(-\pi, \pi)$ scalar product of (83.1) with e_m to give

$$c_m = \frac{(f, e_m)}{(e_m, e_m)}.$$

83.2 Warm Up I: Orthonormal Basis in $\mathbb{C}^n$

To prepare we consider an analogous situation in $\mathbb{C}^n$: We recall that $\mathbb{C}^n$ is the set of ordered n-tuples $x = (x_1, \ldots, x_n)$ with $x_k \in \mathbb{C}$ for $k = 1, \ldots, n$. The scalar product (x, y) of two vectors x and y in $\mathbb{C}^n$ is defined by $x \cdot y = (x, y) = \sum_{j=1}^{n} x_j \overline{y_j}$, with corresponding norm $|x| = (x, x)^{1/2}$

Let now $\{g_1, \ldots, g_n\}$ be a set of n vectors in $\mathbb{C}^n$, that is each $g_k = (g_{k1}, \ldots, g_{kn})$ is a vector in $\mathbb{C}^n$ with components $g_{kj} \in \mathbb{C}$. We recall that the set $\{g_1, \ldots, g_n\}$ is an orthonormal basis in $\mathbb{C}^n$ if the g_k are mutually orthogonal and have norm equal to one, that is

$$(g_k, g_m) = 0 \quad \text{if } k \neq m, \quad \text{and } |g_m| = 1 \quad \text{for } m = 1, \ldots, n.$$

If $\{g_1, \ldots, g_n\}$ is an orthonormal basis, then we can express a given vector $u \in \mathbb{C}^n$ as a linear combination of basis vectors in the form

$$u = \sum_{k=1}^{n} c_m g_m, \quad \text{where } c_m = (u, g_m) \quad \text{for } m = 1, \ldots, n,$$

where the fact that $c_m = (u, g_m)$ follows by taking the scalar product and using the orthonormality.

83.3 Warm Up II: Series

We recall from Chapter *Series* that a series $\sum_{m=1}^{\infty} \alpha_m$ with coefficients $\alpha_m \in \mathbb{C}$, is said to be *convergent* if the sequence $\{s_n\}_{n=1}^{\infty}$ of partial sums $s_n = \sum_{m=1}^{n} \alpha_m$ converges as n tends to infinity. The series is said to be *absolutely convergent* if $\sum_{m=1}^{\infty} |\alpha_m|$ is convergent, which is the same as requiring the sequence of partial sums $\hat{s}_n = \sum_{m=1}^{n} |\alpha_m|$ to be bounded above, that is $\hat{s}_n \leq K$ for $n = 1, 2, \ldots$, where K is a positive constant. For a series with non-negative terms, the concepts of convergence and absolute convergence coincide. A typical example of a positive (absolutely) convergent series is given by $\sum_{m=1}^{\infty} m^{-2}$. To see that $s_n = \sum_{m=1}^{n} m^{-2}$ is bounded above, we use the fact that

$$s_n \leq 1 + \sum_{m=2}^{n} \int_{m-1}^{m} x^{-2}\, dx \leq 1 + \int_{1}^{n} x^{-2}\, dx \leq 1 + [-x^{-1}]_1^n \leq 2.$$

The same argument shows that $\sum_{m=1}^{\infty} m^{-\alpha}$ is convergent if $\alpha > 1$.

We also recall that an alternating series of the form $\sum_{m=1}^{\infty} (-1)^m a_m$, with $\{a_m\}$ a decreasing positive sequence tending to zero, is convergent.

83.4 Complex Fourier Series

A series of the form

$$\sum_{m=-\infty}^{\infty} c_m e^{imx} = \sum_{1}^{\infty} c_{-m} e^{-imx} + c_0 + \sum_{1}^{\infty} c_m e^{imx}, \qquad (83.5)$$

where $x \in \mathbb{R}$, is said to be a *Fourier series* with *Fourier coefficients* $c_m \in \mathbb{C}$, $m = 0, \pm 1, \pm 2, \ldots$. The corresponding *truncated Fourier series*

$$\sum_{m=-n}^{n} c_m e^{imx} = \sum_{m=1}^{n} c_{-m} e^{-imx} + c_0 + \sum_{m=1}^{n} c_m e^{imx}, \qquad (83.6)$$

where $n = 1, 2, \ldots$, may be viewed as a finite linear combination of the set of basis functions

$$\{1, e^{\pm ix}, e^{\pm i2x}, \ldots, e^{\pm inx}\}$$

with coefficients c_m.

The orthogonality of the basis functions $\{e^{imx}\}$ is expressed by:

$$\int_{-\pi}^{\pi} e^{imx} e^{-ikx} dx = \begin{cases} 0 & \text{if } k \neq m, \\ 2\pi & \text{if } k = m, \end{cases} \qquad (83.7)$$

which follows by direct integration.

We shall typically consider cases with the Fourier coefficients c_m satisfying for some positive constant K,

$$|c_m| \leq K m^{-2}, \quad m = \pm 1, \pm 2, \ldots. \qquad (83.8)$$

In this case the series (83.5) converges absolutely for all x, since

$$\sum_{-\infty}^{\infty} |c_m e^{imx}| = \sum_{-\infty}^{\infty} |c_m| \leq |c_0| + 2K \sum_{m=1}^{\infty} m^{-2} < \infty,$$

and thus defines a function $f : \mathbb{R} \to \mathbb{C}$ represented by a converging Fourier series:

$$f(x) = \sum_{m=-\infty}^{\infty} c_m e^{imx}. \qquad (83.9)$$

The series (83.9) gives a *spectral decomposition* of $f(x)$ into harmonics e^{imx} with different amplitudes c_m. The series (83.9) thus gives a description of the function $f(x)$ in terms of amplitudes of different harmonics included in $f(x)$. In musical terms we may think of $f(x)$ as a "chord" built by a number of "tones" $c_m e^{imx}$ of different frequencies m and amplitudes c_m. A spectral decomposition of a "chord" $f(x)$ would display the "tones" building the "chord", try *The Sound of Functions* in the *Mathematics Laboratory* for a direct experience.

We note that the basis functions $\{e^{imx}\}$ have *global support*, that is, each basis function e^{imx} is nonzero for all $x \in \mathbb{R}$. The basis functions $\{e^{imx}\}$ thus combines the following properties: orthogonality and global support. We contrast this to the 'hat functions' which are the basis functions for continuous piecewise linear approximation: the hat functions have local support but are not (quite) orthogonal. The best combination would be orthogonality together with local support. So-called *wavelets* introduced in recent years combine these properties.

Suppose now that $f(x)$ is defined by a converging Fourier series (83.9). Multiplying by e^{-imx} with $m = 0, \pm 1, \pm 2, \ldots$ and integrating over the interval $[-\pi, \pi]$, and using the orthogonality properties (83.7), we find that

$$c_m = c_m(f) = \frac{1}{2\pi} \int_{-\pi}^{\pi} f(x) e^{-imx} \, dx, \qquad (83.10)$$

where we indicated the dependence of the Fourier coefficient $c_m = c_m(f)$ on the function $f(x)$. We thus have the *Fourier series representation*

$$f(x) = \sum_{m=-\infty}^{\infty} c_m(f) e^{imx}, \qquad (83.11)$$

expressing $f(x)$ as a linear combination of different harmonics e^{imx} with different frequencies, where the Fourier coefficients $c_m(f)$ are given by (83.10).

Conversely, if $f : \mathbb{R} \to \mathbb{C}$ is a given 2π-periodic (Lipschitz continuous) function and we define $c_m(f)$ by (83.10), then we may ask if $f(x)$ can be represented by its Fourier series (83.11) for all x. We shall prove below that this is true if $f(x)$ is 2π-periodic and differentiable. This is the basic result of Fourier analysis stating that an arbitrary 2π-periodic differentiable function can be given a spectral decomposition in the form of a Fourier series. This result includes the "completeness" aspect of the basis functions $\{e^{imx}\}$, that is, the fact that *any* differentiable function can be represented as a Fourier series.

83.5 Fourier Series as an Orthonormal Basis Expansion

Normalizing the basis functions e^{imx} we obtain the orthonormal basis functions $e_m(x) = \frac{1}{\sqrt{2\pi}} e^{imx}$ satisfying

$$(e_m, e_k) = 0 \quad \text{if } k \neq m, \quad (e_m, e_m) = 1. \qquad (83.12)$$

A Fourier series representation takes the following form in the normalized basis:

$$f(x) = \sum_{m=-\infty}^{\infty} \tilde{c}_m(f) \frac{1}{\sqrt{2\pi}} e^{imx}, \quad \tilde{c}_m(f) = \frac{1}{\sqrt{2\pi}} \int_{-\pi}^{\pi} f(x) e^{-imx} \, dx.$$

Of course, it would be natural to work with the normalized basis functions $\{\frac{1}{\sqrt{2\pi}} e^{imx}\}$ and the corresponding renormalized Fourier coefficients $\tilde{c}_m(f)$ thus distributing the 2π-factor into two $\sqrt{2\pi}$-factors, but we follow the most common notation and include the 2π-factor in the Fourier coefficient $c_m(f)$ coupled to the basis function e^{imx}, which also simplifies notation somewhat.

83.6 Truncated Fourier Series and Best L_2-Approximation

The *truncated Fourier series*

$$S_n f(x) = \sum_{m=-n}^{n} c_m(f) e^{imx}$$

of a given function $f(x)$ is a *best approximation* of $f(x)$ in the sense that

$$\|f - S_n f\| \le \|f - g_n\|$$

for any function $g_n(x) = \sum_{m=-n}^{n} d_m e^{imx}$ with $d_m \in \mathbb{C}$, $m = 0, \pm 1, \ldots, \pm n$. This is because, by the definition of the Fourier coefficients,

$$(f - S_n f, e_m) = 0 \quad \text{for } m = 0, \pm 1, \ldots, \pm n,$$

and thus $S_n f(x)$ is the best approximation in the $L_2(-\pi, \pi)$ norm of $f(x)$ in the linear space spanned by the functions $\{1, e^{\pm ix}, e^{\pm i2x}, \ldots, e^{\pm inx}\}$, compare Chapter *Piecewise Linear Approximation*.

83.7 Real Fourier Series

Using that $e^{imx} = \cos(mx) + i\sin(mx)$ and $\cos(-mx) = \cos(mx)$ and $\sin(-mx) = -\sin(mx)$, we can write (83.9) in the form

$$\sum_{m=-\infty}^{\infty} c_m e^{imx} = c_0 + \sum_{m=1}^{\infty} a_m \cos(mx) + \sum_{m=1}^{\infty} b_m \sin(mx),$$

where

$$a_m = c_m + c_{-m}, \quad b_m = i(c_m - c_{-m}), \quad m = 1, 2, \ldots .$$

If $f(x)$ is real, that is $f : \mathbb{R} \to \mathbb{R}$, then $\bar{c}_m = c_{-m}$ and thus $a_m = c_m + \bar{c}_m = 2\text{Re}(c_m) \in \mathbb{R}$ and $b_m = i(c_m - \bar{c}_m) = -2\text{Im}(c_m) \in \mathbb{R}$, and

$$c_m = \frac{a_m}{2} - i\frac{b_m}{2}, \quad c_{-m} = \frac{a_m}{2} + i\frac{b_m}{2}, \quad m = 0, 1, 2, \ldots. \tag{83.13}$$

The Fourier series of a real-valued 2π-periodic function $f : \mathbb{R} \to \mathbb{R}$ can thus be written alternatively as a Sine and Cosine series of the form

$$f(x) = \frac{a_0}{2} + \sum_{m=1}^{\infty} a_m \cos(mx) + \sum_{m=1}^{\infty} b_m \sin(mx),$$

where $a_m, b_m \in \mathbb{R}$, are given by

$$a_m = a_m(f) = \frac{1}{\pi} \int_{-\pi}^{\pi} f(x) \cos(mx)\, dx \quad \text{for } m = 0, 1, 2, \ldots,$$

$$b_m = b_m(f) = \frac{1}{\pi} \int_{-\pi}^{\pi} f(x) \sin(mx)\, dx \quad \text{for } m = 1, 2, \ldots.$$

We note that if $f(x)$ is even, that is $f(x) = f(-x)$, then $b_m = 0$ for $m = 1, 2, \ldots$, and thus $f(x)$ has a *Cosine series representation*:

$$f(x) = \frac{a_0(f)}{2} + \sum_{m=1}^{\infty} a_m(f) \cos(mx). \tag{83.14}$$

Correspondingly, if $f(x)$ is odd, that is $f(x) = -f(-x)$, then $a_m = 0$ for $m = 0, 1, \ldots$, and thus $f(x)$ has a *Sine series representation*:

$$f(x) = \sum_{m=1}^{\infty} b_m(f) \sin(mx). \tag{83.15}$$

In the applications below we usually consider Cosine and Sine series for real-valued functions $f : \mathbb{R} \to \mathbb{R}$. The complex Fourier series is useful in the analysis of convergence of Fourier series.

We now present a couple of examples with Fourier coefficients having different rates of convergence to zero (as m^{-2}, m^{-3} and m^{-1}).

Example 83.1. Let $f : \mathbb{R} \to \mathbb{R}$ be a 2π-periodic function given by $f(x) = |x|$ for $-\pi \leq x \leq \pi$. The function $f(x)$ is real-valued and even, and thus has a Cosine series of the form (83.14). We compute using integration by parts if $m > 0$:

$$a_0(f) = \frac{1}{\pi} \int_{-\pi}^{\pi} f(x)\, dx = \frac{2}{\pi} \int_0^{\pi} x\, dx = \pi,$$

$$a_m(f) = \frac{1}{\pi} \int_{-\pi}^{\pi} f(x) \cos(mx)\, dx = \frac{2}{\pi} \int_0^{\pi} x \cos(mx)\, dx$$

$$= \frac{2}{\pi} \left[\frac{x \sin(mx)}{m} \right]_0^{\pi} - \frac{2}{\pi} \int_0^{\pi} \frac{\sin(mx)}{m}\, dx = \frac{2}{\pi} \frac{(-1)^m - 1}{m^2}.$$

Since $(-1)^m - 1 = -2$ if m is odd and $(-1)^m - 1 = 0$ if m is even, the Fourier series representation of $f(x) = |x|$ takes the form

$$|x| = \frac{\pi}{2} - \frac{4}{\pi} \sum_{k=1}^{\infty} \frac{\cos((2k-1)x)}{(2k-1)^2}.$$

We plot the corresponding truncated series with summation over $k = 1, \ldots, n$ for different values of n in Fig. 83.3:

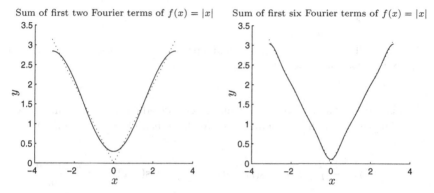

Fig. 83.3. The sum of the first two and first six terms of the fourier series of $|x|$ (*dotted*)

Example 83.2. Let $f : \mathbb{R} \to \mathbb{R}$ be an odd 2π-periodic function given by $f(x) = x(\pi - x)$ for $0 \leq x \leq \pi$. We compute its Sine series coefficients:

$$b_m(f) = \frac{1}{\pi} \int_{-\pi}^{\pi} f(x) \sin(mx)\, dx = \frac{2}{\pi} \int_{0}^{\pi} x(\pi - x) \sin(mx)\, dx$$

$$= -\frac{2}{\pi} \left[\frac{x(\pi - x) \cos(mx)}{m} \right]_{0}^{\pi} + \frac{2}{\pi} \int_{0}^{\pi} \frac{(\pi - 2x) \cos(mx)}{m}\, dx$$

$$= \frac{2}{\pi m} \left[\frac{(\pi - 2x) \sin(mx)}{m} \right]_{0}^{\pi} + \frac{2}{\pi m^2} \int_{0}^{\pi} 2 \sin(mx)\, dx$$

$$= \frac{4}{\pi m^3} (1 - (-1)^m).$$

Example 83.3. Define a 2π-periodic function $f(x)$ by setting

$$f(x) = \begin{cases} 1 & \text{for } |x| < a, \\ 0 & \text{for } a < |x| \leq \pi, \end{cases}$$

where $0 < a < \pi$, see Fig. 83.4.

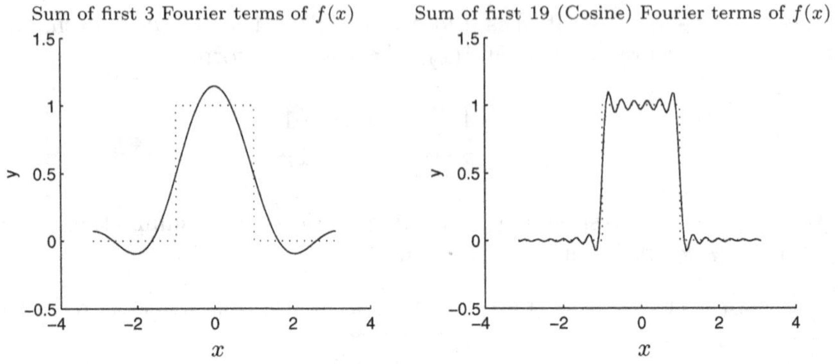

Sum of first 3 Fourier terms of $f(x)$

Sum of first 19 (Cosine) Fourier terms of $f(x)$

Fig. 83.4. The sum of the first 3 and first 19 terms of the fourier series of a piecewise constant function (*dotted*)

This is a piecewise Lipschitz continuous 2π-periodic even function, and we can compute its Fourier coefficients. We have $b_m(f) = 0$ and, for $m > 0$, $2c_m(f)$

$$= a_m(f) = \frac{1}{\pi} \int_{-\pi}^{\pi} f(x) \cos(mx)\, dx = \frac{2}{\pi} \int_0^a \cos(mx)\, dx = \frac{2\sin(ma)}{\pi m},$$
$$(83.16)$$

while $a_0(f) = \frac{2a}{\pi}$. We thus expect that

$$f(x) = \frac{a}{\pi} + \frac{2}{\pi} \sum_{m=1}^{\infty} \frac{\sin(ma)}{m} \cos(mx).$$

We shall return to this equality below, with particular focus on the values $x = \pm a$ where $f(x)$ has jump discontinuities.

83.8   Basic Properties of Fourier Coefficients

We now present some basic properties of the Fourier coefficients

$$c_m(f) = \frac{1}{2\pi} \int_{-\pi}^{\pi} f(x) e^{-imx}\, dx \quad m = 0, \pm1, \pm2, \ldots,$$

of a given 2π-periodic Lipschitz continuous function $f : \mathbb{R} \to \mathbb{C}$.

Linearity

Fourier coefficients satisfy the following obvious linearity properties:

$$c_m(f + g) = c_m(f) + c_m(g), \quad c_m(\alpha f) = \alpha c_m(f),$$

where f and g are two functions with Fourier coefficients $c_m(f)$ and $c_m(g)$, and $\alpha \in \mathbb{C}$.

Fourier Coefficients of the Derivative $Df = f'$

We now couple the Fourier coefficients of the derivative $Df = \frac{df}{dx}$ of a 2π-periodic function $f : \mathbb{R} \to \mathbb{C}$ to the Fourier coefficients of f. The trick is to integrate by parts: Using the periodicity of $f(x)$, we find that

$$c_m(Df) = \frac{1}{2\pi} \int_{-\pi}^{\pi} Df(x)e^{-imx}\, dx = im\frac{1}{2\pi} \int_{-\pi}^{\pi} f(x)e^{-imx}\, dx = im\, c_m(f),$$

and we have thus proved:

Theorem 83.1 *If $f : \mathbb{R} \to \mathbb{C}$ is 2π-periodic and differentiable with derivative Df, then for $m = 0, \pm 1, \pm 2, \ldots$*

$$c_m(Df) = im\, c_m(f). \tag{83.17}$$

This is one of the fundamental results of Fourier analysis, and translates the operation of differentiation $D = \frac{d}{dx}$ with respect to x to multiplication of Fourier coefficients with im where m is the frequency. This opens the way of translating differential equations in the variable x to algebraic equations in the frequency m, which may be very useful and illuminating in certain applications.

We can directly generalize to

Theorem 83.2 *If $f : \mathbb{R} \to \mathbb{C}$ is 2π-periodic and k times differentiable with derivative $D^k f$, then for $m = 0, \pm 1, \pm 2, \ldots$*

$$c_m(D^k f) = (im)^k c_m(f). \tag{83.18}$$

Example 83.4. Consider the differential equation $Du(x) + u(x) = f(x)$, where $f(x)$ a given 2π-periodic function and we seek a 2π-periodic solution $u(x)$. This equation models, for example, a resistor and capacitor in series, with $u(x)$ a primitive function of the current, $f(x)$ an applied voltage, and x representing time, see the Chapter Electrical circuits. Alternatively, $Du(x) + u(x) = f(x)$ models an inductor and resistance in series with $u(x)$ now the current, and again $f(x)$ an applied voltage. For the Fourier coefficients we have using Theorem 83.1

$$im\, c_m(u) + c_m(u) = c_m(f),$$

and thus

$$c_m(u) = \frac{c_m(f)}{1 + im} = \frac{(1 - im)c_m(f)}{1 + m^2}.$$

This shows that the indicated circuits act as so-called *low-pass filters*, with the property of damping high-frequency components: we view $f(x)$ as the input and $u(x)$ as the output and note that the Fourier coefficients of $u(x)$ decay quicker than those of $f(x)$.

Example 83.5. Consider the differential equation $-D^2u(x) + u(x) = f(x)$ with $f(x)$ a given 2π-periodic function and we seek a 2π-periodic solution $u(x)$. Since $c_m(D^2u) = (im)^2 c_m(u)$, we obtain the following algebraic equation for the Fourier coefficients:

$$(m^2 + 1)c_m(u) = c_m(f) \quad \text{for } m \neq 0.$$

We can thus express the solution $u(x)$ of $-D^2u(x) = u(x) = f(x)$ as a Fourier series

$$u(x) = \sum_{-\infty}^{\infty} \frac{c_m(f)}{m^2 + 1} e^{imx},$$

if the data $f(x)$ is given as a Fourier series: $f(x) = \sum_{-\infty}^{\infty} c_m(f)e^{imx}$. Again, we see that the differential equation acts as a low-pass filter with damping of high-frequency components of the data $f(x)$.

Example 83.6. More generally, consider the following differential equation $p(D)u(x) = f(x)$, where $p(D) = \sum_{k=0}^{q} a_k D^k$ is a differential equation with constant coefficients $a_k \in \mathbb{C}$, the data $f(x)$ is 2π-periodic and we seek a 2π-periodic solution $u(x)$. Arguing as above, we get the following equation for the Fourier coefficients:

$$p(im)c_m(u) = \sum_{k=0}^{q} a_k(im)^k c_m(u) = c_m(f),$$

that is, assuming $p(im) \neq 0$ (or $c_m(f) = 0$ if $p(im) = 0$),

$$c_m(u) = \frac{c_m(f)}{p(im)},$$

which gives the Fourier series for the solution, if the Fourier series for the data $f(x)$ is given.

The Fourier coefficients $c_m(f)$ tend to zero as $|m| \to \infty$

As a direct consequence of the preceding result, we conclude that the Fourier coefficients $c_m(f)$ of a 2π-periodic differentiable function $f(x)$ with integrable derivative Df, tend to zero as $|m|$ tends to infinity: Since $|im\,c_m(f)| = |c_m(Df)|$, we have

$$|c_m(f)| = \frac{1}{|m|}|c_m(Df)| \leq \frac{1}{2\pi|m|} \int_{-\pi}^{\pi} |Df| \, dx \to 0 \quad \text{as } |m| \to \infty.$$

Similarly, if $f(x)$ is 2π-periodic with integrable derivative $D^k f$ of order $k > 1$, then for $m = \pm 1, \pm 2, \ldots,$

$$|c_m(f)| \leq \frac{1}{2\pi|m^k|} \int_{-\pi}^{\pi} |D^k f| \, dx.$$

We conclude that the larger k is, the more rapid is the convergence of $c_m(f)$ to zero.

We can also go in the direction of less regularity and ask if we can show that the Fourier coefficients $c_m(f)$ tend to zero as $|m| \to \infty$ under the weaker assumption that f is Lipschitz continuous only. To this end we first note that for any $-\pi < a < b < \pi$, we have

$$\int_a^b e^{-imx} dx = \frac{1}{-im} [e^{-imx}]_a^b \to 0 \quad \text{as } m \to \infty. \tag{83.19}$$

This may be seen as a consequence of the rapid oscillations of e^{-imx} with $|m|$ large, which causes a lot of cancellations in any integral of the form (83.19) with the effect that the integrals decreases to zero as m increases to infinity, see the following figure.

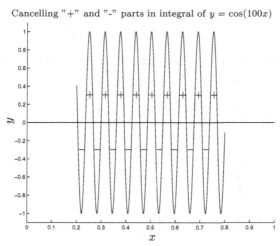

Fig. 83.5. An illustration of the fact that $\int_a^b \cos(mx) dx$ and $\int_a^b \sin(mx) dx$ is small for m large

The estimate (83.19) shows that if $f(x)$ is piecewise constant on $[-\pi, \pi]$, that is a linear combination sum of functions equal to one on a certain interval and zero elsewhere, then $c_m(f) \to 0$ as $|m| \to \infty$.

Finally, a given Lipschitz continuous function $f : [-\pi, \pi] \to \mathbb{C}$ can be approximated by a piecewise constant function $\tilde{f}(x)$, so that

$$\int_{-\pi}^{\pi} |f(x) - \tilde{f}(x)|$$

is as small as we please, which leads to the famous

Theorem 83.3 (Riemann-Lebesgue lemma) *If* $f : [-\pi, \pi]$ *is Lipschitz continuous, then* $c_m(f) \to 0$ *as* $|m| \to \infty$.

The assumption can be relaxed to piecewise Lipschitz continuity.

Convolution

Given two 2π-periodic functions $f(x)$ and $g(x)$, we define a new 2π-periodic function $f * g$ by

$$(f * g)(x) = \int_{-\pi}^{\pi} f(x - y)g(y)\, dy \quad x \in \mathbb{R}.$$

We say that $f * g$ is the *convolution* of f and g. Changing variables, setting $y = x - t$, we find that

$$(f * g)(x) = \int_{-\pi}^{\pi} f(t)g(x - t)\, dt = \int_{-\pi}^{\pi} f(y)g(x - y)\, dy \quad x \in \mathbb{R},$$

and thus the integrand can take the form $f(x - y)g(y)$ or $f(y)g(x - y)$.
    We shall now prove that

$$c_m(f * g) = 2\pi\, c_m(f)c_m(g). \tag{83.20}$$

By direct computation, changing order of integration and using the change of variable $t = x - y$, we have

$$
\begin{aligned}
c_m(f * g) &= \frac{1}{2\pi} \int_{-\pi}^{\pi} (f * g)(x)e^{-imx}\, dx \\
&= \frac{1}{2\pi} \int_{-\pi}^{\pi} \int_{-\pi}^{\pi} f(x - y)g(y)\, dy\, e^{-imx}\, dx \\
&= \int_{-\pi}^{\pi} g(y)e^{-imy} \left(\frac{1}{2\pi} \int_{-\pi}^{\pi} f(x - y)\, e^{-im(x-y)}\, dx \right) dy \\
&= \int_{-\pi}^{\pi} g(y)e^{-imy} \left(\frac{1}{2\pi} \int_{-\pi}^{\pi} f(t)\, e^{-imt}\, dt \right) dy \\
&= c_m(f) \int_{-\pi}^{\pi} g(y)e^{-imy}\, dy = 2\pi c_m(f)c_m(g).
\end{aligned}
$$

Example 83.7. Let $g : \mathbb{R} \to \mathbb{R}$ be a 2π-periodic function defined by

$$g(x) = \frac{1}{2a} \quad \text{for } -a \le x \le a,$$

where $0 < a < \pi$. For a small, we may view $g(x)$ as an approximate delta function. The convolution

$$(f * g)(x) = \int_{-\pi}^{\pi} f(x - y)g(y)\, dy = \frac{1}{2a} \int_{-a}^{a} f(x - y)\, dy$$

is an average of $f(x)$ over the interval $[x - a, x + a]$. Recalling (83.16), and using (83.20), we get

$$c_m(f * g) = c_m(f)\frac{\sin(ma)}{ma}.$$

We conclude that $c_m(f * g)$ is close to $c_m(f)$ if ma is small, and $c_m(f * g)$ is much smaller than $c_m(f)$ if ma is large. The Fourier coefficients of the average $f * g$ thus decay quicker than those of f, and thus $f * g$ is a smoothed version of f: taking the average increases smoothness reflected by quickly decreasing Fourier coefficients.

83.9   The Inversion Formula

We shall now prove that if $f : \mathbb{R} \to \mathbb{C}$ is 2π-periodic and differentiable, then for all $x \in \mathbb{R}$

$$\lim_{n \to \infty} \sum_{-n}^{n} c_m(f) e^{imx} = f(x).$$

In other words, the function $f(x)$ can be represented as a convergent Fourier series:

$$f(x) = \sum_{m=-\infty}^{\infty} c_m(f) e^{imx} \quad \text{for } x \in \mathbb{R}.$$

We have

$$\sum_{-n}^{n} c_m e^{imx} = \sum_{-n}^{n} \frac{1}{2\pi} \int_{-\pi}^{\pi} f(y) e^{-imy} \, dy \, e^{imx}$$

$$= \int_{-\pi}^{\pi} f(y) \frac{1}{2\pi} \sum_{-n}^{n} e^{im(x-y)} \, dy = \int_{-\pi}^{\pi} f(y) D_n(x-y) \, dy,$$

$$(83.21)$$

where, setting $\theta = x - y$,

$$D_n(\theta) = \frac{1}{2\pi} \sum_{-n}^{n} e^{im\theta} = \frac{1}{2\pi} e^{-in\theta} \sum_{m=0}^{2n} e^{im\theta}$$

$$= \frac{1}{2\pi} e^{-in\theta} \frac{1 - e^{i(2n+1)\theta}}{1 - e^{i\theta}} = \frac{1}{2\pi} \frac{e^{-i\frac{\theta}{2}}}{e^{-i\frac{\theta}{2}}} \frac{e^{-in\theta} - e^{i(n+1)\theta}}{1 - e^{i\theta}}$$

$$= \frac{1}{2\pi} \frac{\sin(n\theta + \frac{\theta}{2})}{\sin(\frac{\theta}{2})}$$

is the so-called *Dirichlet kernel*. We here used that $\sum_{m=0}^{2n} e^{im\theta}$ is a finite geometric series with factor $e^{i\theta}$. Using the convolution notation we can write (83.21) in the compact form

$$\sum_{-n}^{n} c_m e^{imx} = f * D_n(x).$$

In order for $f * D_n(x)$ to approximate $f(x)$, we expect D_n to somehow behave like the identity. We look at a plot of $D_n(\theta)$:

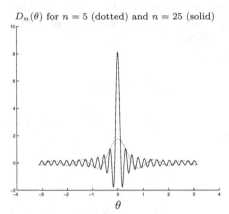

$D_n(\theta)$ for $n = 5$ (dotted) and $n = 25$ (solid)

θ

Fig. 83.6. A plot of $D_n(\theta)$

We see that $D_n(\theta)$ oscillates and has a peak at $\theta = 0$. Integrating $D_n(\theta) = \frac{1}{2\pi} \sum_{-n}^n e^{im\theta}$ term by term over $[-\pi, \pi]$ noting that all integrated terms vanish but one, we see that the total area (with sign) under the graph of D_n is equal to one, that is

$$\int_{-\pi}^{\pi} D_n(\theta) \, d\theta = 1, \tag{83.22}$$

which expresses one aspect of the idea that D_n behaves like the identity. The other aspect of the approximate identity nature of D_n is the increasing focussing of the peak of D_n at 0 as n increases.

Using (83.22), we can write

$$f(x) - f * D_n(x) = \frac{1}{2\pi} \int_{-\pi}^{\pi} (f(x) - f(y)) D_n(x - y) \, dy$$

$$= \frac{1}{2\pi} \int_{-\pi}^{\pi} g(x, y) \sin\left(\left(n + \frac{1}{2}\right)(x - y)\right) dy$$

where

$$g(x, y) = \frac{f(x) - f(y)}{\sin(\frac{x-y}{2})}.$$

Now if $f(x)$ is twice differentiable, then $g(x, y)$ is differentiable with respect to y for all $y \in \mathbb{R}$ with derivative $Dg(x, y)$ (see the corresponding argument in the proof of Cauchy's formula). Integrating by parts we thus have

$$f(x) - D_n * f(x) = -\frac{1}{2\pi} \frac{1}{n + \frac{1}{2}} \int_{-\pi}^{\pi} Dg(x, y) \cos\left(\left(n + \frac{1}{2}\right)(x - y)\right) dy \to 0$$

as $n \to \infty$. In case $f(x)$ is differentiable with piecewise Lipschitz continuous derivative, then $Dg(x,y)$ is Lipschitz continuous in y, and by Riemann-Lebesgue' lemma, we find the same conclusion. We summarize in the following basic theorem:

Theorem 83.4 *If $f : \mathbb{R} \to \mathbb{C}$ is 2π-periodic with piecewise Lipschitz continuous derivative, then $f(x)$ may be represented by a convergent Fourier series:*

$$f(x) = \sum_{m=-\infty}^{\infty} c_m(f)e^{imx} \quad \text{for } x \in \mathbb{R},$$

where the coefficients c_m are given by (83.10).

The assumption on $f(x)$ can be relaxed: it suffices to assume that $f(x)$ is piecewise differentiable with piecewise Lipschitz continuous derivative. At a point x of discontinuity, the Fourier series converges to the mean value of the left hand limit $f^-(x) = \lim_{y \to x, y < x} f(y)$ and the right hand limit $f^+(x) = \lim_{y \to x, y > x} f(y)$:

$$\sum_{m=-\infty}^{\infty} c_m(f)e^{imx} = \frac{f^-(x) + f^+(x)}{2}. \tag{83.23}$$

Example 83.8. We have

$$\sum_{m=1}^{\infty} \frac{\sin(ma)}{\pi m} \cos(mx) = \begin{cases} 1 & \text{if } |x| < a, \\ \frac{1}{2} & \text{if } |x| = a, \\ 0 & \text{if } |x| > a. \end{cases}$$

83.10   Parseval's and Plancherel's Formulas

Suppose $f : \mathbb{R} \to \mathbb{C}$ is 2π-periodic with a convergent Fourier series representation:

$$f(x) = \sum_{m=-\infty}^{\infty} c_m(f)e^{imx}, \tag{83.24}$$

where

$$c_m(f) = \frac{1}{2\pi} \int_{-\pi}^{\pi} f(x)e^{-imx}\,dx.$$

Using the orthogonality (83.7) of the functions $\{e^{imx}\}$, we find that

$$
\int_{-\pi}^{\pi} |f(x)|^2 \, dx = \int_{-\pi}^{\pi} f(x)\overline{f(x)} \, dx
$$

$$
= \int_{-\pi}^{\pi} \left(\sum_{m=-\infty}^{\infty} c_m(f)e^{imx} \right) \left(\sum_{k=-\infty}^{\infty} \overline{c_k(f)}e^{-ikx} \right) dx
$$

$$
\sum_{m,k=-\infty}^{\infty} c_m(f)\overline{c_k(f)} \int_{-\pi}^{\pi} e^{imx}e^{-ikx} \, dx
$$

$$
= 2\pi \sum_{m=-\infty}^{\infty} |c_m(f)|^2.
$$

We have now proved the celebrated:

Theorem 83.5 (Parseval's formula) *If $f(x)$ has a convergent Fourier series representation, then*

$$
\int_{-\pi}^{\pi} |f(x)|^2 \, dx = 2\pi \sum_{m=-\infty}^{\infty} |c_m(f)|^2.
$$

We can in an obvious way generalize to obtain:

Theorem 83.6 (Plancherel's formula) *If $f(x)$ and $g(x)$ have convergent Fourier series representations, then*

$$
\int_{-\pi}^{\pi} f(x)\overline{g(x)} \, dx = 2\pi \sum_{m=-\infty}^{\infty} c_m(f)\overline{c_m(g)}.
$$

83.11   Space Versus Frequency Analysis

We are now ready to lean back and reflect a bit about the nature of Fourier series. Suppose that $f(x)$ is a given 2π-periodic function. If we want to describe the nature of the function $f(x)$, that is the variation of $f(x)$ with x, we can try to give some kind of list of $f(x)$ values for different values of x. We may call this a physical description where we think of x as a space or time variable. Now using Fourier series we can instead express the function $f(x)$ as a Fourier series, determined by the Fourier coefficients $\{c_m(f)\}$. Describing $f(x)$ through its Fourier coefficients, may be viewed as a frequency-description. In the physical description, we describe the function f in terms of its function values $f(x)$ for different values of x. In the frequency description, we describe f in terms of the Fourier coefficients $c_m(f)$ as a sum $f(x) = \sum_m c_m(f)e^{imx}$.

To describe a given function $f(x)$ we may thus look at the variation of $f(x)$ with x, or the variation of $c_m(f)$ with m.

We have noted that the decay of $c_m(f)$ with m couples to the regularity of $f(x)$: if $f(x)$ is highly regular with many derivatives, then the Fourier coefficients $c_m(f)$ decay quickly with increasing m, and vice versa. If the Fourier coefficients decay quickly, then only a few terms in the Fourier series suffices to represent the function to high accuracy.

83.12   Different Periods

Suppose $f : \mathbb{R} \to \mathbb{C}$ is periodic with period $\frac{2\pi}{\omega}$ with $\omega > 0$. We considered above the case $\omega = 1$, and we now generalize to $\omega > 0$. For example: the functions $\sin(\omega x)$, $\sin(2\omega x)$, $\sin(3\omega x)\ldots$, are periodic with period $\frac{2\pi}{\omega}$.

Defining $g(x) = f(\frac{x}{\omega})$, we have that $g(x)$ is 2π-periodic since $g(x+2\pi) = f(\frac{x+2\pi}{\omega}) = f(\frac{x}{\omega} + \frac{2\pi}{\omega}) = f(\frac{x}{\omega}) = g(x)$, and a Fourier series representation of $g(x)$:

$$g(x) = \sum_{m=-\infty}^{\infty} c_m(g)e^{imx}, \quad c_m(g) = \frac{1}{2\pi} \int_{-\pi}^{\pi} g(y)e^{-imy}\, dy$$

translates into the following Fourier series representation of $f(\frac{x}{\omega})$:

$$f\left(\frac{x}{\omega}\right) = \sum_{m=-\infty}^{\infty} c_m(g)e^{imx}, \quad c_m(g) = \frac{1}{2\pi} \int_{-\pi}^{\pi} f\left(\frac{y}{\omega}\right) e^{-imy}\, dy$$

which takes the following form changing variables from $\frac{x}{\omega}$ to x and $\frac{y}{\omega}$ to y:

$$f(x) = \sum_{m=-\infty}^{\infty} c_m(f)e^{im\omega x}, \quad c_m(f) = \frac{\omega}{2\pi} \int_{-\frac{\pi}{\omega}}^{\frac{\pi}{\omega}} f(y)e^{-im\omega y}\, dy. \quad (83.25)$$

83.13   Weierstrass Functions

Consider a series of the form

$$\sum_{m=1}^{\infty} a^{-m} \sin(b^m x), \quad (83.26)$$

where $a > 1$, $b > a$. This type of series was presented by Weierstrass as an example of a Lipschitz continuous function that is not differentiable at any point, see Fig. 83.7, where we plot the corresponding truncated series $\sum_{m=1}^{n} a^{-m} \sin(b^m x)$ with $n = 10$. We see that as n increases the series oscillates increasingly wildly, and gives an irregular "chaotic" impression.

Sum of first 10 terms of Weierstrass function with $a = 2$ and $b = 3$

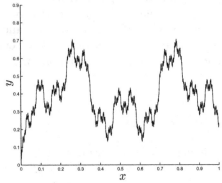

Fig. 83.7. Plots of a truncated Weierstrass function

Since $a > 1$ the series (83.26) is absolutely convergent, and defines a function $f(x) = \sum_{m=1}^{\infty} a^{-m} \sin(b^m x)$, but the series

$$\sum_{m=1}^{\infty} a^{-m} b^m \cos(b^m x)$$

obtained by termwise differentiation, does not converge since $\frac{b}{a} > 1$, which indicates that $f(x)$ is nowhere differentiable. The Weierstrass function, or the corresponding truncated series, is an example of a function with a sequence of "microscales" $\frac{2\pi}{b^m}$, $m = 1, 2, \ldots$ corresponding to the different basis functions $\sin(b^m x)$. The function $f(x)$ thus has the same oscillating nature on all scales and thus has a "fractal" nature. It is believed that phenomena like turbulence also have a fractal nature, which may be useful in attempts to model microscales which are not possible to model numerically.

Choosing $b = 2$ (or b any natural number > 1), gives a series of the form $\sum_{m=1}^{\infty} a^{-m} \sin(2^m x)$, which is an example of a *lacunary Fourier series*, with just very few Fourier coefficients being non-zero. A Weierstrass function with b a natural number is thus a lacunary Fourier series.

83.14 Solving the Heat Equation Using Fourier Series

We consider the 1d homogeneous heat equation:

$$\begin{aligned}
\dot{u}(x,t) - u''(x,t) &= 0 \quad \text{for } 0 < x < \pi,\ t > 0, \\
u(0,t) = u(\pi,t) &= 0 \quad \text{for } t > 0, \\
u(x,0) &= u_0(x) \quad \text{for } 0 < x < \pi,
\end{aligned} \qquad (83.27)$$

where u_0 is a given initial value. We observe that for $m = 1, 2, \ldots$, the function $v(x,t) = v_m(x,t) = e^{-m^2 t} \sin(mx)$ satisfies

$$\dot{v}(x,t) - v''(x,t) = 0 \quad \text{for } 0 < x < \pi, \quad v(0,t) = v(\pi, t) = 0 \quad \text{for } t > 0,$$

and thus any finite linear combination

$$u(x,t) = \sum_{m=1}^{J} b_m e^{-m^2 t} \sin(mx)$$

with coefficients $b_m \in \mathbb{R}$, satisfies (83.27) with corresponding initial data $u_0 = \sum_{m=1}^{J} b_m \sin(mx)$. Each term $e^{-m^2 t} \sin(mx)$ has the form of a product of a function of x only, namely $\sin(mx)$ with *frequency m*, and a function of t only, namely $e^{-m^2 t}$. We see that the factor $e^{-m^2 t}$ decays with increasing t and the rate of decay increases quickly with increasing frequency m. We illustrate this in Fig. 83.8.

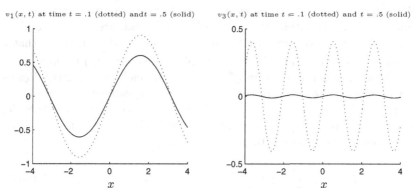

Fig. 83.8. The solutions $v_j(x,t)$ of the heat equation corresponding to frequencies $j = 1$ and $j = 3$

More generally, if the initial data u_0 has a convergent Sine series (with $u_0(x)$ extended as an odd function to $[-\pi, \pi]$)

$$u_0(x) = \sum_{m=1}^{\infty} b_m(u_0) \sin(mx), \tag{83.28}$$

with Fourier coefficients

$$b_m(u_0) = \frac{2}{\pi} \int_0^\pi u_0(x) \sin(mx)\, dx, \tag{83.29}$$

then the function defined by

$$u(x,t) = \sum_{m=1}^{\infty} b_m(u_0) e^{-m^2 t} \sin(mx), \tag{83.30}$$

solves the initial value problem (83.27).

83.15 Computing Fourier Coefficients with Quadrature

To compute the Fourier coefficients

$$c_m(f) = \frac{1}{2\pi} \int_0^{2\pi} f(x)e^{-imx}\, dx, m = 0, \pm 1, \pm 2, \ldots$$

of a given 2π-periodic function $f : \mathbb{R} \to \mathbb{C}$, we will in general have to use quadrature. Using the quadrature points $x_n = \frac{2\pi n}{N}$, $n = 0, \ldots, N-1$, with weights $\omega_n = \frac{2\pi}{N}$, corresponding to a left end-point quadrature formula with N uniformly distributed points, we would approximate $c_m(f)$ by

$$c_m(f) = \frac{1}{2\pi} \int_0^{2\pi} f(x)e^{-imx}\, dx \approx \frac{1}{2\pi} \sum_{n=0}^{N-1} f(x_n)e^{-imx_n}\, \omega_n \equiv \widehat{f}(m),$$

We cannot expect this quadrature formula to be accurate for $m > N$ since then the variation of e^{imx} would not be captured by the quadrature points $\frac{2\pi n}{N}$. We note that $\widehat{f}(m)$ is periodic with period N: that is $\widehat{f}(m) = \widehat{f}(m + N)$, and it is thus natural to consider $\widehat{f}(m)$ for $m = 0, \ldots,$ $N-1$, or equivalently with $|m| \leq (N-1)/2$. We call $(N-1)/2$ the *Nyquist cut-off frequency* which corresponds to at least 2 quadrature points on each period for frequencies m with $|m| \leq (N-1)/2$. According to the inversion formula, we could hope, assuming that the Fourier coefficients $c_m(f)$ are small enough for m larger than cut-off, that

$$f(x_n) \approx \sum_{m=0}^{N-1} \widehat{f}(m)e^{imx_n} \quad \text{for } n = 0, \ldots, N-1, \tag{83.31}$$

which thus would represent an approximate discrete Fourier decomposition for the selected values x_n based on computing the Fourier coefficients $c_m(f)$ by quadrature for $m = 0, \ldots, N-1$. This leads us directly into the *discrete Fourier transform*, which we now discuss.

83.16 The Discrete Fourier Transform

Suppose $\{f_n\}_{n=0}^{N-1}$ is a set of N given complex numbers. We define a corresponding sequence $\{\widehat{f}_m\}_{m=0}^{N-1}$ by

$$\widehat{f}_m = \frac{1}{N} \sum_{n=0}^{N-1} f_n e^{-2\pi imn/N}, \quad \text{for } m = 0, \ldots, N-1.$$

We say that the sequence $\{\widehat{f}_m\}_{m=0}^{N-1}$ is the *discrete Fourier transform* of the sequence $\{f_n\}_{n=0}^{N-1}$. In the setting of the previous section we have $f_n = f(\frac{2\pi n}{N})$ and $\widehat{f}_m \approx c_m(f)$.

We find from the definitions

$$\sum_{m=0}^{N-1} \widehat{f}(m)e^{2\pi imn/N} = \sum_{m=0}^{N-1} \frac{1}{N} \sum_{k=0}^{N-1} f_k e^{-2\pi imk/N} e^{2\pi imn/N}$$

$$= \sum_{k=0}^{N-1} f_k \frac{1}{N} \sum_{m=0}^{N-1} e^{2\pi im(n-k)/N}$$

and using that

$$\frac{1}{N} \sum_{m=0}^{N-1} e^{2\pi im(n-k)/N} = \begin{cases} 1 & \text{if } k = n, \\ 0 & \text{else,} \end{cases}$$

we obtain the following inversion formula, to be compared with (83.31),

$$f_n = \sum_{m=0}^{N-1} \widehat{f}(m)e^{2\pi imn/N}, \quad \text{for } n = 0, \ldots, N-1. \tag{83.32}$$

To compute the discrete Fourier transform of $\{f_n\}_{n=0}^{N-1}$, we would need on the order of N^2 operations (multiplications or additions). If $N = 2^k$ for some natural number k, it is possible to organize the computation of the discrete Fourier transform so that required operations would be of the order N up to a logarithm. The corresponding transform referred to as the *Fast Fourier Transform FFT* developed by Cooley and Tukey in the 1960s, is one of the highlights of applied mathematics of modern time.

Chapter 83 Problems

83.1. Complete the details of the proof of (83.17) and (83.18).

83.2. Prove (83.23).

83.3. Show that the Sine series coefficients for the odd function $f(x) = x^3 - \pi^2 x$ for $-\pi \le x \le \pi$, are given by $b_m(f) = 12\frac{(-1)^m}{m^3}$.

83.4. Show that the Cosine series coefficients for the even function $f(x) = x^4 - 2\pi^2 x^2$ for $-\pi \le x \le \pi$, are given by $a_0 = \frac{14\pi^4}{15}$, $a_m(f) = 48\frac{(-1)^{m+1}}{m^4}$, $m = 1, 2, \ldots$.

83.5. Prove that $\sum_{m=1}^{\infty} \frac{1}{m^4} = \frac{\pi^4}{90}$.

83.6. Define a 2-periodic function $f(x)$ by $f(x) = (x+1)^2$ for $-1 < x < 1$. Expand $f(x)$ in a complex Fourier seriea. Find a 2-periodic solution to the differential equation $2y'' - y' - y = f$.

83.7. Expand the function $\cos x$ as a π-periodic Fourier sine series on the interval $(0, \frac{\pi}{2})$. Use the result to compute $\sum_{n=1}^{\infty} \frac{n^2}{(4n^2-1)^2}$.

83.8. Determine the discrete Fourier transform $\hat{f}_m$ of

$$f_n = \left\{ \begin{array}{ll} 1, & 0 \le n \le k-1, \\ 0, & k \le n \le N-1. \end{array} \right\},$$

and use a Parseval formula to compute

$$\sum_{\mu=1}^{N-1} \frac{1 - \cos \frac{2\pi\mu k}{N}}{1 - \cos \frac{2\pi\mu}{N}}.$$

83.9. Determine the discrete Fourier transform of $f_n = \sin \frac{n\pi}{N}, n = 0, \ldots, N-1$.

84
Fourier Transforms

As the natural ideas of equality developed it was possible to conceive the sublime hope of establishing among us a free government exempt from kings and priests, and to free from this double yoke the long-usurped soil of Europe. I readily became enamoured of this cause, in my opinion the greatest and most beautiful which any nation has ever undertaken. (Fourier 1793, joining a Revolutionary Committee of the French Revolution)

Fourier series concern function $f : \mathbb{R} \to \mathbb{C}$ which are periodic. We now consider functions $f : \mathbb{R} \to \mathbb{C}$ which are non-periodic and the analogous concept is then the *Fourier transform*, which we will study in this chapter. For a given (piecewise Lipschitz continuous) function $f : \mathbb{R} \to \mathbb{C}$ such that $f(x)$ is integrable over $\mathbb{R}$, that is,

$$\int_{\mathbb{R}} |f(x)|\, dx < \infty, \qquad (84.1)$$

we define for $\xi \in \mathbb{R}$

$$\widehat{f}(\xi) = \frac{1}{2\pi} \int_{-\infty}^{\infty} f(x) e^{-i\xi x}\, dx, \qquad (84.2)$$

noting that the integral is absolutely convergent and thus well defined under the assumption (84.1). We say that the function $\widehat{f} : \mathbb{R} \to \mathbb{C}$ defined by (84.2) is the *Fourier transform* of $f(x)$.

We shall now develop a calculus for the Fourier transform which is analogous to that developed for Fourier series in the previous chapter. In particular we shall prove the inversion formula:

$$f(x) = \int_{-\infty}^{\infty} \widehat{f}(\xi) e^{i\xi x} \, d\xi \quad \text{for } x \in \mathbb{R},$$

under the assumption that $f(x)$ is differentiable on $\mathbb{R}$. As we go along the analogy between Fourier series and Fourier transforms will be uncovered.

We compute the Fourier transform of some basic functions.

Example 84.1. If $f(x) = e^{-|x|}$ for $x \in \mathbb{R}$, then

$$\begin{aligned}
\widehat{f}(\xi) &= \frac{1}{2\pi} \int_{-\infty}^{\infty} e^{-|x| - i\xi x} \, dx \\
&= \frac{1}{2\pi} \int_{-\infty}^{0} e^{x - i\xi x} \, dx + \frac{1}{2\pi} \int_{0}^{\infty} e^{-x - i\xi x} \, dx \\
&= \frac{1}{2\pi} \frac{1}{1 - i\xi} + \frac{1}{2\pi} \frac{1}{1 + i\xi} = \frac{1}{\pi} \frac{1}{1 + \xi^2}.
\end{aligned}$$

Example 84.2. If $f(x) = e^{-\frac{ax^2}{2}}$ for $x \in \mathbb{R}$, where $a > 0$ is a constant then

$$\widehat{f}(\xi) = \frac{1}{2\pi} \int_{-\infty}^{\infty} e^{-\frac{ax^2}{2}} e^{-i\xi x} \, dx = \frac{1}{2\pi} e^{-\frac{\xi^2}{2a}} \int_{-\infty}^{\infty} e^{-\frac{1}{2}(\sqrt{a}x + i\frac{\xi}{\sqrt{a}})^2} \, dx.$$

To evaluate the integral, we shall use Cauchy's theorem for analytic functions as follows: We note that the function $g(z) = e^{-\frac{1}{2}(\sqrt{a}z + i\frac{\xi}{\sqrt{a}})^2}$ is analytic in z, and we may thus shift the line of integration to obtain

$$\widehat{f}(\xi) = \frac{1}{2\pi} e^{-\frac{\xi^2}{2a}} \int_{-\infty}^{\infty} e^{-\frac{1}{2}(\sqrt{a}x)^2} \, dx,$$

and recalling (64.25) we thus have

$$\widehat{f}(\xi) = \frac{1}{2\pi} e^{-\frac{\xi^2}{2a}} \int_{-\infty}^{\infty} e^{-\frac{ax^2}{2}} \, dx = \frac{1}{\sqrt{2\pi a}} e^{-\frac{\xi^2}{2a}}.$$

We note that as a tends to zero, the function $f(x)$ tends to 1 for all x, and $\widehat{f}(\xi)$ tends to $\delta(0)$, the delta function at 0, see Fig. 84.1

Example 84.3. Defining $f(x)$ by

$$f(x) = 1 \quad \text{for } -a \le x \le a$$

where $a > 0$, we obtain (see Fig. 84.2)

$$\widehat{f}(\xi) = \frac{1}{2\pi} \int_{-a}^{a} e^{-i\xi x} \, dx = \frac{1}{\pi} \frac{\sin(\xi a)}{\xi}.$$

exp($-ax^2/2$) for $a = 0.2$ (dotted) and $a = 5$ (solid)   Corresponding Fouriertransforms

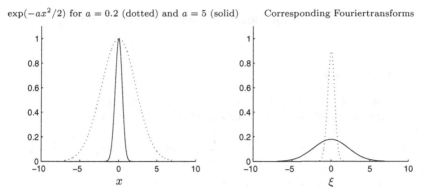

Fig. 84.1. The functions $e^{-\frac{ax^2}{2}}$ and its Fourier transform $\frac{1}{\sqrt{2\pi a}}e^{-\frac{\xi^2}{2a}}$ for different $a > 0$

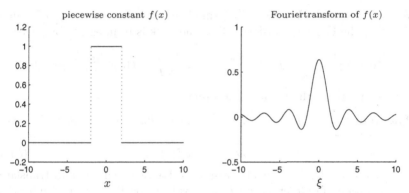

Fig. 84.2. The above piecewise constant function and its Fourier transform $\frac{1}{\pi}\frac{\sin(\xi a)}{\xi}$

84.1   Basic Properties of the Fourier Transform

We now present some basic properties of the Fourier transform

$$\widehat{f}(\xi) = \frac{1}{2\pi}\int_{-\infty}^{\infty} f(x)e^{-i\xi x}\,dx, \quad \xi \in \mathbb{R},$$

of given functions $f : \mathbb{R} \to \mathbb{R}$ which are integrable over $\mathbb{R}$.

Linearity

The Fourier transform satisfies the following obvious linearity properties:

$$\widehat{(f+g)}(\xi) = \widehat{f}(\xi) + \widehat{g}(\xi), \quad \widehat{(\alpha f)}(\xi) = \alpha\widehat{f}(\xi).$$

where f and g are two functions with Fourier transforms $\hat{f}$ and $\hat{g}$, and $\alpha \in \mathbb{C}$.

84.1.1 Scaling

Let $g : \mathbb{R} \to \mathbb{R}$ be integrable and define $f(x) = g(ax)$, where $a > 0$ is a constant. Then, changing variables setting $y = ax$, we have

$$\hat{f}(\xi) = \frac{1}{2\pi} \int_{-\infty}^{\infty} g(ax)e^{-i\frac{\xi}{a}ax}\, dx = \frac{1}{2\pi} \int_{-\infty}^{\infty} g(y)e^{-i\frac{\xi}{a}y}\frac{1}{a}\, dy = \frac{1}{a}\,\hat{g}\left(\frac{\xi}{a}\right).$$

We conclude that if $f(x) = g(ax)$, then $\hat{f}(\xi) = \frac{1}{a}\hat{g}(\frac{\xi}{a})$.

The Fourier Transform of the Derivative $Df = \frac{df}{dx}$

We now couple the Fourier transform of the derivative $Df = \frac{df}{dx}$ of a function f to the Fourier transform of f. The trick is to integrate by parts:

$$\widehat{Df}(\xi) = \frac{1}{2\pi} \int_{-\infty}^{\infty} Df(x)e^{-i\xi x}\, dx = i\xi \frac{1}{2\pi} \int_{-\infty}^{\infty} f(x)e^{-i\xi x}\, dx = i\xi \hat{f}(\xi).$$

We summarize in the following theorem:

Theorem 84.1 *If $f : \mathbb{R} \to \mathbb{C}$ is integrable with integrable derivative Df, then for $\xi \in \mathbb{R}$,*

$$\widehat{Df}(\xi) = i\xi \hat{f}(\xi) \tag{84.3}$$

This is one of the fundamental results of Fourier analysis, and translates the operation of differentiation $D = \frac{d}{dx}$ with respect to x to multiplication of Fourier transforms with $i\xi$ where ξ is the frequency. More generally we have

$$\widehat{D^k f}\xi = (i\xi)^k \hat{f}(\xi). \tag{84.4}$$

This opens the way of translating differential equations in the variable x to algebraic equations in the frequency ξ, which may be very useful and illuminating in certain applications.

Example 84.4. Consider the differential equation $-D^2u(x) + u(x) = f(x)$ on $\mathbb{R}$ with $f(x)$ a given integrable function and seeking an integrable solution $u(x)$. Since $\widehat{D^2u}(\xi) = (i\xi)^2\hat{u}(\xi)$, we obtain the algebraic equation

$$(\xi^2 + 1)\hat{u}(\xi) = \hat{f}(\xi) \quad \text{for } \xi \neq 0$$

and we can thus express the solution $u(x)$ as a Fourier integral

$$u(x) = \int_{\mathbb{R}} \frac{\hat{f}(\xi)}{\xi^2 + 1} e^{i\xi x}\, d\xi,$$

in terms of the Fourier transform $\hat{f}(\xi)$ of the data $f(x)$.

84.2   The Fourier Transform $\widehat{f}(\xi)$ Tends to 0 as $|\xi| \to \infty$

As a direct consequence of the preceding result, we conclude that the Fourier transform $\widehat{f}(\xi)$ of a differentiable function $f(x)$ with integrable derivative $Df(x)$, tends to zero as $|\xi|$ tends to infinity. This is simply because

$$|\widehat{f}(\xi)| = \frac{1}{|\xi|}\widehat{Df}(\xi) \le \frac{1}{2\pi|\xi|}\int_{-\infty}^{\infty}|Df|\,dx \to 0 \quad \text{as } |\xi| \to \infty.$$

This result can be extended to the case of $f(x)$ being integrable, as in the corresponding case of Fourier series.

84.3   Convolution

Given two functions $f : \mathbb{R} \to \mathbb{R}$ and $g : \mathbb{R} \to \mathbb{R}$, we define a new function $f * g : \mathbb{R} \to \mathbb{R}$ by

$$(f * g)(x) = \int_{-\infty}^{\infty} f(x - y)g(y)\,dy.$$

We say that $f * g$ is the *convolution* of f and g. We shall prove that

$$\widehat{f * g} = 2\pi \widehat{f}\,\widehat{g}$$

By direct computation, changing order of integration and using the change of variable $t = x - y$, we have

$$\widehat{f * g}(\xi) = \frac{1}{2\pi}\int_{-\infty}^{\infty}(f * g)(x)e^{-i\xi x}\,dx = \frac{1}{2\pi}\int_{-\infty}^{\infty}\int_{-\infty}^{\infty} f(x - y)g(y)\,dy\,e^{-i\xi x}\,dx$$

$$= \int_{-\infty}^{\infty} g(y)e^{-i\xi y}\left(\frac{1}{2\pi}\int_{-\infty}^{\infty} f(x - y)\,e^{-i\xi(x-y)}\,dx\right)dy$$

$$= \int_{-\infty}^{\infty} g(y)e^{-i\xi y}\left(\frac{1}{2\pi}\int_{-\infty}^{\infty} f(t)\,e^{-i\xi t}\,dt\right)dy$$

$$= \widehat{f}(\xi)\int_{-\infty}^{\infty} g(y)e^{-i\xi y}\,dy = 2\pi\widehat{f}(\xi)\widehat{g}(\xi).$$

We summarize:

Theorem 84.2 *We have $\widehat{f * g}(\xi) = 2\pi\widehat{f}(\xi)\widehat{g}(\xi)$ for $\xi \in \mathbb{R}$.*

84.4   The Inversion Formula

We shall now prove that if $f(x)$ is differentiable, then for all $x \in \mathbb{R}$,

$$\lim_{n\to\infty} f_n(x) = f(x),$$

where

$$f_n(x) = \int_{-n}^{n} \widehat{f}(\xi) e^{i\xi x}\, d\xi$$

and thus for all $x \in \mathbb{R}$, the function $f(x)$ can be represented as a convergent Fourier integral:

$$f(x) = \int_{-\infty}^{\infty} \widehat{f}(\xi) e^{i\xi x}\, d\xi.$$

We have

$$\begin{aligned} f_n(x) = \int_{-n}^{n} \widehat{f}(\xi) e^{i\xi x}\, d\xi &= \frac{1}{2\pi} \int_{-\infty}^{\infty} f(y) \int_{-n}^{n} e^{i\xi(x-y)}\, d\xi dy \\ &= \int_{-\infty}^{\infty} f(y) D_n(x-y)\, dy, \end{aligned} \tag{84.5}$$

where, setting $\theta = x - y$,

$$D_n(\theta) = \frac{1}{2\pi} \int_{-n}^{n} e^{i\xi\theta}\, d\xi = \frac{1}{\pi} \frac{\sin(n\theta)}{\theta}$$

is the *Dirichlet kernel* for the Fourier transform. Using the convolution notation we can write (84.5) in the compact form

$$f_n(x) = f * D_n(x).$$

With experience from the Dirichlet kernel for Fourier series, we expect D_n to be an approximate identity. Looking at a plot of $D_n(\theta)$ in Fig. 84.3, we see that $D_n(\theta)$ oscillates with a peak at $\theta = 0$, which gets sharper with increasing n. One can prove that, see Problem 84.5,

$$\int_{-\infty}^{\infty} D_n(\theta)\, d\theta = 1 \tag{84.6}$$

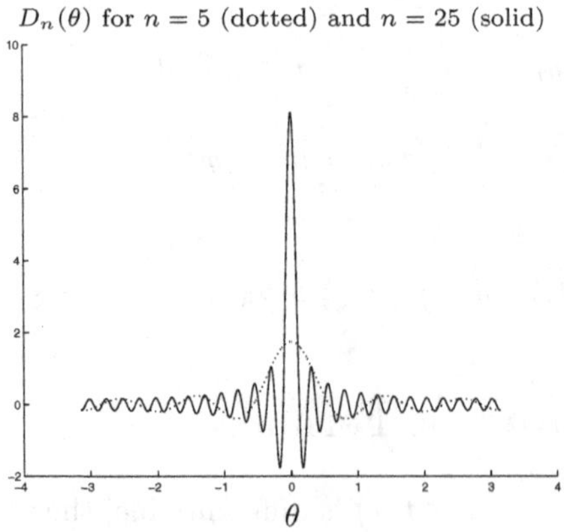
$D_n(\theta)$ for $n = 5$ (dotted) and $n = 25$ (solid)
10
8
6
4
2
0
-2
-4
-3
-2
-1
0
1
2
3
4
θ

Fig. 84.3. The function $D_n(\theta)$

and we can thus write

$$f(x) - f_n(x) = \int_{-\infty}^{\infty} (f(x) - f(y)) D_n(x-y)\, dy$$

$$= \frac{1}{\pi} \int_{-\infty}^{\infty} g(y) \sin((n(x-y))\, dy$$

where

$$g(y) = \frac{f(x) - f(y)}{x - y}.$$

Now if $f(x)$ is differentiable with integrable derivative, it follows with an argument similar to that used in the case of Fourier series (integrating by parts), that

$$f(x) - f_n(x) \to 0$$

as $n \to \infty$, which proves the following basic theorem.

Theorem 84.3 *If $f(x)$ is differentiable, then $f(x)$ is given by a convergent Fourier integral:*

$$f(x) = \int_{-\infty}^{\infty} \widehat{f}(\xi) e^{i\xi x}\, d\xi \quad \text{for } x \in \mathbb{R}.$$

84.5   Parseval's Formula

Parseval's formula takes the following form for the Fourier transform (for a proof see Problem 84.3):

Theorem 84.4 *If $f(x)$ has a convergent Fourier series representation, then*

$$\int_{-\infty}^{\infty} |f(x)|^2\, dx = 2\pi \int_{-\infty}^{\infty} |\widehat{f}(\xi)|^2\, d\xi.$$

84.6   Solving the Heat Equation Using the Fourier Transform

We consider the 1d homogeneous heat equation on $\mathbb{R}$

$$\dot{u}(x,t) - u''(x,t) = 0 \quad \text{for } x \in \mathbb{R},\, t > 0,$$
$$u(x,0) = u_0(x) \quad \text{for } x \in \mathbb{R}, \tag{84.7}$$

where the initial value $u_0(x)$ is integrable over $\mathbb{R}$ and we seek a solution $u(x, t)$ which is integrable over $\mathbb{R}$ for all $t > 0$. Taking Fourier transforms with respect to x, we are led to the following initial value problem for each $\xi \in \mathbb{R}$

$$\frac{d}{dt}\widehat{u}(\xi, t) + \xi^2 \widehat{u}(\xi, t) = 0 \quad \text{for } t > 0, \quad \widehat{u}(\xi, 0) = \widehat{u}_0(\xi)$$

with solution

$$\widehat{u}(\xi, t) = e^{-t\xi^2}\widehat{u}_0(\xi).$$

We thus obtain the following solution formula

$$u(x, t) = \frac{1}{\sqrt{4\pi t}} \int_{-\infty}^{\infty} e^{-\frac{(x-y)^2}{4t}} u_0(y)\, dy.$$

Here we used that $\widehat{u}$ is the product of the Fourier transforms $e^{-t\xi^2}$ of $\sqrt{\pi/t}\, e^{-\frac{x^2}{4t}}$ and $\widehat{u}_0$ of u_0, the inverse transform of which thus is the convolution of $\frac{1}{4\pi t}e^{-\frac{x^2}{4t}}$ and u_0.

84.7 Fourier Series and Fourier Transforms

Suppose $f : \mathbb{R} \to \mathbb{C}$ is periodic with period $\frac{2\pi}{\omega}$ with $\omega > 0$ and has the Fourier series representation

$$f(x) = \sum_{m=-\infty}^{\infty} c_m(f)e^{im\omega x}, \quad c_m(f) = \frac{\omega}{2\pi} \int_{-\frac{\pi}{\omega}}^{\frac{\pi}{\omega}} f(y)e^{-im\omega y}\, dy,$$

which we write in the form

$$f(x) = \sum_{m=-\infty}^{\infty} \frac{1}{2\pi} \left(\int_{-\frac{\pi}{\omega}}^{\frac{\pi}{\omega}} f(y)e^{-im\omega y}\, dy \right) e^{im\omega x}\omega. \tag{84.8}$$

We now compare with a Fourier transform representation of a non-periodic function $f : \mathbb{R} \to \mathbb{C}$ according to the previous section:

$$f(x) = \int_{-\infty}^{\infty} \frac{1}{2\pi} \left(\int_{-\infty}^{\infty} f(y)e^{-i\xi y}\, dy \right) e^{i\xi x}\, d\xi. \tag{84.9}$$

We formally obtain (84.9) from (84.8) by replacing $m\omega$ by ξ and ω by $d\xi$ viewing the sum over m as a Riemann sum and letting ω tend to 0.

Note the normalization used in the definition of the Fourier transform $\widehat{f}(\xi)$ with the factor $\frac{1}{2\pi}$, and the factor $\frac{\omega}{2\pi}$ in the definition of the Fourier coefficients $c_m(f)$ of a function f with period $\frac{2\pi}{\omega}$.

84.8   The Sampling Theorem

Let $f : \mathbb{R} \to \mathbb{C}$ be a given function with Fourier transform $\widehat{f}(\xi)$, and suppose that $\hat{f}(\xi) = 0$ for $|\xi| \geq \pi$. By Fourier's inversion formula, we have

$$f(x) = \int_{-\pi}^{\pi} \widehat{f}(\xi) e^{ix\xi} \, d\xi.$$

We now expand $\widehat{f}(\xi)$ in a Fourier series:

$$\widehat{f}(\xi) = \sum_{m=-\infty}^{\infty} c_m(\widehat{f}) e^{im\xi}$$

with Fourier coefficients

$$c_m(\widehat{f}) = \frac{1}{2\pi} \int_{-\pi}^{\pi} \widehat{f}(\eta) e^{-im\eta} \, d\eta.$$

Using that $\widehat{f}(\eta) = 0$ for $|\eta| \geq \pi$, we can write

$$c_m(\widehat{f}) = \frac{1}{2\pi} \int_{-\infty}^{\infty} \widehat{f}(\eta) e^{-im\eta} \, d\eta = \frac{1}{2\pi} f(-m)$$

where we used Fourier's inversion formula. We thus obtain the representation formula:

$$f(x) = \int_{-\pi}^{\pi} \frac{1}{2\pi} \sum_{m=-\infty}^{\infty} f(-m) e^{im\xi} e^{ix\xi} \, d\xi$$

$$= \frac{1}{2\pi} \sum_{m=-\infty}^{\infty} f(-m) \int_{-\pi}^{\pi} e^{im\xi} e^{ix\xi} \, d\xi$$

$$= \sum_{m=-\infty}^{\infty} f(-m) \frac{\sin(x+m)}{\pi(x+m)} = \sum_{m=-\infty}^{\infty} f(m) \frac{\sin(x-m)}{\pi(x-m)}$$

which gives a representation of $f(x)$ for any value of x in terms of the values $\{f(m)\}$ with m integer. We have now prove the famous:

Theorem 84.5 (Sampling theorem) *If $f : \mathbb{R} \to \mathbb{C}$ has a Fourier transform $\widehat{f}(\xi)$ such that $\hat{f}(\xi) = 0$ for $|\xi| \geq \pi$, then*

$$f(x) = \sum_{m=-\infty}^{\infty} f(m) \frac{\sin(x-m)}{\pi(x-m)}$$

We conclude that *sampling* the values $f(m)$ of the function $f(x)$ for the integer values $m = 0, \pm 1, \pm 2, \ldots$, gives information of all the values of $f(x)$ for any $x \in \mathbb{R}$, under the assumption that $\hat{f}(\xi) = 0$ for $|\xi| \geq \pi$.

Example 84.5. The Sampling theorem takes the following form for a function $f(x)$ such that $\hat{f}(\xi) = 0$ for $|\xi| \geq a\pi$, where $a > 0$ is a constant:

$$f(x) = \sum_{m=-\infty}^{\infty} f\left(\frac{m}{a}\right) \frac{\sin(ax - m)}{\pi(ax - m)}.$$

This follows by applying the Sampling theorem to $g(x) = f(\frac{x}{a})$, recalling that $\hat{g}(\xi) = a\hat{f}(a\xi)$ and noting that $\hat{g}(\xi) = 0$ if $|\xi| \geq \pi$ since $\hat{f}(\xi) = 0$ for $|\xi| \geq a\pi$. We see that the larger the factor a gets, the closer the sampling points $\frac{m}{a}$ will be distributed. Of course this couples to the Nyquist cut-off frequency.

84.9   The Laplace Transform

We give a brief account of the *Laplace transform*, which is closely related to the Fourier transform. The Laplace transform is useful in solving certain constant-coefficient linear initial value problems analytically with classical applications in e.g. control theory.

For a given function $f : [0, \infty) \rightarrow \mathbb{R}$, we define the *Laplace transform* $Lf : [0, \infty)$ by

$$Lf(s) = \int_0^\infty e^{-st} f(t)\, dt \quad \text{for } s \in [0, \infty).$$

We denote here the independent variable by t indicating typical applications with t representing time.

Example 84.6. If $f(t) = e^{-at}$, then $Lf(s) = \frac{1}{s+a}$.

Example 84.7. If $f(t) = \frac{t^n}{n!}$ then $Lf(s) = \frac{1}{s^{n+1}}$. This follows by repeated integration by parts.

Example 84.8. If $f(t) = \sin(mt)$ then $Lf(s) = \frac{m}{m^2+s^2}$. If $f(t) = \cos(mt)$ then $Lf(s) = \frac{s}{m^2+s^2}$.

We note the following connection between the Laplace transform of $Df = f'$ and f:

$$Lf'(s) = sLf(s) - f(0) \tag{84.10}$$

which follows by integration by parts.

Laplace Transforms and Constant-Coefficient Linear Initial Value Problems

The typical application goes as follows: Consider the initial value problem $u'(t) + u(t) = f(t)$ for $t > 0$ with $u(0) = 0$. Taking Laplace transforms of both sides we get

$$sLu(s) + Lu(s) = Lf(s), \quad \text{or } Lu(s) = \frac{Lf(s)}{s+1}$$

For example, if $f(t) = 1$, then $Lf(s) = \frac{1}{s}$ and thus $Lu(s) = \frac{1}{s(s+1)} = \frac{1}{s} - \frac{1}{s+1}$ and we conclude that $u(s) = 1 - e^{-t}$. Having a catalogue of Laplace transforms we may expect to be able to solve constant-coefficient linear initial value problems.

84.10   Wavelets and the Haar Basis

We give a short introduction to wavelets in the simplest setting of 1d piece-wise constant approximation using the *Haar basis*, which combines the features of orthogonality and local support. We thus consider functions defined on the unit interval $[0, 1]$ and we let $0 = x_0 < x_1 < \ldots < x_N = 1$ be a uniform subdivision with $x_j = jh_n$, $h_n = 2^{-n}$ and $N = 2^n$ for some natural number n. A natural orthogonal basis for the space V_n of piecewise constant functions on the subdivision $0 = x_0 < x_1 < \ldots < x_N = 1$ consists of the set of functions $\{\varphi_{n,k}\}_{k=0}^{N}$, where $\varphi_{n,k}(x) = 1$ for $x \in I_{n,k} = (kh_n, (k+1)h_n)$ and $\varphi_{n,k}(x) = 0$ else, that is, each basis function $\varphi_{n,k}(x)$ is equal to 1 on the subinterval $I_{n,k}$ and vanishes elsewhere. We can express these functions through scaling and translation of one single function in the form

$$\varphi_{n,k} = \varphi(2^n x - k) \quad \text{for}, \quad k = 0, \ldots, N - 1,$$

where $\varphi(x) = 1$ for $x \in (0, 1)$, and $\varphi(x) = 0$ else. We note that V_{n-1} is a subspace of V_n since the space V_n is built on a finer subdivision than V_{n-1}.

We shall now present a different orthogonal basis for V_n which displays the "difference" between V_n and V_{n-1}, and which carries useful information on the various scales in V_n. More precisely, we shall express each $u \in V_n$ in the form $u = v + w$ with $v \in V_{n-1}$ and $w \in W_{n-1}$, where W_{n-1} is spanned by the functions $\psi_{n-1,k} = \psi(2^{n-1}x - k)$ for $k = 1, \ldots, 2^{n-1}$, expressed through scaling and translation of the single function $\psi(x)$ given by

$$\psi(x) = \begin{cases} 1 & \text{for } 0 < x < \frac{1}{2}, \\ -1 & \text{for } \frac{1}{2} < x < 1, \end{cases}$$

and $\psi(x) = 0$ else. We note that $(v, w) = \int_0^1 v(x)w(x)\, dx = 0$ if $v \in V_{n-1}$ and $w \in W_{n-1}$. Further, the two functions $\varphi_{n-1,k}$ and $\psi_{n-1,k}$ obviously span the two-dimensional space of functions on the interval $I_{n,k}$ which are piecewise constant on the two subintervals $kh_n < x < kh_n + h_{n+1}$ and $kh_n + h_{n+1} < x < kh_n + h_n$ of $I_{n,k}$. We thus have the following orthogonal decomposition

$$V_n = V_{n-1} \oplus W_{n-1},$$

stating that each function $u \in V_n$ can be expressed in the form $u = v + w$ with $v \in V_{n-1}$, $w \in W_{n-1}$ and $(v, w) = 0$, see Fig. 84.4.

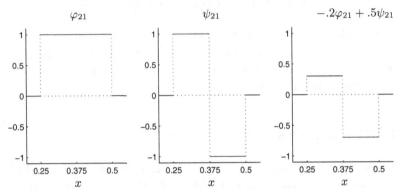

Fig. 84.4. Illustration of the orthogonal decomposition $V_n = V_{n-1} \oplus W_{n-1}$

We can thus express V_n as an orthogonal sum:

$$V_n = V_0 \oplus W_0 \oplus W_1 \oplus \ldots \oplus W_{n-1}$$

where each space $\oplus W_j$ measures variations on the scale 2^{-j}. The corresponding basis functions comprise the so-called *Haar basis* for V_n:

$$\{\varphi_0 = \varphi, \psi_{1,1} = \psi, \psi_{2,1}, \psi_{2,2}, \psi_{3,1}, \psi_{3,2}, \psi_{3,3}, \psi_{3,4}, \ldots, \psi_{n-1,1}, \ldots, \psi_{n-1,2^{n-1}}\}$$

combining orthogonality and local support.

Chapter 84   Problems

84.1. Solve using Fourier series the differential equation $-D^2 u(x) = f(x)$ with $f(x)$ a given 2π-periodic function with zero mean value and we seek a 2π-periodic solution $u(x)$ with zero mean value.

84.2. Model the following electrical circuits: (i) resistor 1 and inductor in series over applied voltage (ii) resistor 1 and capacitor in series over applied voltage (iii) resistor 2 coupled in series with resistor 1 and inductor in parallel over applied voltage. Output voltage drop over resistor 1. Solve using Fourier series. Show that (i) and (ii) correspond to low-pass filters and (iii) to high-pass filter.

84.3. Prove Parseval's formula for the Fourier transform. Hint: Set $\hat{g}(\xi) = \overline{\hat{f}(\xi)}$, which is the same as setting $g(-x) = \overline{f}(x)$, and integrate over ξ:

$$\int_{-\infty}^{\infty} f(x)\overline{f}(x)\, dx = f * g(0) = \int_{-\infty}^{\infty} \widehat{f * g}(\xi)\, d\xi = 2\pi \int_{-\infty}^{\infty} |\hat{f}(\xi)|^2\, d\xi,$$

and use Theorem 84.2.

84.4. Prove that for $a \in \mathbb{R}$, we have (i) $\hat{g}(\xi) = e^{-ia\xi}\hat{f}(\xi)$ if $g(x) = f(x-a)$, (ii) $\hat{g}(\xi) = \hat{f}(\xi - a)$ if $g(x) = e^{iax}f(x)$.

84.5. Prove (84.6).

84.6. Compute the Fourier transform of the functions a) $\frac{x}{(x^2+a^2)^2}$, b) $\frac{1}{(x^2+a^2)^2}$, c) $\frac{x}{(x^2+1)(x^2+2x+5)}$, d) $e^{-a|x|}\sin xt$ $(a > 0, b > 0)$.

84.7. The function $f(x)$ has the Fourier transform $\frac{1-i\xi}{1+i\xi}\frac{\sin\xi}{\xi}$. Compute $\int_{-\infty}^{\infty}|f(x)|^2 dx$.

84.8. Compute $\int_{-\infty}^{\infty}\frac{\sin x}{x(x^2+1)}dx$ using the Fourier transform.

84.9. A function $f(x)$ has the Fourier transform $\frac{1}{|\xi|^3+1}$. Compute $\int_{-\infty}^{\infty}|f * f'|^2 dx$.

84.10. Compute the Fourier transform of the function $f(x) = \int_0^2 \frac{\sqrt{\xi}}{1+\xi}e^{i\xi x}d\xi$. Then compute a) $\int_{-\infty}^{\infty} f(x)\cos x dx$, b) $\int_{-\infty}^{\infty}|f(x)|^2 dx$.

84.11. Determine the solution $f(t)$, $t > 0$, to the initial value problem

$$f''(t) - f'(t) + f(t) + 6\int_0^t f(\tau)d\tau = 2e^t \quad \text{for } t > 0,$$

with initial values $f(0) = 1$, $f'(0) = 0$.

85

Analytic Functions Tool Bag

85.1 Differentiability and Analyticity

A function $f : \Omega \to \mathbb{C}$ is *differentiable* at $z_0 \in \Omega$ with derivative $f'(z_0) \in \mathbb{C}$, if for z close to z_0, we have

$$|f(z) - f(z_0) - f'(z_0)(z - z_0)| \le K_f(z_0)|z - z_0|^2,$$

where $K_f(z_0)$ is a non-negative real constant depending on f and z_0.

A function $f : \Omega \to \mathbb{C}$ is *analytic* in the open domain Ω of the complex plane if $f(z)$ is differentiable at all $z_0 \in \Omega$ with derivative $f'(z_0)$. If $f : \Omega \to \mathbb{C}$ is analytic, then also $f' : \Omega \to \mathbb{C}$ is analytic with derivative $f'' : \Omega \to \mathbb{C}$, which is also analytic, and so on. An analytic function $f : \Omega \to \mathbb{C}$ thus has derivatives of all orders $f^{(n)} : \Omega \to \mathbb{C}$, $n = 1, 2, \ldots$, which are all analytic.

The usual rules for differentiation of sums, products and quotients valid for functions $f : \mathbb{R} \to \mathbb{R}$ extend to functions $f : \mathbb{C} \to \mathbb{C}$.

The function $f(z) = z^n$ is analytic in $\mathbb{C}$ for $n = 1, 2, \ldots$

The function $f(z) = z^{-n}$ is analytic for $z \ne 0$ if $n = 1, 2, \ldots$.

85.2 The Cauchy-Riemann Equations

If $f(z) = u(x, y) + iv(x, y)$ is analytic in the open domain Ω of the complex plane, then the real and imaginary parts $u(x, y)$ and $v(x, y)$ satisfy the

Cauchy-Riemann equations in Ω:

$$\frac{\partial u}{\partial x} = \frac{\partial v}{\partial y}, \qquad \frac{\partial u}{\partial y} = -\frac{\partial v}{\partial x},$$

or in polar coordinates $z = re^{i\theta}$:

$$\frac{\partial u}{\partial r} = \frac{1}{r}\frac{\partial v}{\partial \theta}, \qquad \frac{\partial v}{\partial r} = -\frac{1}{r}\frac{\partial u}{\partial \theta}.$$

85.3   The Real and Imaginary Parts of an Analytic Function

If $f : \Omega \to \mathbb{C}$ is analytic, where Ω is an open domain of the complex plane $\mathbb{C}$, then the real part $u(x,y) = \operatorname{Re} f(z)$ and the imaginary part $v(x,y) = \operatorname{Im} f(z)$ are harmonic in Ω.

85.4   Conjugate Harmonic Functions

If $u(x,y)$ is harmonic in a simply connected domain Ω in $\mathbb{R}^2$, then there exists a harmonic function $v(x,y)$, uniquely determined up to a constant, such that $f(z) = u(x,y) + iv(x,y)$ is analytic in Ω. The function $v(x,y)$ is *conjugate* to $u(x,y)$.

85.5   Curves in the Complex Plane

A set $\Gamma = \text{Range of } \gamma = \{\gamma(t) : t \in I\}$ in an open domain Ω in the complex plane $\mathbb{C}$ parameterized by a Lipschitz continuous mapping $\gamma : I \to \Omega$, where $I = [a,b]$ is an interval of $\mathbb{R}$, is said to be a curve. The unit circle is a curve parameterized by the function $\gamma(t) = \exp(it)$ with $0 \le t < 2\pi$. Γ is a *differentiable curve* if the corresponding parametrization $\gamma : I \to \mathbb{C}$ is differentiable on I, that is, decomposing $\gamma(t) = x(t) + iy(t)$ into real and imaginary parts,that is, if $x : I \to \mathbb{R}$ if $x(t)$ and $y(t)$ are differentiable on I.

A curve Γ with parametrization $\gamma : [a,b] \to \mathbb{C}$ is said to *closed and simple* if $\gamma(a) = \gamma(b)$ and $\gamma(s) = \gamma(t)$ only if $a = b$. A domain Ω in $\mathbb{C}$ which is bounded by a simple closed curve, is *simply connected*. A simply connected domain does not have any "holes".

85.6 An Analytic Function Defines a Conformal Mapping

An analytic function $f : \Omega \to \mathbb{C}$, where Ω is an open domain in $\mathbb{C}$, is conformal in Ω in the sense that angles are preserved under the mapping $w = f(z)$.

85.7 Complex Integrals

We define

$$\int_\Gamma f(z)\, dz = \int_a^b \big(u(x(t), y(t)) + iv(x(t), y(t))\big)(\dot{x}(t) + i\dot{y}(t))\, dt,$$

where Ω is an open domain in the complex plane, Γ is a differentiable curve in $\mathbb{C}$ parameterized by $\gamma = (x, y) : [a, b] \to \mathbb{C}$, and $f = u + iv : \Gamma \to \mathbb{C}$ is Lipschitz continuous. Formally we have $dz = dx + idy = \dot{x}dt + i\dot{y}dt = (\dot{x} + i\dot{y})\, dt$.

85.8 Cauchy's Theorem

If $f(z)$ is analytic in Ω and Γ is a simple closed curve in Ω enclosing a domain contained in Ω, then

$$\int_\Gamma f(z)\, dz = 0.$$

85.9 Cauchy's Representation Formula

If $f(z)$ is analytic in an open domain Ω, and Γ is a simple closed curve in Ω oriented counter-clockwise and enclosing the open domain Ω_Γ contained in Ω, then for $z_0 \in \Omega_\Gamma$,

$$f(z_0) = \frac{1}{2\pi i} \int_\Gamma \frac{f(z)}{z - z_0}\, dz,$$

and for $n = 1, 2, \ldots,$

$$f^{(n)}(z_0) = \frac{n!}{2\pi i} \int_\Gamma \frac{f(z)}{(z - z_0)^{n+1}}\, dz.$$

85.10   Taylor's Formula

If $f(z)$ is analytic in a neighborhood Ω of a $z_0 \in \mathbb{C}$, then

$$f(z) = f(z_0) + f'(z_0)(z - z_0) + \ldots + \frac{f^{(n)}(z_0)}{n!}(z - z_0)^n + R_n(z),$$

where

$$R_n(z) = \frac{(z - z_0)^{n+1}}{2\pi i} \int_\Gamma \frac{f(\zeta)}{(\zeta - z_0)^{n+1}(\zeta - z)} \, d\zeta.$$

85.11   The Residue Theorem

If $f(z)$ is analytic in a simply connected open domain Ω, except at finitely many isolated points $z_1, z_1, \ldots, z_n$ in Ω, where $f(z)$ has simple or multiple poles, and Γ is a simple closed curve in Ω surrounding all the z_m counterclockwise, then

$$\int_\Gamma f(z) \, dz = \sum_{m=1}^{n} 2\pi i \operatorname{Res} f(z_m).$$

86

Fourier Analysis Tool Bag

86.1 Properties of Fourier Coefficients

The Fourier coefficients $c_m(f)$ of a given 2π-periodic Lipschitz function $f : \mathbb{R} \to \mathbb{C}$ are defined by

$$c_m(f) = \frac{1}{2\pi} \int_{-\pi}^{\pi} f(x)e^{-imx}\, dx \quad m = 0, \pm 1, \pm 2, \ldots,$$

and satisfy

$$c_m(f + g) = c_m(f) + c_m(g), \quad c_m(\alpha f) = \alpha c_m(f), \quad \text{for } \alpha \in \mathbb{C},$$
$$c_m(D^k f) = (im)^k c_m(f) \quad \text{for } k = 0, 1, 2, \ldots,$$

If $f : [-\pi, \pi]$ is Lipschitz continuous, then $c_m(f) \to 0$ as $|m| \to \infty$ (Riemann-Lebesgue' Lemma).

86.2 Convolution

Defining for 2π-periodic functions $f(x)$ and $g(x)$, the convolution $f * g$ by

$$(f * g)(x) = \int_{-\pi}^{\pi} f(x - y)g(y)\, dy \quad x \in \mathbb{R},$$

we have

$$c_m(f * g) = 2\pi\, c_m(f)c_m(g).$$

86.3   Fourier Series Representation

If $f : \mathbb{R} \to \mathbb{C}$ is 2π-periodic with piecewise Lipschitz continuous derivative, then $f(x)$ may be represented by a convergent Fourier series:

$$f(x) = \sum_{m=-\infty}^{\infty} c_m(f)e^{imx} \quad \text{for } x \in \mathbb{R}.$$

86.4   Parseval's Formula

If $f(x)$ has a convergent Fourier series representation, then

$$\int_{-\pi}^{\pi} |f(x)|^2 \, dx = 2\pi \sum_{m=-\infty}^{\infty} |c_m(f)|^2.$$

86.5   Discrete Fourier Transforms

If $\{f_n\}_{n=0}^{N-1}$ is a sequence of N given complex numbers, then we may define a corresponding sequence $\{\widehat{f}_m\}_{m=0}^{N-1}$ by

$$\widehat{f}_m = \frac{1}{N} \sum_{n=0}^{N-1} f_n e^{-2\pi imn/N}, \quad \text{for } m = 0, \ldots, N-1,$$

and we say that the sequence $\{\widehat{f}_m\}_{m=0}^{N-1}$ is the *discrete Fourier transform* of the sequence $\{f_n\}_{n=0}^{N-1}$. We have the following inversion formula:

$$f_n = \sum_{m=0}^{N-1} \widehat{f}(m)e^{2\pi imn/N}, \quad \text{for } n = 0, \ldots, N-1.$$

86.6   Fourier Transforms

For $f : \mathbb{R} \to \mathbb{C}$ piecewise Lipschitz continuous and integrable over $\mathbb{R}$, we define the Fourier transform of $f(x)$ for $\xi \in \mathbb{R}$ by

$$\widehat{f}(\xi) = \frac{1}{2\pi} \int_{-\infty}^{\infty} f(x)e^{-i\xi x} \, dx.$$

We have the inversion formula:

$$f(x) = \int_{-\infty}^{\infty} \widehat{f}(\xi)e^{i\xi x} \, d\xi \quad \text{for } x \in \mathbb{R},$$

under the assumption that $f(x)$ is differentiable on $\mathbb{R}$ with integrable derivative. If $f(x) = e^{-|x|}$ for $x \in \mathbb{R}$, then

$$\widehat{f}(\xi) = \frac{1}{\pi}\frac{1}{1+\xi^2}.$$

If $f(x) = e^{-\frac{ax^2}{2}}$ for $x \in \mathbb{R}$, where $a > 0$ is a constant, then

$$\widehat{f}(\xi) = \frac{1}{2\sqrt{a}}e^{-\frac{\xi^2}{2a}}.$$

If $f(x) = 1$ for $-a \le x \le a$ and $f(x) = 0$ else, where $a > 0$, then

$$\widehat{f}(\xi) = \frac{\sin(\xi a)}{\xi}.$$

86.7 Properties of Fourier Transforms

If f and g are two functions with Fourier transforms $\widehat{f}$ and $\widehat{g}$, and $\alpha \in \mathbb{C}$, then

$$\widehat{(f+g)}(\xi) = \widehat{f}(\xi) + \widehat{g}(\xi), \quad \widehat{(\alpha f)}(\xi) = \alpha\widehat{f}(\xi).$$

If $g : \mathbb{R} \to \mathbb{R}$ is integrable and $f(x) = g(ax)$, then $\widehat{f}(\xi) = \frac{1}{a}\widehat{g}(\frac{\xi}{a})$.
If $f : \mathbb{R} \to \mathbb{C}$ is integrable with integrable derivative, then

$$\widehat{Df}(\xi) = i\xi\widehat{f}(\xi).$$

Defining for two integrable functions $f : \mathbb{R} \to \mathbb{R}$ and $g : \mathbb{R} \to \mathbb{R}$, the convolution $f * g$ by

$$(f * g)(x) = \int_{-\infty}^{\infty} f(x-y)g(y)\,dy,$$

we have

$$\widehat{f * g}(\xi) = 2\pi\widehat{f}(\xi)\widehat{g}(\xi) \quad \text{for } \xi \in \mathbb{R}.$$

Parseval's formula:

$$\int_{-\infty}^{\infty} |f(x)|^2\,dx = 2\pi\int_{-\infty}^{\infty} |\widehat{f}(\xi)|^2\,d\xi.$$

86.8 The Sampling Theorem

If $f : \mathbb{R} \to \mathbb{C}$ has a Fourier transform $\widehat{f}(\xi)$ such that $\widehat{f}(\xi) = 0$ for $|\xi| \ge a\pi$, where $a > 0$ is a constant, then

$$f(x) = \sum_{m=-\infty}^{\infty} f\left(\frac{m}{a}\right)\frac{\sin(ax-m)}{\pi(ax-m)}.$$

87

Incompressible Navier-Stokes: Quick and Easy

My attention was drawn to various mechanical phenomena, for the explanation of which I discovered that a knowledge of mathematics was essential. (Reynolds)

By this research it is shown that there is one, and only one, conceivable purely mechanical system capable of accounting for all the physical evidence, as we know it in the Universe. (Reynolds)

87.1   Introduction

The Navier-Stokes equations is the basic model for fluid flow and describe a variety of phenomena in hydro and aero-dynamics, processing industry, biology, oceanography, geophysics, meteorology and astrophysics. Fluid flow in all these applications usually contains features of both *turbulent* and *laminar* flow, with turbulent flow being irregular with rapid fluctuations in space and time and laminar flow being more organized. The basic question of *Computational Fluid Dynamics* CFD is how to efficiently and reliably solve the Navier-Stokes equations numerically for both laminar and turbulent flow.

The Navier-Stokes equations is a system of nonlinear differential equations coupling the phenomena of convection and diffusion. Traditionally, the study of the Navier-Stokes equations is separated into *incompressible* and *compressible* flow, using different dependent variables: *primitive variables* (velocity, pressure, temperature) for incompressible flow and *conservation variables* (density, momentum, energy) for compressible flow.

We focus in this chapter on the incompressible Navier-Stokes equations in the case of constant density, viscosity and temperature, with the velocity and pressure as variables. We present the cG(1)dG(0) finite element method with cG(1) in space and dG(0) in time, and follow up with the corresponding cG(1)dG(1) and cG(1)cG(1) methods. In Fig. 87.2 and Fig. 87.3 below we show results from computations of two time-dependent bench-marks: flow around a bluff body and flow in a channel with a back-ward facing step.

87.2   The Incompressible Navier-Stokes Equations

The Navier-Stokes equations for an incompressible Newtonian fluid with constant kinematic viscosity $\nu > 0$, unit density and constant temperature enclosed in a volume Ω in $\mathbb{R}^3$ with boundary Γ, take the form: find the velocity/pressure (u, p) such that

$$
\begin{aligned}
\frac{\partial u}{\partial t} + (u \cdot \nabla)u - \nu \Delta u + \nabla p &= f & \text{in } \Omega \times I, \\
\nabla \cdot u &= 0 & \text{in } \Omega \times I, \\
u &= w & \text{on } \Gamma \times I, \\
u(\cdot, 0) &= u^0 & \text{in } \Omega,
\end{aligned}
\tag{87.1}
$$

where $u = (u_1, u_2, u_3)$ is the velocity and p the pressure of the fluid and f, w, u^0, $I = (0, T)$, is a given driving force, boundary data, initial data and time interval, respectively. Recall that

$$
\frac{\partial v}{\partial t} + (u \cdot \nabla)v = \frac{\partial v}{\partial t} + \sum_{i=1}^{3} u_i \frac{\partial v}{\partial x_i}
\tag{87.2}
$$

is the *particle derivative* of a quantity $v(x, t)$ measuring the rate of change of $v(x(t), t)$ with respect to time, that is the rate of change of v along a trajectory $x(t)$ of a fluid particle with velocity $u(x, t)$, satisfying $\frac{dx}{dt} = u(x(t), t)$. In particular, $\frac{\partial u}{\partial t} + (u \cdot \nabla)u$ is the acceleration (rate of change of velocity) of a fluid particle. The expression $\nu \Delta u - \nabla p$ represents the total force on a fluid particle resulting from of viscous shear force and an isotropic pressure. The first equation of (87.1), which is a vector equation

$$
\frac{\partial u_i}{\partial t} + (u \cdot \nabla)u_i - \nu \Delta u_i + \frac{\partial p}{\partial x_i} = f_i, \qquad i = 1, 2, 3,
$$

is the *momentum equation* expressing Newton's second law stating that the acceleration is proportional to the force, and the second equation expresses the incompressibility condition. We consider here the case of Dirichlet boundary conditions with the velocity u being prescribed on the boundary Γ. Below we consider Neumann and Robin boundary conditions. Below we will often write for short $(u \cdot \nabla)u = u \cdot \nabla u$.

The linear *Stokes equations* are obtained omitting the nonlinear term $u \cdot \nabla u$, which is possible if the velocity u is small, corresponding to *creeping flow*.

The *Reynolds number Re* is defined by $Re = \frac{uL}{\nu}$, where u represents a velocity and and L a length scale characteristic of the flow. The size of the Reynolds number is decisive. If $Re \sim 1$, then the flow is very viscous, a situation met in e.g. polymer flow or forming processes. In most applications in areo/hydro-dynamics, Re is much larger than 1, often very large up to 10^6 or even larger. In these cases with small viscosity, the flow may be very complex or turbulent.

There is a stationary analog of (87.1) assuming the solution to be independent of time along with the driving force and boundary data. A stationary solution normally arises as a limit of a time-dependent solution as time tends to infinity, and this is often reflected in the computation of a stationary solution through some kind of time-stepping until convergence. For larger Reynolds numbers, stable stationary solutions in general do not not exist.

87.3   The Basic Energy Estimate for Navier-Stokes

We now derive a basic stability estimate of energy type for the velocity u of a (u, p) of Navier-Stokes equation (87.1) assuming for simplicity that $f = 0$ and $w = 0$. Scalar multiplication of the momentum equation by u and integration with respect to x gives

$$\frac{1}{2} \frac{d}{dt} \int_\Omega |u|^2 \, dx + \nu \sum_{i=1}^3 \int_\Omega |\nabla u_i|^2 \, dx = 0,$$

because by partial integration (with boundary terms vanishing),

$$\int_\Omega \nabla p \cdot u \, dx = - \int_\Omega p \nabla \cdot u \, dx = 0$$

and

$$\int_\Omega (u \cdot \nabla) u \cdot u \, dx = - \int_\Omega (u \cdot \nabla) u \cdot u \, dx - \int_\Omega \nabla \cdot u |u|^2 \, dx$$

so that

$$\int_\Omega (u \cdot \nabla) u \cdot u \, dx = 0.$$

Integrating next with respect to time, we thus obtain the following basic stability estimate for any time $T > 0$:

$$\|u(\cdot, T)\|^2 + 2\nu \sum_{i=1}^3 \int_0^T \|\nabla u_i\|^2 \, dt = \|u^0\|^2, \qquad (87.3)$$

where $\| \cdot \|$ denotes the $L_2(\Omega)$-norm. This estimate gives a bound on the velocity with the second term on the left representing the dissipation from the viscosity of the fluid. We see that the growth of this term over time corresponds to a decrease of the velocity (momentum) of the flow.

The case of large Reynold's number corresponding to small ν, with a normalization of velocity and typical length scale to unit size, is of particular interest with typically turbulent flows occuring. In laminar flow with small viscosity the dissipation is small because velocity gradients are not large, while in turbulent flow the dissipation is significant because the velocity gradients are large corresponding to a decay of velocities in the case of no driving forces.

Fig. 87.1. Jacques-Louis Lions (1928–2001), founder of the French School of Numerical Analysis: "... optimal control problems for distributed parameter systems modeled by partial differential equations obviously connect to fundamental aspects of Body & Soul..."

87.4   Lions and his School

Jacques-Louis Lions (1929–2001), see Fig. 87.1, carried the strong French mathematical tradition coupled to physics and mechanics through the second half of the 20th century with important contributions to the theory and practice of partial differential equations using tools from Functional Analysis in the spirit of Sobolev. He created the French School of Numerical Analysis, which boomed with the development of the finite element method starting in the 1960s. Among many other things, Lions proved existence and uniqueness of solutions to the Navier-Stokes equations with a regularizing viscosity modification as indicated below.

87.5   Turbulence: Lipschitz with Exponent 1/3?

The mathematical modeling and simulation of turbulent flow represents one of the open problems of classical mechanics and physics, where today computational methods open new possibilities in the form of *Large Eddy Simulation* LES with *subgrid modeling*. Turbulent flow has features (vortices) on a range of scale from largest macroscopic of diameter of order one to smallest of order $\nu^{3/4}$, with ν the viscosity, assuming normalization to characteristic macroscopic velocity and length scale of order one, so that the macroscopic Reynolds number Re equals $1/\nu$. I typical applications Re may be of size 10^8 in which case the smallest length scale may be roughly of order 10^{-6} requiring of the order of 10^{18} degrees of freedom in a *Direct Numerical Simulation* DNS with resolution of all scales. This is way beyond the capacity of any computer within sight, with the present limit being set for DNS with a smallest scale of size 10^{-3} corresponding to Reynolds number roughly of order 10^4. To simulate flows with larger Reynolds number we may seek a *subgrid model* with the objective of modeling the effect on resolvables scales of unresolved scales. This may be possible using features of *scale similarity* of turbulent flow reflecting a certain repetition of flow features in a cascade from coarser to finer scales down to the smallest vortices where significant dissipation occurs. In Fig. 87.4 we show a jet undergoing transition from laminar to turbulent flow on a $128 \times 32 \times 32$ mesh.

Let us give an argument indicating a feature of scale similarity first presented by the Russian mathematician Kolmogorov 1941: Let then h be the smallest scale, that is the diameter of the smallest vorticity, and let $\bar{u}$ be the corresponding velocity of the smallest vorticity. We may then argue that we should have $\bar{u}h \sim \nu$, since the break up of larger vortices into to smaller should continue until the local Reynolds number becomes small enough (of size 50–100). Further, turbulent dissipation on the smallest scale of order one would mean that $\nu(\frac{\bar{u}}{h})^2 \sim 1$. From these two relations, we conclude that $h \sim \nu^{3/4}$ as anticipated and also that $\bar{u} \sim \nu^{1/4}$. We conclude that

$$|u(x) - u(y)| \sim |x - y|^{1/3}$$

for $y = x + h$, and by scale similarity we may expect this relation to hold for general x and y, that is, that the turbulent velocity should be Lipschitz (Hölder) continuous with exponent 1/3.

Does the above derivation have any to do with reality? Yes, both physical experiments and DNS indicate that turbulent flow indeed has features of scale similarity with Lipschitz (Hölder) continuity with exponent 1/3. This gives hope that subgrid modeling may be feasible for turbulent flow and thus that computational simulation of turbulent flow would be possible, and more and more so as the computational power increases.

Summing up, it thus appears that computational simulation of turbulent flow may be possible, and this would in a way settle most questions from a practical point of view: we would be able to simulate and predict turbulent flow. However, we would still lack a mathematical model of turbulence more tractable than simply the Navier-Stokes equations in DNS. So, as human beings we may not be able to "understand turbulence" in the same way as we can understand e.g. the fundamental solution of the Laplacian $(\frac{1}{4\pi|x|})$, but we would be able to computationally simulate turbulent flow. Maybe this is the most we can ask for?

87.6   Existence and Uniqueness of Solutions

The question of existence and uniqueness of solutions to the Navier-Stokes equations is one of the unsolved problems of mathematics. If we change the viscosity from a Newtonian constant viscosity ν to a non-Newtonian solution dependent viscosity $\hat{\nu} = \nu + Ch^2|\nabla u|$, where h is a parameter corresponding to a smallest scale, then, existence and uniqueness is possible to prove using standard methods as shown by Lions. Since with h small the modification will be small, except where ∇u is very large, the modification may be viewed as a regularization eliminating certain extreme situations with very large velocity gradients, where at any rate the Newtonian property of constant viscosity may be questioned. This directly couples to subgrid modeling of turbulent flow, where $\hat{\nu}$ corresponds to a so called *turbulent viscosity*, with the constant C to be modeled computationally.

87.7   Numerical Methods

Trying to solve the incompressible Navier-Stokes equations numerically, we meet the following difficulties:

- instabilities from discretization of convection terms,

- pressure instabilities in equal order interpolation of velocity and pressure.

The simplest cure to convection instability is to increase the viscosity ν in the computation so that $\nu \geq uh$, where u is the local fluid velocity and h is the local mesh size. The simplest stabilization of the pressure p, is to modify the incompressibility equation $\nabla \cdot u = 0$ to $-\nabla \cdot (\delta \nabla p) + \nabla \cdot u = 0$, with $\delta \approx h^2$ with $h(x)$ the local mesh size.

In Galerkin methods the stabilization can be achieved in higher-order consistent form by adding least-squares control of residuals. We present this approach below in the context of the cG(1)dG(0) method with cG(1) in space and dG(0) in time. We also present corresponding cG(1)cG(1) and cG(1)dG(1) methods.

87.8   The Stabilized cG(1)dG(0) Method

We now present the cG(1)dG(0) method for (87.1) starting with the case of homogeneous Dirichlet boundary conditions. Let $0 = t_0 < t_1 < \ldots < t_N = T$ be a sequence of discrete time levels with associated time steps $k_n = t_n - t_{n-1}$. Let W_h be the usual finite element space of continuous piecewise linear functions on a triangulation $\mathcal{T}_h = \{K\}$ of Ω with mesh function $h(x)$. Let W_h^0 be the space of functions in W_h vanishing on Γ. We shall seek an approximate velocity $U(x, t)$ such that $U(x, t)$ is continuous and piecewise linear in x for each t, and $U(x, t)$ is piecewise constant in t for each x. Similarly, we shall seek an approximate pressure $P(x, t)$ which is continuous piecewise linear in x and piecewise constant in t. More precisely, we shall seek $U^n \in V_h^0$ with $V_h^0 = W_h^0 \times W_h^0 \times W_h^0$ and $P^n \in W_h$ for $n = 1, \ldots, N$, and we shall set

$$
\begin{aligned}
U(x, t) &= U^n(x) \quad x \in \Omega, \quad t \in (t_{n-1}, t_n], \\
P(x, t) &= P^n(x) \quad x \in \Omega, \quad t \in (t_{n-1}, t_n].
\end{aligned}
\tag{87.4}
$$

Further we write for velocities $v = (v_i)$ and $w = (w_i)$

$$
(v, w) = \int_\Omega v \cdot w \, dx, \qquad (\nabla v, \nabla w) = \int_\Omega \sum_i^3 \nabla v_i \cdot \nabla w_i \, dx,
$$

and similarly for scalar functions p and q defined on Ω:

$$
(p, q) = \int_\Omega pq \, dx.
$$

We now formulate the cG(1)dG(0) method without stabilization as follows: For $n = 1, \ldots, N$, find $(U^n, P^n) \in V_h^0 \times W_h$ such that

$$
\left(\frac{U^n - U^{n-1}}{k_n}, v \right) + (U^n \cdot \nabla U^n + \nabla P^n, v) + (\nu \nabla U^n, \nabla v) = (f^n, v)
$$

$$
\forall v \in V_h^0,
$$
$$
(\nabla \cdot U^n, q) = 0 \quad \forall q \in W_h,
\tag{87.5}
$$

where $U^0 = u^0$, and we set $f^n(x) = f(x, t_n)$. We see that the discrete equations result from multiplication of the momentum equation with $v \in V_h^0$ and the incompressibility equation by $q \in W_h$, followed by integration over Ω including integration by parts in the term $(-\nu \Delta U, v)$.

We can write the cG(1)dG(0) method without stabilization alternatively as follows: For $n = 1, \ldots, N$, find $(U^n, P^n) \in V_h^0 \times W_h$ such that

$$
\left(\frac{U^n - U^{n-1}}{k_n}, v \right) + (U^n \cdot \nabla U^n + \nabla P^n, v) + (\nabla \cdot U^n, q)
$$

$$
+ (\nu \nabla U^n, \nabla v) = (f^n, v) \quad \forall (v, q) \in V_h^0 \times W_h,
\tag{87.6}
$$

where we simply added the equations in 87.5.

The cG(1)dG(0) method with stabilization takes the form: For $n = 1, \ldots, N$, find $(U^n, P^n) \in V_h^0 \times W_h$ such that

$$
\left(\frac{U^n - U^{n-1}}{k_n}, v\right) + (U^n \cdot \nabla U^n + \nabla P^n, v + \delta(U^n \cdot \nabla v + \nabla q)) + (\nabla \cdot U^n, q)
$$
$$
+ (\nu \nabla U^n, \nabla v) = (f^n, v + \delta(U^n \cdot \nabla v + \nabla q)) \quad \forall (v, q) \in V_h^0 \times W_h, \qquad (87.7)
$$

where δ is a stabilization parameter defined as follows: $\delta(x) = h^2(x)$ in the case of *diffusion-dominated* flow with $\nu \geq Uh$, and

$$
\delta = \left(\frac{1}{k} + \frac{U}{h}\right)^{-1} \qquad (87.8)
$$

in the case of *convection dominated* flow with $\nu < Uh$. Note that if $k \approx \frac{h}{U}$, which is a natural choice of time step in the convection-dominated case, then $\delta \approx \frac{1}{2}\frac{h}{U}$. Note further that the stabilized form (87.7) of the cG(1)dG(0) method is obtained by replacing v by $v + \delta(U^n \cdot \nabla v + \nabla q)$ in the terms $(U^n \cdot \nabla U^n + \nabla P^n, v)$ and (f^n, v). In principle, we should make the replacement throughout, but in the present case of the cG(1)dG(0), only the indicated terms get involved because of the low order of the approximations. The perturbation in the stabilized method is of size δ, and thus the stabilized method has the same order as the original method (first order in h if $k \sim h$).

Letting v vary in (87.7) while choosing $q = 0$, we get the following equation (the discrete momentum equation):

$$
\left(\frac{U^n - U^{n-1}}{k_n}, v\right) + (U^n \cdot \nabla U^n + \nabla P^n, v + \delta U^n \cdot \nabla v)
$$
$$
+ (\nu \nabla U^n, \nabla v) = (f^n, v + \delta U^n \cdot \nabla v) \quad \forall v \in V_h^0, \qquad (87.9)
$$

and letting q vary while setting $v = 0$, we get the following discrete pressure equation:

$$
(\delta \nabla P^n, \nabla q) = -(\delta U^n \cdot \nabla U^n, \nabla q) - (\nabla \cdot U^n, q) + (\delta f^n, \nabla q) \quad \forall q \in W_h. \qquad (87.10)
$$

We normally seek to solve the system (87.7) iteratively alternatively solving the velocity equation (87.9) for U^n with P^n given, and the pressure equation (87.10) for P^n with U^n given.

87.9   The cG(1)cG(1) Method

We present the following cG(1)cG(1) variant of the cG(1)dG(0) method with cG(1) in time instead of dG(0): For $n = 1, \ldots, N$, find

$(U^n, P^n) \in V_h^0 \times W_h$ such that

$$\left(\frac{U^n - U^{n-1}}{k_n}, v\right) + (\hat{U}^n \cdot \nabla \hat{U}^n + \nabla P^n, v + \delta(\hat{U}^n \cdot \nabla v + \nabla q)) + (\nabla \cdot \hat{U}^n, q)$$

$$+ (\nu \nabla \hat{U}^n, \nabla v) = (f^n, v + \delta(\hat{U}^n \cdot \nabla v + \nabla q)) \quad \forall (v, q) \in V_h^0 \times W_h, \qquad (87.11)$$

where $\hat{U}^n = \frac{1}{2}(U^n + U^{n-1})$. Evidently, we obtained the cG(1) version by changing from U^n to $\hat{U}^n$ in all terms but the first in the cG(1)dG(0) method.

87.10 The cG(1)dG(1) Method

We shall now formulate the cG(1)dG(1) method obtained by replacing dG(0) by dG(1) in the cG(1)dG(0) method. In this method the discrete velocity $U(x, t)$ is piecewise linear linear in time on each time interval I_n, with possibly discontinuities at the discrete time levels t_n. More precisely, we make the Ansatz:

$$U^n(x, t) = \frac{t_n - t}{k_n} U_+^{n-1}(x) + \frac{t - t_{n-1}}{k_n} U_-^n(x), \qquad \text{for } t_{n-1} < t < t_n, \qquad (87.12)$$

where U_+^{n-1} and U_-^n belong to V_h^0. We note that

$$U_{\pm}^n(x) = \lim_{s \to 0^+} U(x, t_n \pm s)$$

is the limit of $U(x, t)$ as t approaches t_n from below $(-)$, or above $(+)$. The cG(1)dG(1) method takes the form: For $n = 1, \ldots, N$, find U^n of the form (87.12) and $P^n \in W_h$, such that for all $v(x, t) = w_1(x, t) + (t - t_{n-1}) w_2(x, t)$ with $w_1, w_2 \in V_h^0$ and $q \in W_h$,

$$(U_+^{n-1} - U_-^{n-1}, v) + \int_{t_{n-1}}^{t_n} ((\dot{U}^n + U^n \cdot \nabla U^n$$

$$+ \nabla P^n, v + \delta(\dot{U}^n + U^n \cdot \nabla v + \nabla q)) + (\nabla \cdot U^n, q)) \, dt$$

$$+ \int_{t_{n-1}}^{t_n} (\nu \nabla U^n, \nabla v) \, dt = \int_{t_{n-1}}^{t_n} (f^n, v + \delta(\dot{U} + U^n \cdot \nabla v + \nabla q)). \qquad (87.13)$$

We may similarly let P be piecewise linear discontinuous in time.

87.11 Neumann Boundary Conditions

To properly model Neumann boundary conditions, we first need to recall that the components σ_{ij} of the *total stress tensor* $\sigma = (\sigma_{ij})$ acting on a fluid element, are given by

$$\sigma_{ij} = \bar{\sigma}_{ij} - p\delta_{ij}, \quad i, j = 1, 2, 3,$$

where the *stress deviatoric* $\bar{\sigma} = (\bar{\sigma}_{ij})$ is coupled to the *strain tensor* $\epsilon(u) = (\epsilon_{ij}(u))$ with components

$$\epsilon_{ij}(u) = (\partial u_i / \partial x_j + \partial u_j / \partial x_i)/2, \quad i, j = 1, 2, 3,$$

through the constitutive relation of a *Newtonian fluid*:

$$\bar{\sigma}_{ij} = 2\nu\epsilon_{ij}(u), \quad i, j = 1, 2, 3,$$

where ν is the constant viscosity, and $\delta_{ij} = 1$ if $i = j$ and $\delta_{ij} = 0$ if $i \neq j$. We observe that the trace of the stress deviatoric is zero, that is,

$$\sum_{i=1}^{3} \bar{\sigma}_{ii} = 2\nu \sum_{i=1}^{3} \epsilon_{ii}(u) = 2\nu\nabla \cdot u = 0,$$

and thus the total stress σ is decomposed into a stress deviatoric $\bar{\sigma}$ with zero trace and an isotropic pressure p. Further, a direct computation shows that

$$\nu\Delta u - \nabla p = \nabla \cdot \sigma, \tag{87.14}$$

where $\nabla \cdot \sigma$ is a vector with components $(\nabla \cdot \sigma)_i$ given by

$$(\nabla \cdot \sigma)_i = \sum_{j=1}^{3} \frac{\partial \sigma_{ij}}{\partial x_j}.$$

Multiplying 87.14 by $v = (v_i)$ with $v = 0$ on Γ and integrating by parts, we find that

$$\nu(\nabla u, \nabla v) + (\nabla p, v) = 2\nu(\epsilon(u), \epsilon(v)) + (\nabla p, v),$$

where

$$(\epsilon(u), \epsilon(v)) = \sum_{i,j=1}^{3} \int_{\Omega} \epsilon_{ij}(u)\epsilon_{ij}(v)\, dx.$$

We are thus led to replace the term $(\nu\nabla u, \nabla v)$ by the term $(2\nu\epsilon(u), \epsilon(v))$ in variational formulations of the Navier-Stokes equations. In the case of Dirichlet boundary conditions for the velocity the two expressions are equal, since the test velocity v vanishes on Γ, but in the case of Neumann type boundary conditions the replacement opens the possibility of enforcing in variational form a Neumann boundary condition of the form

$$\sum_{j=1}^{3} \sigma_{ij}n_j = \sum_{j=1}^{3} \bar{\sigma}_{ij}n_j - pn_i = \sum_{j=1}^{3} 2\nu\epsilon_{ij}(u)n_j - pn_i = g_i \text{ on } \Gamma_2, \ i = 1, 2, 3,$$

$$\tag{87.15}$$

which expresses that the total force on the boundary part Γ_2 is equal to the given force $g = (g_i)$. For example, if $g = 0$, then this condition expresses that the total force is zero on Γ_2, which we may use as an outflow boundary condition simulating that the fluid freely flows out into a large reservoir. More precisely, the presence of the terms

$$-(p, \nabla \cdot v) + (2\nu\epsilon(u), \epsilon(v))$$

in a variational formulation with v varying freely on Γ_2, will enforce a homogeneous Neumann boundary condition 87.15 upon integration by parts.

We now consider a typical situation with the boundary Γ decomposed into two parts Γ_1 an Γ_2 with the velocity being equal to a given velocity w on Γ_1 and imposing the homogeneous Neumann condition 87.15 on Γ_2. For simplicity, we assume that w is independent of time, the extension to time dependence of w being evident. Typically, w will be zero on a part of Γ_1 and will be directed into Ω on the remaining part corresponding to a given inflow.

We let V_h be the space of continuous piecewise linear velocities v on a triangulation $\mathcal{T}_h = \{K\}$ of Ω with mesh function $h(x)$, satisfying the boundary condition $v = w$ on Γ_1, and let V_h^0 be the corresponding test space of functions with $v = 0$ on Γ_1. Let W_h be the space of continuous piecewise linear pressures p on $\mathcal{T}_h = \{K\}$, and W_h^0 the corresponding test space of pressures q such that $q = 0$ on Γ_2.

The stabilized cG(1)dG(0) method can be formulated as follows: For $n = 1, \ldots, N$ seek $U^n \in V_h$ and $P^n \in W_h$ such that

$$\left(\frac{U^n - U^{n-1}}{k_n}, v \right) + (U^n \cdot \nabla U^n, v + \delta U^n \cdot \nabla v) - (P^n, \nabla \cdot v)$$
$$+ (2\nu\epsilon(U^n), \epsilon(v)) = (f^n, v + \delta U^n \cdot \nabla v) \quad \forall v \in V_h^0, \tag{87.16}$$

$$(\delta \nabla P^n, \nabla q) = -(\delta U^n \cdot \nabla U^n, \nabla q) - (\nabla \cdot U^n, q) + (\delta f^n, \nabla q) \quad \forall q \in W_h^0, \tag{87.17}$$

where we choose P^n on Γ_2 according to 87.15 with $g = 0$ and u replaced by U. Again we seek to solve the system iteratively alternatively solving the velocity equation (87.16) for U^n with P^n given, and the pressure equation (87.17) for P^n with U^n given.

87.12   Computational Examples

We now present some computational examples of 3d time dependent flows, using the stabilized cG(1)cG(1) method on a mesh with meshsize $h = 1/32$.

In Fig. 87.2 we present the solution of a bluff body problem: a flow in a channel with 1x1 square cross section and length 4, with a square obstacle

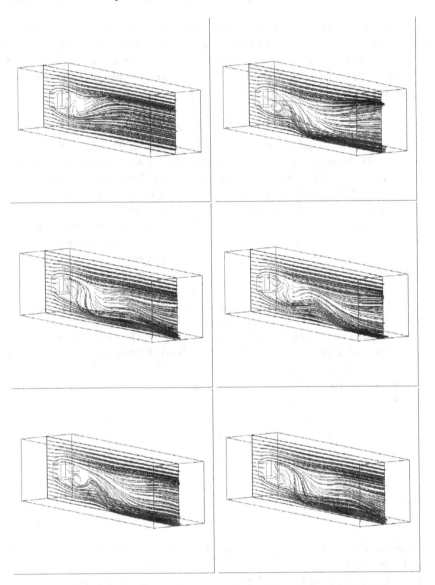

Fig. 87.2. Bluff body flow computations for $t = 2, 4, 6, 8, 10, 12$

with side length 0.25 centered at $(0.5, 0.5, 0.5)$. We have used zero Dirichlet boundary condition for the velocity on the side walls and Neumann outflow boundary conditions on the outflow boundary. On the inflow a parabolic velocity is prescribed.

In Fig. 87.3 we present the solution of a step down problem in a similar channel with a step down of height and length 0.5.

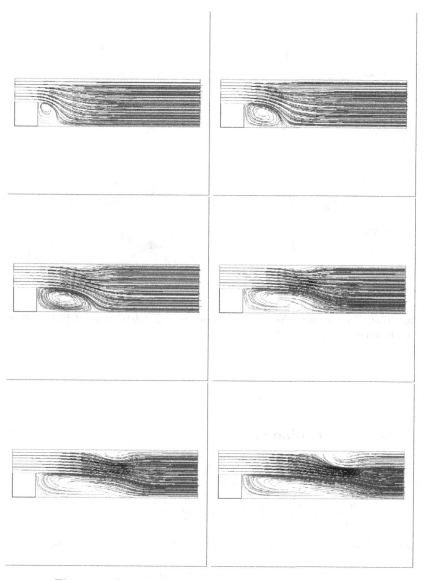

Fig. 87.3. Step down flow computations for $t = 1, 2, 3, 4, 5, 6$

Finally in Figure Fig. 87.4 we present computations of transition to turbulence in a circular jet flow with streamwise velocity 1 in the jet and zero outside the jet, where we apply a small random perturbation. Here we have used periodic boundary conditions in all directions.

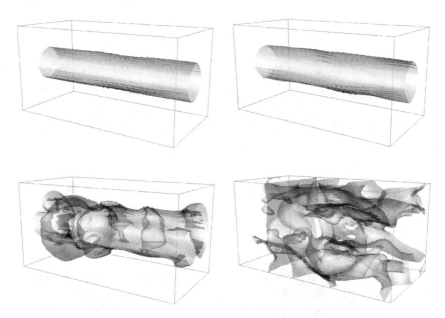

Fig. 87.4. Streamwise velocity isosurfaces for $|u_1| = 0.02$ in a jet in transition from laminar to turbulent flow, for $t = 5, 7, 10, 15$

Chapter 87 Problems

87.1. Prove for that a solution (u, p) of (87.1) with $f = 0$ and $w = 0$ satisfies the following energy estimate for $t > 0$:

$$\int_\Omega |u(x, t)|^2 + 2\nu \int_0^t \int_\Omega |\nabla u(x, s)|^2 \, dx \, ds = \int_\Omega |u^0(x)|^2 \, dx.$$

Hint: Multiply the momentum equation by u and use that if $\nabla \cdot u = 0$, then

$$\int_\Omega (u \cdot \nabla) u \cdot u \, dx = 0,$$

which follows by integration by parts.

87.2. Prove a basic stability estimate for (87.7) by choosing $(v, q) = (U, P)$.

Thus the methods of Lagrange and Hamilton are undoubtedly *useful* in helping us to carry out the primary task of dynamics - namely, to find out how systems move. But it would be wrong to think that this is the sole purpose of these general methods or even their main purpose. They do much more. In fact, they teach us what dynamics

really is : It is the study of certain types of differential equations.
(Synge and Griffiths, Principles of Mechanics, 1959)

I sing the body electric,
The armies of those I love engirth me and I engirth them,
They will not let me off till I go with them, respond to them,
And discorrupt them, and charge them full
with the charge of the soul.
Was it doubted that those who corrupt
their own bodies conceal themselves?
And if those who defile the living are as bad as
they who defile the dead?
And if the body does not do fully as much as the soul?
And if the body were not the soul, what is the soul?
(Walt Whitman).

References

[1] L. AHLFORS, *Complex Analysis*, McGraw-Hill Book Company, New York, 1979.

[2] K. ATKINSON, *An Introduction to Numerical Analysis*, John Wiley and Sons, New York, 1989.

[3] L. BERS, *Calculus*, Holt, Rinehart, and Winston, New York, 1976.

[4] M. BRAUN, *Differential Equations and their Applications*, Springer-Verlag, New York, 1984.

[5] R. COOKE, *The History of Mathematics. A Brief Course*, John Wiley and Sons, New York, 1997.

[6] R. COURANT AND F. JOHN, *Introduction to Calculus and Analysis*, vol. 1, Springer-Verlag, New York, 1989.

[7] R. COURANT AND H. ROBBINS, *What is Mathematics?*, Oxford University Press, New York, 1969.

[8] P. DAVIS AND R. HERSH, *The Mathematical Experience*, Houghton Mifflin, New York, 1998.

[9] J. DENNIS AND R. SCHNABEL, *Numerical Methods for Unconstrained Optimization and Nonlinear Equations*, Prentice-Hall, New Jersey, 1983.

[10] K. ERIKSSON, D. ESTEP, P. HANSBO, AND C. JOHNSON, *Computational Differential Equations*, Cambridge University Press, New York, 1996.

[11] I. GRATTAN-GUINESS, *The Norton History of the Mathematical Sciences*, W.W. Norton and Company, New York, 1997.

[12] P. HENRICI, *Discrete Variable Methods in Ordinary Differential Equations*, John Wiley and Sons, New York, 1962.

[13] E. ISAACSON AND H. KELLER, *Analysis of Numerical Methods*, John Wiley and Sons, New York, 1966.

[14] M. KLINE, *Mathematical Thought from Ancient to Modern Times*, vol. I, II, III, Oxford University Press, New York, 1972.

[15] J. O'CONNOR AND E. ROBERTSON, *The MacTutor History of Mathematics Archive*, School of Mathematics and Statistics, University of Saint Andrews, Scotland, 2001. http://www-groups.dcs.st-and.ac.uk/~history/.

[16] W. RUDIN, *Principles of Mathematical Analysis*, McGraw–Hill Book Company, New York, 1976.

[17] T. YPMA, *Historical development of the Newton-Raphson method*, SIAM Review, 37 (1995), pp. 531–551.

Index